Hematologic Malignancies

Manorama Bhargava

Hematologic Malignancies

Case Studies in Cytogenetic and Molecular Genetics

Springer

Manorama Bhargava
Former Senior Advisor and Head Hematopathology
Medanta-The Medicity
New Delhi
India

Former Professor and Head Hematology
All India Institute of Medical Sciences
and Sir Ganga Ram Hospital
New Delhi
India

ISBN 978-981-33-4798-4 ISBN 978-981-33-4799-1 (eBook)
https://doi.org/10.1007/978-981-33-4799-1

This Springer imprint is published by the registered company Springer Nature Singapore Pte Ltd.
The registered company address is: 152 Beach Road, #21-01/04 Gateway East, Singapore 189721, Singapore

To my Late Parents
and to my Family

Foreword

> Progress in science depends on new techniques, new discoveries and new ideas. Probably in that order **(Prof. Sydney Brenner 1927-2019; Recipient of the Noble Prize in Physiology or Medicine 2002)**.

Haematology has the unique and proud distinction of being at the forefront of the application of novel techniques and successfully translating the resulting knowledge in improving both the understanding and treatments of blood diseases. This is particularly true of particular haematological malignancies, where researchers and clinicians continue to work in close partnership and successfully carry out an assault on the disease mechanisms. This has led to an unprecedented accumulation of knowledge as well as diagnostic and therapeutic innovations. Specific and easily available dyes used for staining of peripheral blood (PB) and bone marrow (BM) films meant for identification of different cell types in healthy and diseased subjects with haematological cancers became the global norm. The discovery of Philadelphia chromosome in 1959 by David Hungerford and Peter Nowell, as a consistent abnormality in chronic myeloid leukaemia patients, ushered in an era of molecular and cellular haemato-oncology, leading to a vast array of therapeutic agents that specifically target and alter the behaviour of leukaemia clone(s).

Furthermore, immunophenotypic profiling of malignant cells has also become a given in the diagnostic work-ups and assignment of specific therapies. Application of the molecular techniques such as PCR has become a routine in haematology laboratories globally. More recently, in many departments worldwide results obtained from next-generation sequencing of the malignant cells have identified the genetic/epigenetics of characteristics of haematological cancers. Thus, it has become almost mandatory for accurate diagnosis, prognosis and treatment selection to combine information obtained from morphology, immunophenotyping, cytogenetics and molecular genetics. Prof. Manorama Bhargava is a distinguished haematologist and clinician scientist, who like me, has witnessed the genesis of our discipline and the importance of multidisciplinary laboratory investigations. She firmly believes in ensuring that the next generation is knowledgeable in the application and interpretation of the tests performed in the laboratories. Although there is vast literature in each of diagnostic disciplines of haematological malignancies, but books that bring together the 'real world', clinical cases and results from specialised diagnostics are lacking. In this book, Dr. Bhargava has carefully chosen 107 case studies of haemo-oncology patients from India seen at a single institution. The book has a logical layout with an introductory chapter that describes step-by-step details as well as tips in cytogenetic, fluorescent in situ hybridisation (FISH), immunophenotyping and some molecular techniques such as polymerase chain reaction (PCR).

Chapters follow a uniform, easy to digest logical presentations with a brief clinical history, full blood count, bone marrow features, immunophenotype, cytogenetics and key molecular characteristics. Each case study has a commentary that brings together the results of the aforementioned investigations and a final diagnosis with key references regarding the final diagnosis. Clear emphasis of the book is the integration of results from cytogenetics/FISH studies, immunophenotyping and in some instances the results of molecular genetic typing. There are high-quality photomicrographs of various chromosomal abnormalities with excellent quality FISH studies in acute and chronic myeloid/lymphoid leukaemia. Throughout the text, there are excellent examples that illustrate the complementary value of the various diagnostic techniques. For example, case studies 9.2 and 9.4 clearly illustrate that t (12; 21) involving ETV6 and RUNX1 genes were not detected by cytogenetics techniques, but could be detected by using relevant molecular probes. The book is punctuated with examples faced in routine clinical laboratory practice in the interpretation of the laboratory tests and bringing together of the clinical phenotype with the results from the specialist laboratory investigations. Haematologists (clinical and laboratory), those in training and technical and scientific trainees in haematology will find the case studies extremely educational and helpful in understanding the value of integration of results from morphology, immunophenotyping, cytogenetics and molecular testing.

The 107 case studies cover the full spectrum of common haematological malignancies encountered in clinical practice.

I hope clinicians/scientists will be sufficiently impressed by the examples and the huge research opportunities that diagnostic and follow-up clinical material provides in haematological cancers in India. This book is a reminder of this untapped resource. In this regard, Prof. Bhargava has added further to her massive contributions to the field of haematology in India. Although next-generation sequencing (NGS) is not the main focus of the book, but reference to this increasing application is made in 'Chapter 8'. Several of the case studies deal with the key results from NGS, but interested readers will be well advised to consult recent reviews on NGS in haematological cancers.

The use of the word peripheral blood smear and BM 'smear' in this book reminds me of a conversation that I had almost three decades ago with the late Prof. David Galton (Director of MRC Leukaemia Unit at the Hammersmith Hospital, Royal Postgraduate School). He explained to me that microbiology colleagues *smear* a bit of fluid (sputum, stool, etc.) on a slide and then examine it under the microscope. By contrast, haematologists make a 'film', by carefully spreading a film of blood and/or BM on a glass slide. Thus, he emphasised to me relatively sternly for a mild-mannered gentle giant of haematology that we (haematologists) make a 'film' and *not* a 'smear'. Semantics I know!

In common with our American practitioners, this book has preferred smear over film. I personally prefer the David Galton explanation.

I wish you all the young and not so young budding haematologists and scientists a happy and inquisitive reading. It would be wonderful to see 'Volume 2' of this book with specific emphasis on molecular genetics diagnostics.

Ghulam J. Mufti, OBE, DM, FRCP, FRCPath
Professor of Haemato-oncology Haematological Medicine
King's College Hospital NHS Foundation Trust
London, UK
ghulam.mufti@nhs.net

Preface

In recent years, there has been a prolific growth in the understanding of the genetic profile of Hematologic Malignancies. Malignant hematologic disorders are often driven by genetic mutations and epigenetic alterations. They are heterogeneous both in biological and clinical aspects. The alterations of genomic profile associated with these diseases are complex and variable including mutations, translocations, karyotypic rearrangements, and post-translocation modifications. In some cases, several genetic changes are required to induce the onset of the disease. The development of new methods in molecular biology has allowed not only the individualized molecular diagnosis of a disease, its biological behavior, and prognosis but has also led to effective clinical decision making. The discovery of genetic profile has made possible targeted therapy as an important additional option to chemotherapy and radiotherapy used currently to treat most hematologic malignancies.

Several methods are available for studying the genetic profile of these disorders. The PCR, first described in 1983, remains the most popular for DNA analysis. Cytogenetics (karyotyping and FISH) provide valuable information on gene mutations, rearrangements, translocations, inversions as well as several numerical and structural alterations. Next Generation Sequencing and Proteomics are other advanced techniques available for molecular characterization of tumors.

All classifications for hematologic malignancies today are based on their molecular genetic characteristics and form an integral part of therapeutic management of patients. While many advanced centers for cancer care, hematology, and medical oncology departments incorporate such data in their clinical protocols, many others have no access to such specialized tests.

This book was conceived keeping these changes in mind. The primary objective was to put on record the correlations between molecular genetics of hematologic malignancies and the clinical diseases resulting in a better understanding of its pathogenesis, its biological behavior and prognosis leading to evidence-based and more rational approach to the management of patients.

The other, and an equally important objective, was to provide a source of learning. The book is primarily targeted to students/trainees/examinees who are engaged in pursuing superspeciality degree (DM, DNB) in Hematology, Medical Oncology, and Pediatric Hematology—Oncology. It would also benefit students who are pursuing MD in Pathology, Internal Medicine, or

Pediatrics. Besides, it would be a source of knowledge for all those interested in molecular genetics of hematologic malignancies.

This book is not a textbook in molecular hematology. It is a compendium of several case studies in hematologic malignancies with particular reference to cytogenetics (karyotyping and FISH) and molecular genetic changes through real-time PCR technology. The reason why the format of case studies is chosen is that, in my opinion, case studies offer an excellent platform for teaching, for rapid learning, clinicopathologic understanding of a disease, and a brief but comprehensive source of information and encourage and enthuse a reader to look for more.

The hematologic malignancies covered are the myeloid malignancies, Acute Myeloid Leukemia (AML), Myelodysplastic Syndrome (MDS), Myelodysplastic Myeloproliferative Neoplasms (MDS-MPN), Myeloproliferative Neoplasms (MPN), Chronic Myeloid Leukemia (CML), and Chronic Myelomonocytic Leukemia (CMML). The lymphoid malignancies included are Acute Lymphoblastic Leukemia (ALL), Chronic Lymphocytic Leukemia (CLL, Lymphomas), and Multiple Myeloma.

A total of 107 case studies have been selected for the uniqueness of the findings they represent, each different from the other, even though they may have had a similar morphologic picture. Molecular genetic findings have been combined with cytogenetic findings in each case since they support and complement each other to give a final diagnosis.

Each case study describes in detail the clinical history of the patient, findings of peripheral blood, bone marrow aspirate, and bone biopsy morphological details. This is followed by flow cytometric immunophenotyping, cytogenetics (karyotyping and FISH), and molecular findings on real-time PCR. All this data collectively leads to a final diagnosis. A brief discussion follows with particular reference to clinical implications, prognosis, treatment response, and survival. Relevant references, pertinent to the discussion, are included in each case. Each case study is accompanied by black and white or color images as appropriate.

The uniqueness of this book lies in providing a complete picture in each case to give a diagnostic basis for treatment. Another distinguishing feature is to put on record the cytogenetic and molecular findings in a spectrum of hematologic malignancies in Indian Patients. It is well known that different ethnic and population groups show different disease characteristics and may have a bearing on treatment adopted and response to treatment.

It is my fervent hope that those who read this book would find it of benefit to them in their areas of work to gain a better understanding of a malignant hematologic disorder and its rational management.

New Delhi, India Manorama Bhargava

Acknowledgments

In the year 2018, the two speciality laboratories of Cytogenetics and Molecular Hematology became functional at Medanta, The Medicity. The idea for establishing these laboratories had its origin in my discussions with Dr. Ashok Vaid, Chief of Medical Oncology and Hematology at Medanta, when I joined Medanta in late 2016. The idea found support from Dr. Naresh Trehan, Chairman and Managing Director of Medanta, and Pankaj Sahni, the CEO. To these three individuals, I owe a deep sense of gratitude for their continued support towards the completion of this project. At the heart of this endeavor was to provide clinicians with state of the art information in cytogenetics and molecular hematology for rational management and decision making for their patients with hematologic malignancies.

The case studies selected for presentation in this book were chosen from the data collected in these two laboratories from April 2018 to March 2020 in the division of Hematopathology, Department of Laboratory Medicine at Medanta. I would like to express my deep appreciation to the consultants in the lab who were associated with the entire hematologic workup on these patients, Dr. Shalini Goel, Dr. Ritu Chadha, Dr. Bhavna Jha, and Dr. Uday Kumar; Dr. Renu Saxena, Director, Laboratory Medicine, encouraged me to push ahead with the publication of this data. I am particularly grateful for the assistance provided by my clinical colleagues. Dr. Ashok Vaid, Dr. Nitin Sood, Dr. Amrita, and Dr. Roshan Dixit in Medical Oncology and Clinical Hematology and Dr. S P Yadav and his colleagues in Pediatric Hematology—Oncology for their interactions and discussions leading to additions and improvements in these two labs.

My very special thanks and gratitude are due to my three young scientists who generated all this data, without whom this book would not have been possible. Dr. Nitin Mathur for Molecular Hematology and Mr. Anil Kumar Yadav for cytogenetics virtually established these labs with me. Mr. Akshay Gore, who joined later, contributed immensely to cytogenetics. All of them, with years of experience behind them, were instrumental in the success of these laboratories. I would also like to thank the technical staff in the hematology lab for their willingness to help at all times.

I would also like to thank the staff in the administrative and financial wings of Medanta in helping remove the roadblocks in procurement of essentials for these labs. I thank the companies (Thermo-Fischer, Quiagen, Promega, 3B Bio Black, BioRad, Zytovision, and Metasystems) for their support in maintaining the equipment and for the timely supply of reagents, etc.

Last but not least, my family, my emotional anchor, for their unstinted support all through the journey of establishing these labs, taking them forward, and eventually writing this book. My husband Dr. Santosh Bhargava has been my source of immense strength, and my children Sumit, Shibani, Anna, and Hiren and grandchildren Alex and Arjun were a source of great joy through difficult times.

Contents

1 Techniques in Cytogenetics 1

2 Techniques in Molecular Hematology 17
2.1 Polymerase Chain Reaction 17

Part I Case Studies in Myeloid Malignancies

3 Acute Myeloid Leukemia 39
3.1 Case Study 3.1: Acute Myeloid Leukemia with t(8;21)(q22;q22.1);RUNX1-RUNX1T1 and KIT Gene Mutation 39
3.2 Case Study 3.2: Acute Myeloid Leukemia with a Variant Three-Way Translocation t(8;17;21)(q22;q23;22.1);RUNX1-RUNX1T1 41
3.3 Case Study 3.3: Acute Myeloid Leukemia, t(8;21)(q22;q22) with Rare Duplication of der (21) 42
3.4 Case Study 3.4: AML with inv(16)(p13;q22)/t(16;16)p(13;q22);CBFB–MYH11 44
3.5 Case Study 3.5: Acute Myeloid Leukemia with t(9;11)(p21;q23.3)(KMT2A-MLL3) with Trisomies of Chromosomes 8 and 21 46
3.6 Case Study 3.6: Acute Promyelocytic Leukemia with PML-RARA, t(15;17) with Mutated FLT3-TKD (D835) 47
3.7 Case Study 3.7: Acute Promyelocytic Leukemia with Variant (11;17)(q23;q21.1)(PLZF/RARA) 49
3.8 Case Study 3.8: Acute Promyelocytic Leukemia with t(15;17) and t(9;22) Translocations, Trisomy 8 and Multiple Molecular Abnormalities 51
3.9 Case Study 3.9: Acute Promyelocytic Leukemia with PML-RARA t(15;17) with Mutated FLT3-ITD. 52
3.10 Case Study 3.10: Acute Myeloid Leukemia with Normal Karyotype, Mutated NPM1 and FLT3-ITD . . . 54
3.11 Case Study 3.11: Acute Myeloid Leukemia with normal Karyotype and NPM 1 Gene Mutation 56
3.12 Case Study 3.12: Acute Myeloid Leukemia with Normal Karyotype and FL3-ITD Gene Mutation 58

3.13 Case Study 3.13: Acute Myeloid Leukemia with t(11;19) and FLT3-ITD Mutation 59
3.14 Case Study 3.14: t-Acute Myeloid Leukemia with Myelodysplasia-Related Changes (AML-MRC) with Monosomy 7 Transformed from Aplastic Anemia 60
3.15 Case Study 3.15: Acute Myeloid Leukemia with Myelodysplasia-Related Changes with Monosomy 7 62
3.16 Case Study 3.16: Acute Myeloid Leukemia with Myelodysplasia-Related Changes with Monosomal Karyotype 63
3.17 Case Study 3.17: Acute Myeloid Leukemia (s-AML) from (MDS-EB2) with Myelodysplasia-Related Changes with Complex Karyotype............................ 65
3.18 Case Study 3.18: Acute Myeloid Leukemia (t-AML) with Myelodysplasia-Related Changes with Monosomy 7 67
3.19 Case Study 3.19: Acute Myeloid Leukemia with Myelodysplasia-Related Changes with Complex Karyotype............................ 68
3.20 Case Study 3.20: Acute Myeloid Leukemia with Myelodysplasia-Related Changes (MDS-EB1) with Complex Karyotype 70
3.21 Case Study 3.21: Acute Myeloid Leukemia (t-AML) with t(11;19)(q23;p13.1)(KMT2A-ELL) 73
3.22 Case Study 3.22: Therapy-Related Acute Myeloid Leukemia (t-AML) with Hypodiploid Complex Karyotype 74
3.23 Case Study 3.23: s-AML with Monocytic Differentiation with der(1) t(1;15)(q11;q13) Evolving from Polycythemia Vera..................... 77
3.24 Case Study 3.24: Acute Myeloid Leukemia (s-AML) from Primary Myelofibrosis with Complex Karyotype............................ 78
3.25 Case Study 3.25: Acute Megakaryoblastic Leukemia with Isolated Deletion 9q 82
3.26 Case Study 3.26: Acute Myeloid Leukemia, NOS, with Isolated Trisomy 11 83
3.27 Case Study 3.27: Acute Myeloid Leukemia, NOS with Isolated Trisomy 14 85
3.28 Case Study 3.28: Acute Myeloid Leukemia, NOS with Sole Abnormality of Trisomy 8 86
3.29 Case Study 3.29: Acute Myeloid Leukemia: AML-M7 (NOS) with Complex Karyotype 88
3.30 Case Study 3.30: Acute Myeloid Leukemia with Lineage Switch from ALL with TP53 alterations and Complex Karyotype.................... 89

3.31 Case Study 3.31: Acute Myeloid Leukemia with Normal Karyotype and NPM1 Mutation on (RT-PCR) with NPM1 and FLT3-ITD Mutation on NGS . . . 91
3.32 Case Study 3.32: Acute Myeloid Leukemia with t(8;21) and RUNX1-RUNX1T1 Gene Rearrangement on RT-PCR and NGS . . . 93
3.33 Case Study 3.33: Acute Myeloid Leukemia with Normal Karyotype and NPM1 Mutation on RT-PCR and NGS . . . 95
Annexure . . . 96
References . . . 97

4 Myelodysplastic Syndrome . . . 101
4.1 Case Study 4.1: Myelodysplastic Syndrome (MDS-EB2) with Rare Interstitial Deletion of the Long Arm of Chromosome 9 . . . 101
4.2 Case Study 4.2: Myelodysplastic Syndrome with Excess Blasts (MDS-EB2) and Trisomy 8 . . . 102
4.3 Case Study 4.3: Myelodysplastic Syndrome with Multilineage Dysplasia (MDS-MLD) with Monosomal Karyotype . . . 104
4.4 Case Study 4.4: Myelodysplastic Syndrome with Multilineage Dysplasia (MDS-MLD) with Deletion 13q . . . 106
4.5 Case Study 4.5: Therapy Related Myelodysplastic Syndrome with Complex Karyotype . . . 108
4.6 Case Study 4.6: Myelodysplastic Syndrome with Excess Blasts (MDS-EB-2) with Dicentric Chromosome . . . 110
4.7 Case Study 4.7: Myelodysplastic Syndrome with Excess Blasts-EB1 (MDS-EB-1) with Composite Karyotype . . . 111
4.8 Case Study 4.8: Myelodysplastic Syndrome (MDS-EB-2) with Monosomal Karyotype . . . 114
4.9 Case Study 4.9: Myelodyaplastic Syndrome with Isolated 5q Deletion (5q-Syndrome) . . . 116
Annexure . . . 117
References . . . 117

5 Myelodysplastic/Myeloproliferative Neoplasms (MDS/MPN) . . . 119
5.1 Case Study 5.1: Myelodysplastic Syndrome: Myeloproliferative Neoplasm (CMML-1-RS) . . . 119
5.2 Case Study 5.2: Myelodysplastic Syndrome/ Myeloproliferative Neoplasm (CMML-2) with del (20q) . . . 122
5.3 Case Study 5.3: Myelodysplastic: Myeloproliferative Neoplasm MDS-MPN-CMML-O with Rare t(2;11) . . . 124

5.4 Case Study 5.4: Myelodysplastic Syndrome: Myeloproliferative Neoplasm CMML-1 with Isolated Trisomy 8 126
5.5 Case Study 5.5: Myelodysplastic Syndrome: Myeloproliferative Neoplasm Atypical CML (a CML) with i(17q). 128
5.6 Case Study 5.6: Myelodysplastic Syndrome Myeloproliferative Neoplasm (CMML-1) with t(1;15) with Derivative Chromosome 15 130
Annexure. 131
References. 131

6 Myeloproliferative Neoplasms (MPNs). 133
6.1 Case Study 6.1: Myeloproliferative Neoplasm: Primary Myelofibrosis with Trisomy 9 133
6.2 Case Study 6.2: Myeloproliferative Neoplasm: Polycythemia Vera with JAK 2 Mutation. 135
6.3 Case Study 6.3: Myeloproliferative Neoplasm: Primary Myelofibrosis with JAK2 Mutation 135
6.4 Case Study 6.4: Myeloproliferative Neoplasm: Essential Thrombocythemia with CALR Mutation 136
6.5 Case Study 6.5: Myeloproliferative Neoplasm: Essential Thrombocythemia with JAK2 Mutation. 137
Annexure. 138
References. 138

7 Chronic Myeloid Leukemia 139
7.1 Case Study 7.1: Chronic Myeloid Leukemia with Lymphoid Blast Crisis with Rare dic (7;9) 139
7.2 Case Study 7.2: Chronic Myeloid Leukemia in Myeloid Blast Crisis with Hyperdiploid Complex Karyotype. 140
7.3 Case Study 7.3: Chronic Myeloid Leukemia in Myeloid Blast Crisis with t(9;22) and inv(16). 143
7.4 Case Study 7.4: Chronic Myeloid Leukemia in Myeloid Blast Crisis with a Novel dic(16;17) and TP53 Gene Deletion 144
Annexure. 145
References. 146

8 Next Generation Sequencing for Diagnostics of Myeloid Malignancies for Routine Clinical Use: Pitfalls and Advantages 147
Reference 147

Part: II Case Studies in Lymphoid Malignancies

9 Acute Lymphoblastic Leukemia 151
9.1 Case Study 9.1: Acute Lymphoblastic Leukemia with High Hyperdiploidy 151
9.2 Case Study 9.2: Acute Lymphoblastic Leukemia with Complex Karyotype with t(12;21) (ETV6/RUNX1) ... 152
9.3 Case Study 9.3: Philadelphia Chromosome Positive Acute Lymphoblastic Leukemia with Hypodiploidy, Monosomy 7, and der 6 t(2;6) 154
9.4 Case Study 9.4 Acute Lymphoblastic Leukemia with normal Karyotype with t(12;21) on FISH 156
9.5 Case Study 9.5: Acute Lymphoblastic Leukemia with t (1;19) (q23;p13.3) and i(1)(q10) 158
9.6 Case Study 9.6: Acute Lymphoblastic Leukemia with del(9p) and dup (1q) 160
9.7 Case Study 9.7: Relapsed Precursor B Acute Lymphoblastic Leukemia with t(12;17)(p13;q11.2) 161
9.8 Case Study 9.8: Philadelphia Chromosome Positive Acute Lymphoblastic Leukemia with High Hypodiploidy with Monosomy 7 163
9.9 Case Study 9.9: Therapy Related Acute Lymphoblastic Leukemia with Complex Karyotype 164
9.10 Case Study 9.10: Early T Cell Precursor Acute Lymphoblastic Leukemia (ETP-ALL) with Normal Karyotype and FLT3-ITD Mutation Positive 166
9.11 Case Study 9.11: Acute Lymphoblastic Leukemia with del (9p) 167
9.12 Case Study 9.12: Acute Lymphoblastic Leukemia with High Hyperdiploidy with Complex Karyotype and TP53 Gene Deletion 169
9.13 Case Study 9.13: Acute Lymphoblastic Leukemia with Complex Karyotype with t(4;11) (q21;q23) 171
9.14 Case Study 9.14: Acute Lymphoblastic Leukemia with both Hyperdiploid and Hypodiploid Clones 173
9.15 Case Study 9.15: Philadelphia Chromosome Positive Acute Lymphoblastic Leukemia with additional Cytogenetic Abnormalities 175
9.16 Case Study 9.16: Acute Lymphoblastic Leukemia with Masked t(9;22) 177
9.17 Case Study 9.17: Philadelphia Chromosome Positive Acute Lymphoblastic Leukemia 179

9.18 Case Study 9.18: Acute Lymphoblastic Leukemia with der 19 t(1;19) (q23;p13). 180
9.19 Case Study 9.19: Acute Lymphoblastic Leukemia with Philadelphia Chromosome t(9;22)(q34;q11.2) with i(8)(q10)and der (5) t (5;8). 181
9.20 Case Study 9.20: Acute Lymphoblastic Leukemia/Non-Hodgkin's Lymphoma with a Variant t(8;22), MYC/IGL. 184
9.21 Case Study 9.21: Relapsed Philadelphia Chromosome Positive Acute Lymphoblastic Leukemia with Composite Karyotype and del(7p) 186
9.22 Case Study 9.22: Acute Lymphoblastic Leukemia with Complex Karyotype. 187
9.23 Case Study 9.23: Acute Lymphoblastic Leukemia with Eosinophilia 189
Annexure. 190
References. 191

10 Chronic Lymphocytic Leukemia and Non-Hodgkin's Lymphomas. 195
10.1 Case Study 10.1: Chronic Lymphocytic Leukemia with ATM Gene (11q22) Gene Deletion and Trisomy 12 195
10.2 Case Study 10.2: Chronic Lymphocytic Leukemia with Isolated Trisomy 12 197
10.3 Case Study 10.3: Chronic Lymphocytic Leukemia with TP53 (17p13) and RB 1 (13q14) Gene Deletions. 198
10.4 Case Study 10.4: Chronic Lymphocytic Leukemia with ATM (11q22) and RB 1 (13q14) Gene Deletions. 200
10.5 Case Study 10.5: Chronic Lymphocytic Leukemia with IgH and Gene Rearrangement and Trisomy 12 202
10.6 Case Study 10.6: Chronic Lymphocytic Leukemia with Isolated RB 1 Gene Deletion (13q14). 204
10.7 Case Study 10.7: Chronic Lymphocytic Leukemia with Partial deletion of IgH (14q32) Gene Rearrangement 205
10.8 Case Study 10.8: Chronic Lymphocytic Leukemia with Biallelic and Monoallelic RB 1 Gene Deletion (13q14) 206
10.9 Case Study 10.9: Chronic Lymphocytic Leukemia with Sole Abnormality of ATM Gene Deletion (11q22) 207
10.10 Case Study 10.10: Leukemic Mantle Cell Lymphoma with Rearranged IgH gene, t(11;14) and Trisomy 12 208

10.11 Case Study 10.11: Chronic Lymphocytic Leukemia with Trisomy 12 and RB1 (13q14) Gene Deletion 210
10.12 Case Study 10.12: Non-Hodgkin's lymphoma (Follicular Lymphoma)—Double-Hit Extra Copy Lymphoma (Atypical Double-Hit Lymphoma) 211
10.13 Case Study 10.13: Non-Hodgkin's Lymphoma—Double-Hit Diffuse Large B Cell Lymphoma 214
10.14 Case Study 10.14: Non-Hodgkin's Lymphoma—Splenic Marginal Zone Lymphoma with Extra Copies of MYC gene (NOS) 216
10.15 Case Study 10.15: Non-Hodgkin's Lymphoma—Anaplastic Large Cells Lymphoma 218
10.16 Case Study 10.16: Non-Hodgkin's Lymphoma—Mantle Cell Lymphoma with t(11;14). 219
10.17 Case Study 10.17: Non-Hodgkin's Lymphoma—High-Grade B Cell Lymphoma (NOS) with BCL-2 Extra Copies. 221
Annexure. 222
References. 223

Part III Case Studies in Multiple Myeloma

11 Multiple Myeloma. 227
11.1 Case Study 11.1: Multiple Myeloma with IgH Gene Rearrangement, t(11;14) 227
11.2 Case Study 11.2: Multiple Myeloma with Partial Deletion of IgH Gene Rearrangement, Monosomy Chromosome 13, Trisomy Chromosome 11 and Gain of Chromosome 1q 228
11.3 Case Study 11.3: Multiple Myeloma with IgH Gene Rearrangement with t(14;16) and Monosomies of Chromosomes 13 and 16 230
11.4 Case Study 11.4: Multiple Myeloma with Biallelic IgH Gene Rearrangement, t(4;14). 231
11.5 Case Study 11.5: Multiple Myeloma with del(17) (p13.1) and Trisomy of Chromosome 11 233
11.6 Case Study 11.6: Multiple Myeloma with Trisomy Chromosome 17 and 1q Amplification 234
11.7 Case Study 11.7: Multiple Myeloma (Plasmablastic) with TP53 Gene Deletion and Tetrasomies of Chromosomes 4, 11, 13, 14, and 1 235
11.8 Case Study 11.8: Double-Hit Multiple Myeloma with TP53 Gene Deletion, Monosomy 13 and Gain of 1 q 238

11.9 Case Study 11.9: Triple-Hit Multiple Myeloma with TP53 Gene Deletion, t(4;14) with Gain of 1q and Deletion 1p 240
11.10 Case Study 11.10: Triple-Hit Multiple Myeloma with IgH Gene Rearrangement t(4;14), TP53 Gene Deletion and Monosomy Chromosome 17, Monosomy 13 and Gain of 1q 242
Annexure..... 245
References..... 245

About the Author

Manorama Bhargava has served as a Professor and Head in the Department of Hematology at the prestigious All India Institute of Medical Sciences until September 2001, Sir Ganga Ram Hospital till 2016, and as Senior Advisor and Head of Hematopathology at Medanta, The Medicity until May 2020.

Dr. Bhargava is a pioneer in hematology in India and is widely recognized for her contributions to hematology both nationally and internationally. In a career spanning almost 50 years, she advocated and brought awareness for hematology as a discipline separate from internal medicine and pathology where it constituted a minor part. She has been the recipient of prestigious fellowships such as Alexander Von Humboldt, Fogarty, and INSA fellowships. She has been the principal investigator on several research grants, notably Indo-US, Indo-French, Indo-German, and Indo-Japan grants, and has published more than 300 articles in national and international journals.

In recognition of her contributions to hematology, she was awarded the Dr. B.C. Roy National Award and the highest award of the Indian Council of Medical Research, the Basanti Devi Amirchand prize. An acclaimed teacher, she established the first -ever superspeciality degrees of DM in Clinical Hematology and DM in Hematopathology in the year 2000 at the All India Institute of Medical Sciences and the Diplomate National Board in Hematology in 2008.

Abbreviations

ALCL	Anaplastic Large Cell Lymphoma
ALL	Acute Lymphoblastic Leukemia
AML	Acute Myeloid Leukemia
AML-MRC	Acute Myeloid Leukemia with Myelodysplasia-Related Changes
ANC	Absolute Neutrophil Count
APML	Acute Promyelocytic Leukemia
BM	Bone Marrow
CBC	Complete Blood Count
Chr	Chromosome
CLL	Chronic Lymphocytic Leukemia
CLPD	Chronic Lymphoproliferative Disease
CML	Chronic Myeloid Leukemia
CML-BC	Chronic Myeloid Leukemia—Blast Crisis
CML-CP	Chronic Myeloid Leukemia—Chronic Phase
CMML	Chronic Myelomonocytic Leukemia
COO	Cell of Origin
DFS	Disease-Free Survival
DLBCLL	Diffuse Large B cell Lymphoblastic lymphoma
DLC	Differential Leucocyte count
ET	Essential Thrombocythemia
FCM	Flow cytometry
Hb	Hemoglobin
HGBL	High Grade B Cell Lymphoma
IHC	Immunohistochemistry
MCL	Mantle Cell Lymphoma
MDS	Myelodysplastic Syndrome
MGUS	Monoclonal Gammopathy of Unspecified Significance
MK	Monosomal Karyotype
MM	Multiple Myeloma
MPN	Myeloproliferative Neoplasm
MRD	Minimal Residual Disease
NDMM	Newly Diagnosed Multiple Myeloma
NGS	Next Generation Sequencing
NHL	Non-Hodgkin's Lymphoma
OS	Overall Survival
PBS	Peripheral Blood Smear

PCR	Polymerase Chain Reaction
PMF	Primary Myelofibrosis
PV	Polycythemia Vera
SLL	Small Lymphocytic Lymphoma
SMM	Smoldering Multiple Myeloma
SMZL	Splenic Marginal Zone Lymphoma
SVT	Splanchnic Vein Thrombosis
TIFT	Time to First Treatment
TLC	Total Leucocyte Count
TTP	Time to Progression
WHO	World Health Organization

1 Techniques in Cytogenetics

Cytogenetic studies have become a critical factor in the workup of hematologic neoplasms. They provide invaluable information for diagnosis, prognosis, and clinical decision-making for the disease.

This chapter describes in detail the infrastructure required for a cytogenetics laboratory, the equipment, the kits and reagents, probes, and procedures based on standard operating procedures used universally. It is aimed at familiarizing the reader with the principles involved, the terminology employed and the multiple steps that need to be followed sequentially to analyze the preparation for interpretation of karyotyping and FISH. Both karyotyping and FISH have been used extensively in the following case studies in the book.

The chapter is presented in a format that facilitates the establishment of a cytogenetics laboratory if required/desired.

The following figures (Figs. 1.1–1.35) give in a sequential manner the theoretical background for karyotyping and FISH together with the steps involved for both.

M. Bhargava, *Hematologic Malignancies*, https://doi.org/10.1007/978-981-33-4799-1_1

Cytogenetics can be classified in two categories:

- Conventional cytogenetics: Karyotyping
- Molecular cytogenetics: FISH, CGH, M-FISH, Spectral karyotyping, WCP.

Fig. 1.1 Shows the two types of cytogenetic techniques commonly used for analysis

Overview

- Specimen collection and handling
- Culture initiation
- Culture maintenance
- Cell Harvest

Fig. 1.2 Provides an overview of the steps to be followed beginning from sample collection of blood and bone marrow material

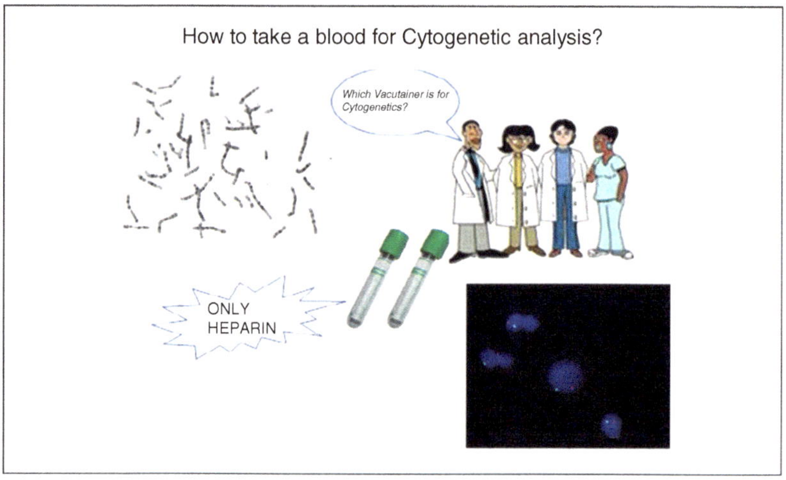

Fig. 1.3 Shows graphically the importance of the collection of samples only in heparin a prerequisite for further processing

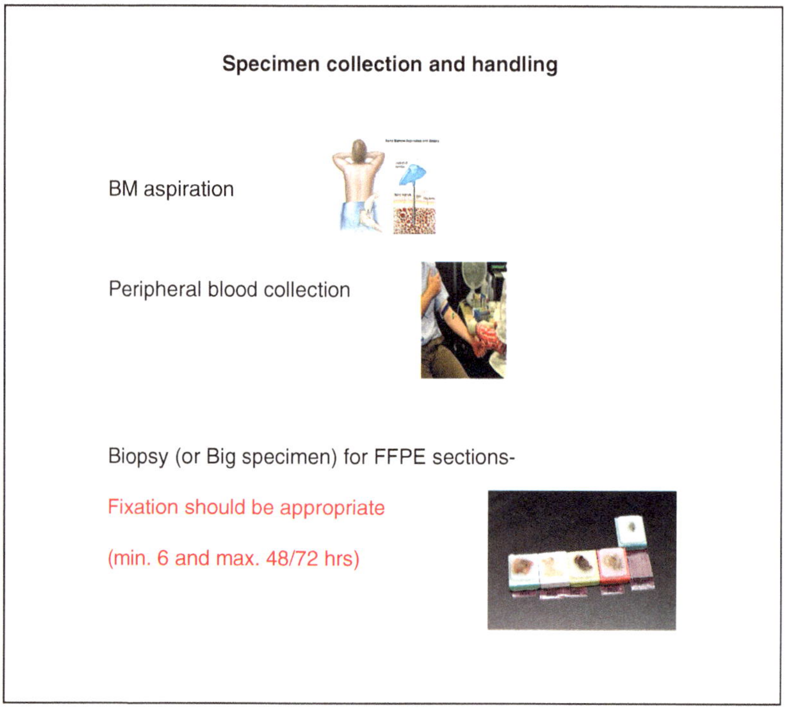

Fig. 1.4 and 1.5 Signify the importance of handling of samples to avoid errors, a commonly faced problem in laboratories. It includes handling of FFPE tissues for FISH

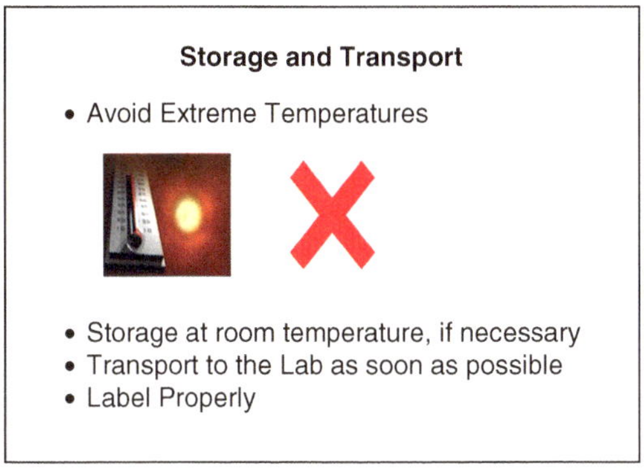

Fig. 1.4 and 1.5 (continued)

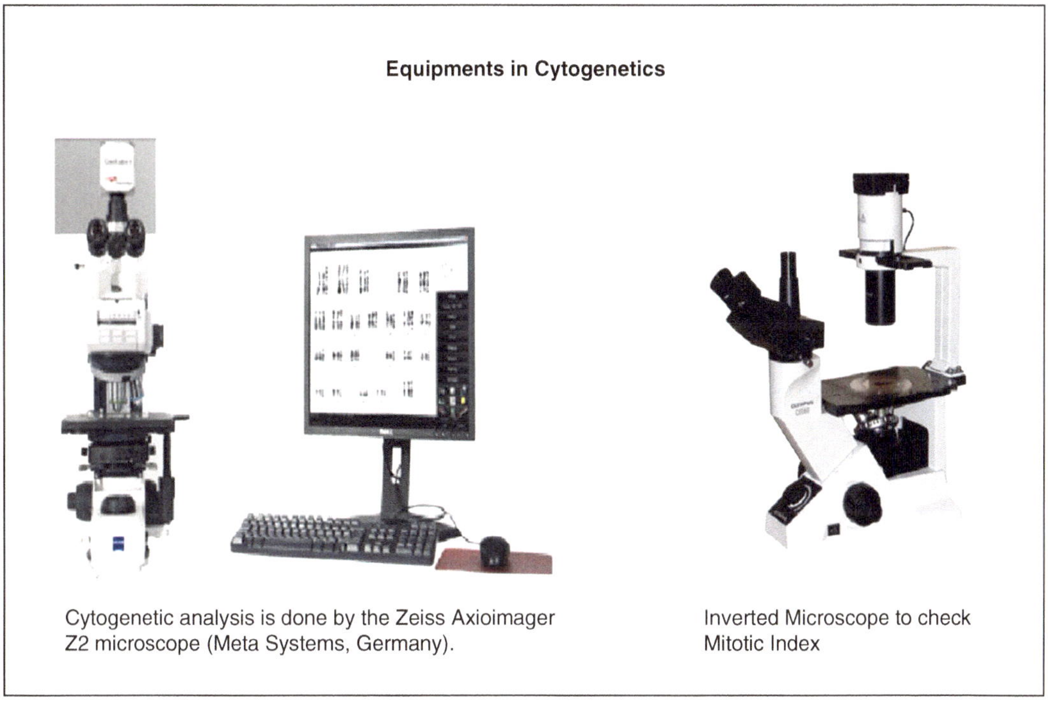

Fig. 1.6 and 1.7 Show the main equipment required for analysis. The main stay of analysis is the microscope on which depends the quality of images obtained for karyotyping and FISH analysis

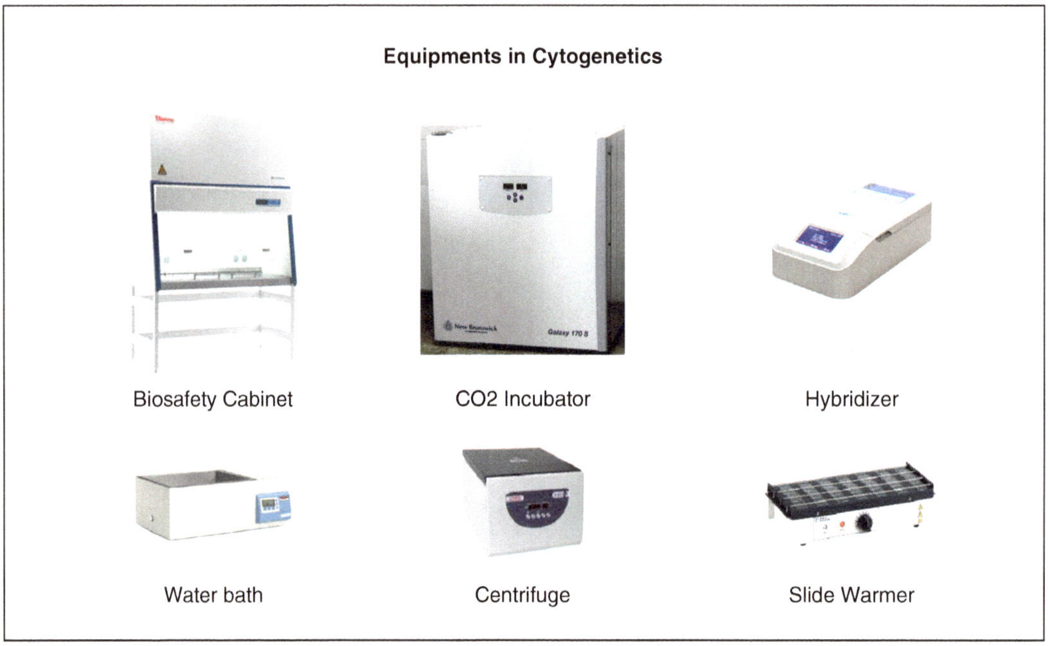

Fig. 1.6 and 1.7 (continued)

Fig. 1.8, 1.9, and 1.10 Show the various types of culture vessels used, the maintenance of the culture, and culture initiation

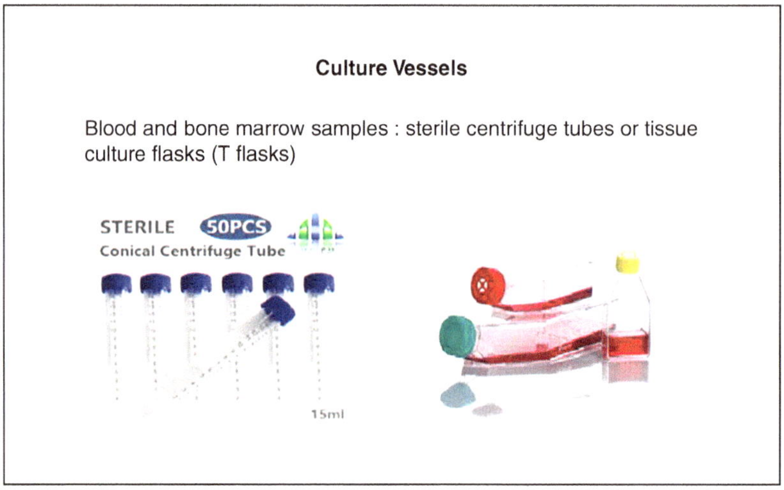

Culture Maintenance and Growth Interval

Temp: 37°C, humidity: 97%, pH: 7.2-7.4

CO_2 incubator: exchange of gas

Peripheral Blood: 69-72 hours

Bone Marrow: 16-18 hours

Culture Initiation

Growth Media : Balanced salt solutions : salts, glucose, buffering system, microbial inhibitors, FBS

25% Culture medium (RPMI 1640 + FBS)

Mitotic stimulants: PHA stimulates T lymphocytes; LPS, IL2 : B lymphoid cells

Fig. 1.8, 1.9, and 1.10 (continued)

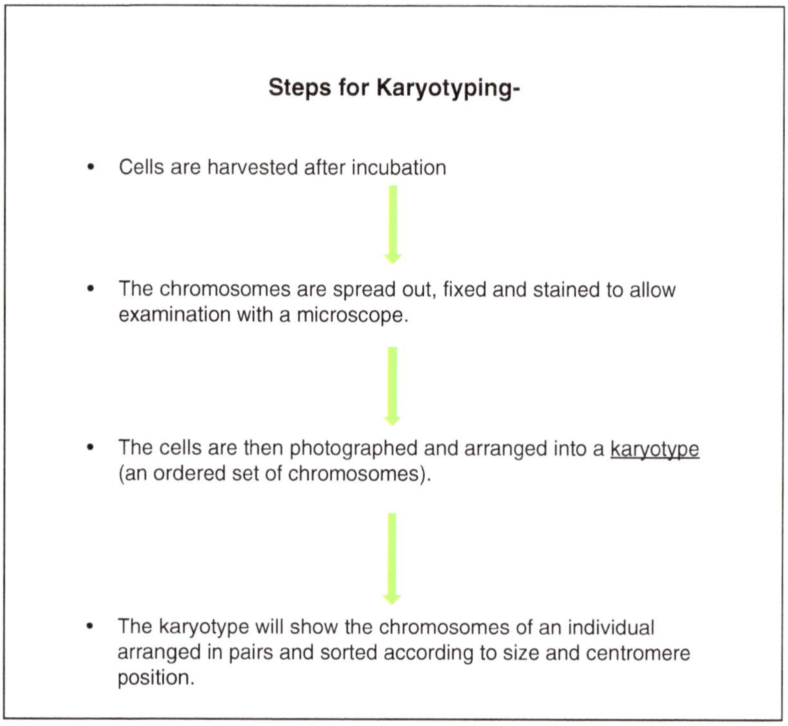

Fig. 1.11 Shows the actual steps for karyotyping to obtain a karyotype based on the arrangement of chromosomes

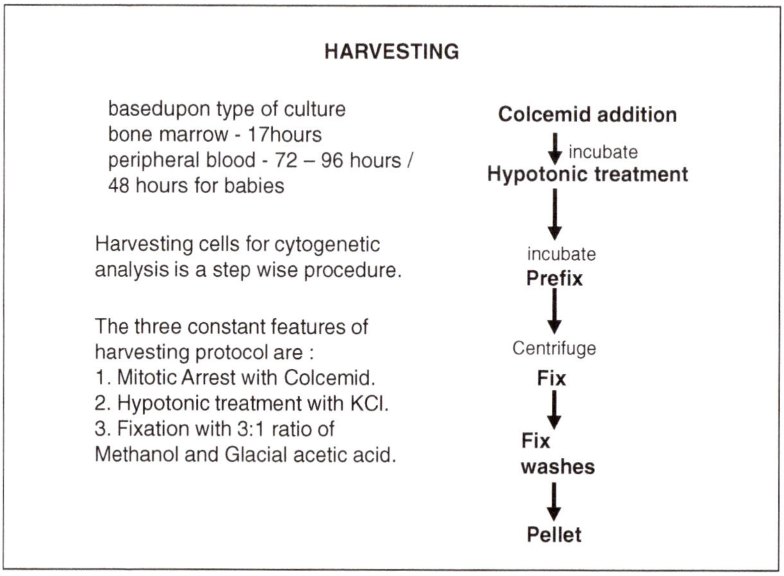

Fig. 1.12 Shows the details of harvesting of the cells after incubation prior to preparing a karyotype

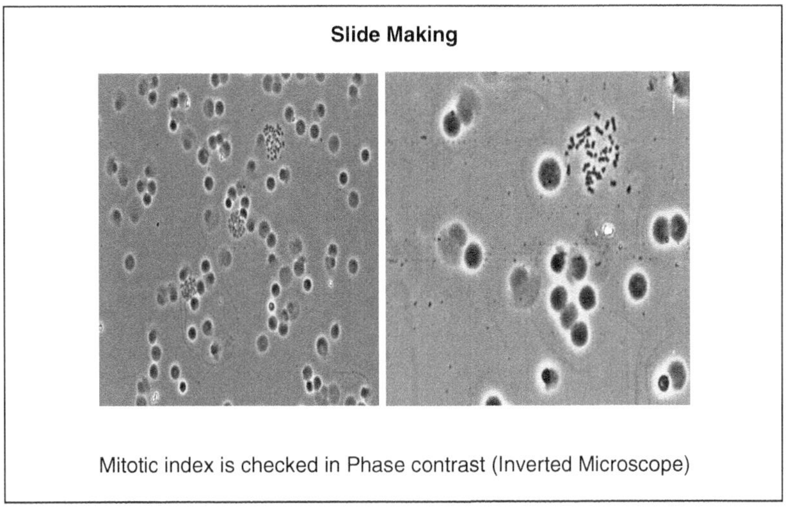

Fig. 1.13 Shows the checking of the mitotic index on slides in a phase contrast (inverter and microscope)

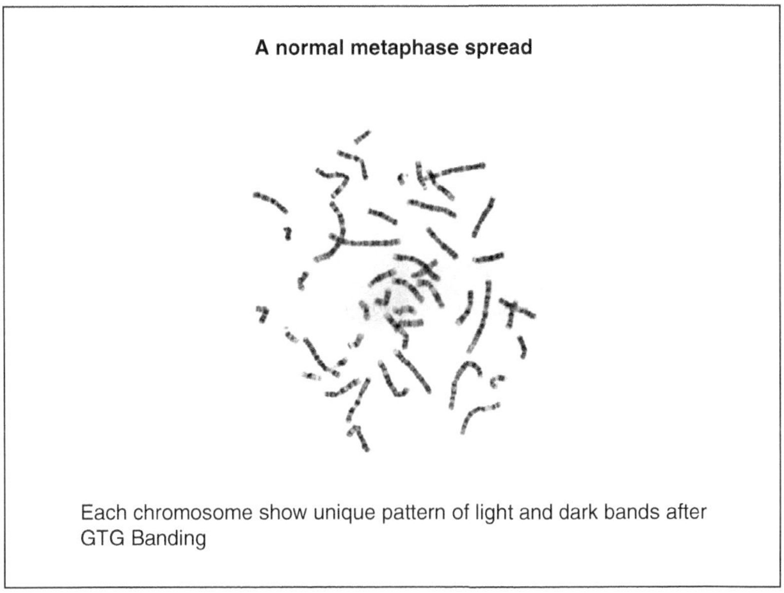

Fig. 1.14 Shows how the chromosomes appear after GTG banding

Fig. 1.15 Chromosome analysis is based on the actual structure of a chromosome as seen in this figure

Techniques of Analysis

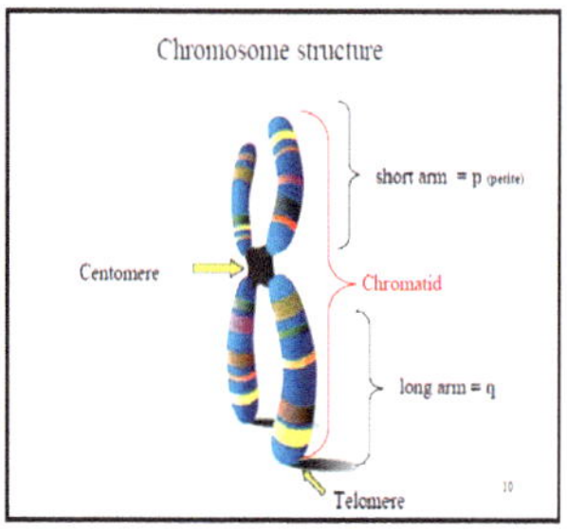

- Routine chromosome analysis refers to analysis of chromosomes which have been banded using trypsin (a serine proteases) followed by Giemsa, Leishmanns, or a mixture of the two.
- This creates a unique banding pattern on the chromosomes.
- Generally 20 cells are analysed to rule out chromosome abnormalities.

Chromosome shapes

Chromosomes are said to be **Metacentric** (centromere in the middle with equal length long and short arms
Submetacentric (centromere placed above the middle with a short arm and a long arm
Acrocentric (centromere almost at the end of the chromosome with short arms and usually with satellites

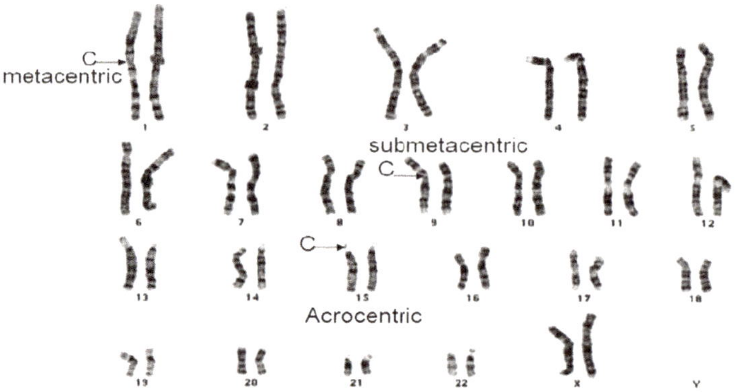

Fig. 1.16 Describes the various chromosome shapes based on where the centromere is placed in a chromosome. Therefore, terms such as metacentric, submetacentric, and acrocentric are used

ISCN

- del - deletion
- dic - dicentric
- fra - fragile site
- i - isochromosome
- inv - inversion
- p - short arm
- r - ring

- der - derivative
- dup - duplication
- h - heterochromatin
- ins - insertion
- mat - maternal origin
- q - long arm
- t - translocation

- Normal male
  - 46,XY
- Normal female
  - 46,XX

Fig. 1.17 Gives the ISCN terminology followed universally for expressing the alterations seen in chromosome number and structure on analysis of a karyotype

Types of chromosome abnormalities

- Numerical
  - Aneuploidy (monosomy, trisomy, tetrasomy)
  - Polyploidy (triploidy, tetraploidy)
- Structural
  - Translocations
  - Inversions
  - Insertions
  - Deletions
  - Rings
  - Isochromosomes

Fig. 1.18 Shows a description of chromosome abnormalities, both numerical and structural. On this is based the entire interpretation of abnormalities seen in a given hematologic disease

Philadelphia Chromosome (Ph)

CML is an acquired cytogenetic abnormality that is characterized by the presence of the Philadelphia Chromosome (Ph)

The Ph chromosome is a result of an exchange of material (translocation) between the long arms of chromosomes 9 and 22 e.g. t(9;22)

- This translocation brings together the BCR gene on chromosome 22 and the ABL gene on chromosome 9
- The resulting hybrid gene BCR/ABL causes uncontrolled cell growth

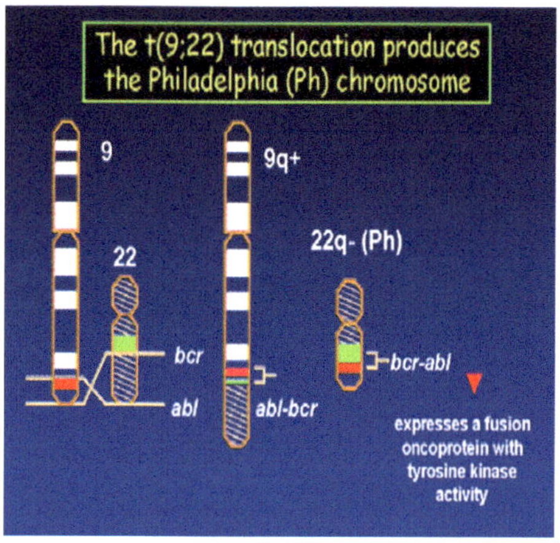

Fig. 1.19 Shows a classic Philadelphia chromosome-positive image in chronic myeloid leukemia and its description

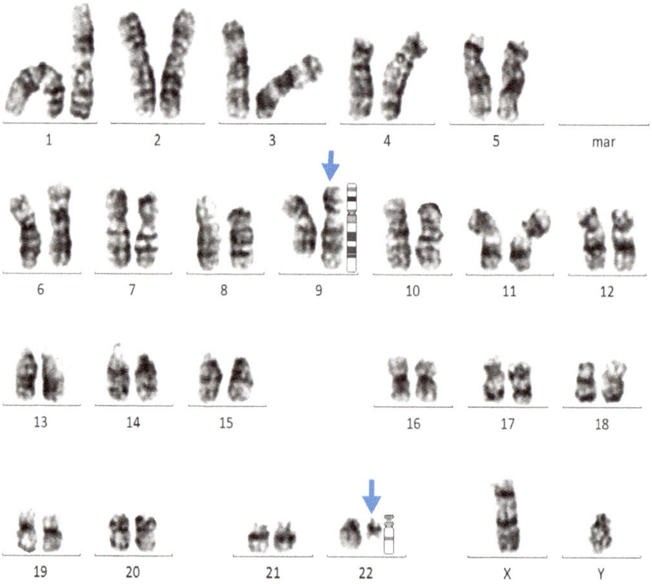

Karyotype-46,XY,t(9;22)(q34;q11.2)

Fig. 1.20 Shows the karyotype seen in a case of chronic myeloid leukemia

Fig. 1.21 An isochrome is seen in certain instances of chromosomal abnormalities. It shows the picture of isochrome 17, i(17q)

Isochromosome

- Loss of a complete arm, "replaced" by the duplication of the other arm.
- Recorded as i, followed by a bracket with the number of the chromosome and the arm (e.gi(17q) or i(17)(q10): duplication of the q arm loss of the p arm)

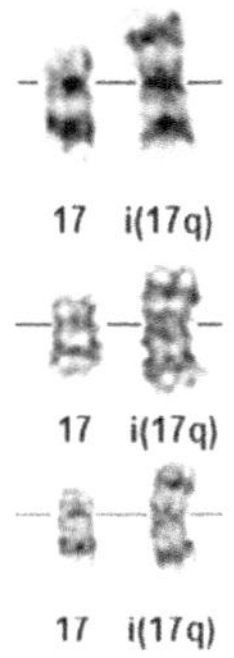

Fig. 1.22 Shows the variations in chromosome number which carry an important diagnostic significance apart from structural abnormalities

Variations in Chromosomal Number

- Euploidy – the normal number and sets of chromosomes
- Polyploidy – the presence of three or more complete sets of chromosomes
- Aneuploidy – the presence of additional or missing individual chromosomes

- **Types of Polyploidy**
- Triploidy – three sets of chromosomes 23 x 3 = 69
- Tetraploidy – four sets of chromosomes 23 x 4 = 92
- **Types of Aneuploidy**
- Monosomy – one less chromosome (23 x 2) –1 = 45
- Trisomy – one additional chromosome (23 x 2) + 1 = 47

Fig. 1.23 Shows the principles of the FISH procedure used either independently or together with karyotyping to localize the presence or absence of specific DNA sequences on chromosomes employing sequence-specific probes

Fluorescent *in situ* hybridization (FISH)

- Fluorescent *in situ* hybridization (FISH) is a cytogenetic technique which can be used to detect and localize the presence or absence of specific DNA sequences on chromosomes.
- FISH uses fluorescent probes which bind only to those parts of the chromosome with which they show a high degree of sequence similarity.
- Generally between 200 cells are counted.

Fig. 1.24, 1.25, 1.26, and 1.27 It give the details of the various steps required for completing the FISH procedure

FISH procedure

- Sample collection - Na-Heparin vaccutainers
- Planting
- Harvesting - Hypotonic and Fixation
- Slide preparation
- Ageing and Pretreatment

- The first step is to break apart (denature) the double strands of DNA in both the probe DNA and the chr DNA so they can bind to each other.
- This is done by heating the DNA.

- Next, the probe is placed on the slide and a glass coverslipis placed on top.
- To prevent evaporation, the edges of the coverslipare sealed with rubber cement.
- The slide is then placed onto "Hybridizer" for overnight for the probe to hybridize with the target chromosome.
- Overnight, the probe DNA seeks out its target sequence on the specific chromosome and binds to it. The strands slowly rejoin (re-anneal).

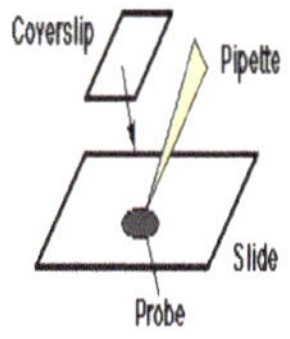

Hybridization

- Nucleic acid hybridization is the formation of a duplex between two complementary sequences.
- Annealing is another term used.
- Bound sequences are available for hybridization with an added nucleic acid (probe).
- Post-hybridisation washing and observe under fluorescence microscope.

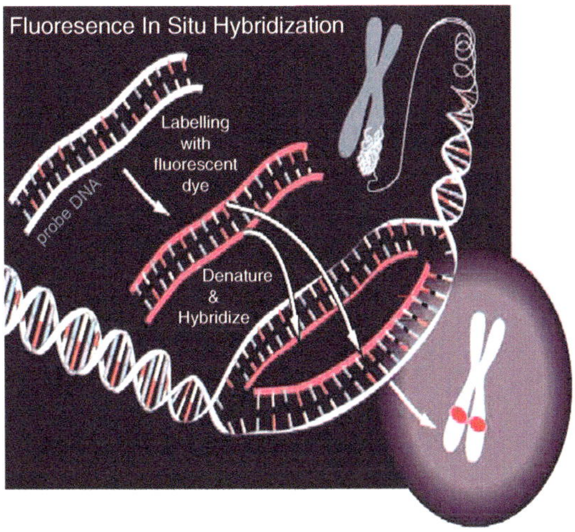

Fig. 1.28 It is a pictorial depiction of the FISH procedure

FISH Probes

- Probe is a segment of DNA that:
  - is labeled with a marker which allows identification and quantitation
  - will hybridize to another nucleic acid on the basis of base complementarity
- Types of labels:
  - Fluorescent -FISH (fluorescence in situ hybridization)
  - Radioactive (^{32}P, ^{35}S, ^{14}C, ^{3}H)
  - Biotinylated (avidin-streptavidin)

  There are 2 basic types of FISH probes:

  1. Chromosome Enumeration Probes (CEP)-CEP probes attach to the centromeres of chromosome and are used to determine how many copies of a chromosome are present in a cell.
  2. Locus Specific Identifier (LSI) probes are designed to attach to genes of interest, for example the HER2 oncogene, and are used to determine if a cell shows deletion or amplification of those genes.

Fig. 1.29 It provides an extensive description of the FISH probes used. A probe is a segment of DNA labeled with a marker for identification which hybridizes to another nucleic acid based on base complementarity. Both Chromosome Enumeration probes (CEP) and Locus Specific Identifier (LSI) probes are used for specific purposes of determining the copy number of chromosomes (CEP) or deletion or amplification of genes (LSI)

General cut off for FISH positivity is >5%.

Fusion probesare used to detect common translocations involving gene regions which, when rearranged and joined, result in leukemia and/or cancer.

1. Dual fusion probes (LSI probes)

The two genes involved in the translocation are labeled in different colors (ex: the ABL gene on chromosome 9 is red and the BCR gene on chromosome 22 is green), a nearby gene on each chromosome is also labeled in the same color.

In normal cells we see two signals for each color (cell on the left).

The piece of chromosome 9 moves (translocates) to the chromosome 22, the piece of chromosome 22 moves to the chromosome 9; one chromosome has a fusion of the ABL and BCR genes, the other chromosome has a fusion of the two nearby genes (two red-green fused signals, cells on the right).

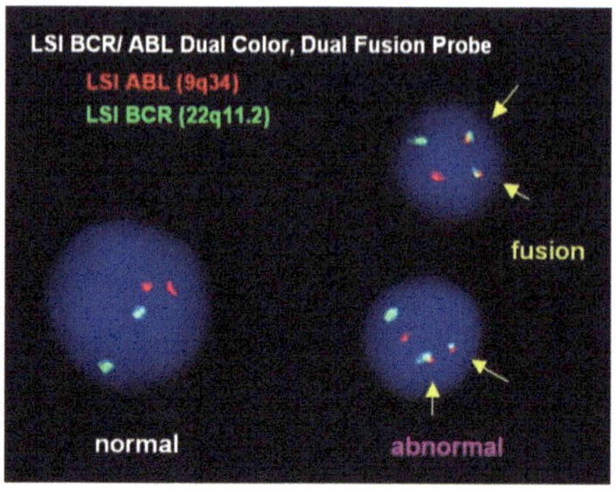

Fig. 1.30 This figure details the information on Fusion probes. Dual color Fusion probes. (LSI) is shown here to depict BCR-ABL fusion, and how a translocation t(9;22) takes place and appears on FISH

2. Break Apart (LSI Probe)

Some genes are involved in many different translocations associated with cancer, so instead of using a specific dual fusion probe to test for each possible translocation, a break apart probe is used to test whether the gene of interest is involved in a rearrangement. One end of the gene is labeled red and the other end of the gene is labeled green. Normal copies of the gene are seen as a red-green fused signal (cell on the right). If there is a translocation or rearrangement, the gene breaks into separate red and green signals.

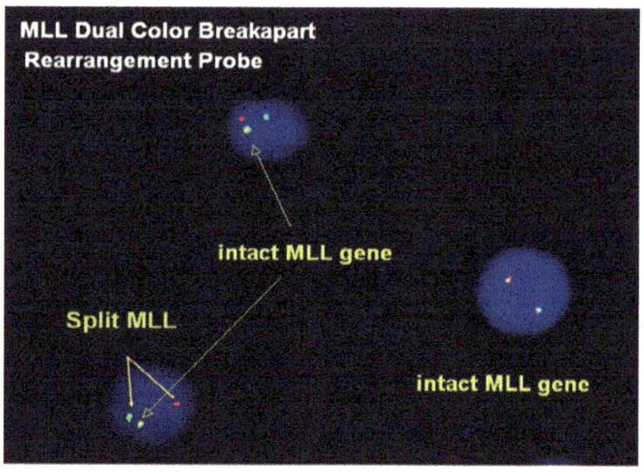

Fig. 1.31 Shows an example of a Break Apart Probe (LSI) in MLL gene rearrangement. These probes are used to identify whether the gene of interest is involved in a rearrangement. One end of the gene is labeled red and the other end is labeled green normal copies are seen as red green fused signal and separate red green signals are seen if there is translocation or rearrangement. It shows an MML gene rearrangement

Fig. 1.32 Shows the ISCN FISH nomenclature used universally expressed as nuc ish to explain the location and number of signals seen in standard 200 interphases nuclei on FISH

ISCN FISH nomenclature

To indicate the number of signals, the symbols nuc ish are followed immediately in parentheses by the location designation, a multiplication sign (x), and the number of signals seen.

nuc ish (DXZ1 x 2)[200]

Two copies of the locus DXZ1 alpha satellite probe on the X chromosome, observed in 200 interphase nuclei.

Minimum of 200 nucleus is counted/examined.

Fig. 1.33 Shows images in CLL using common probes for identification of gene deletions and number. It shows the most important gene deletions of RB1, TP53, and ATM as well as trisomy 12 (CLL images)

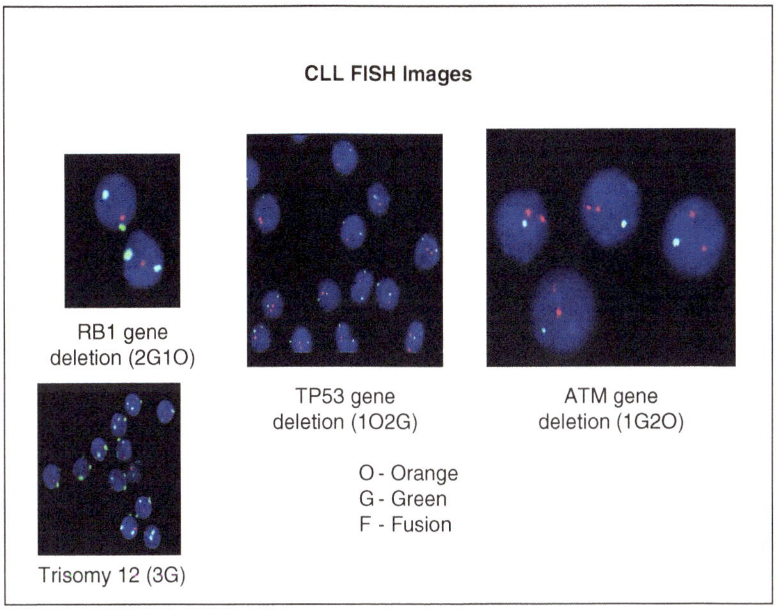

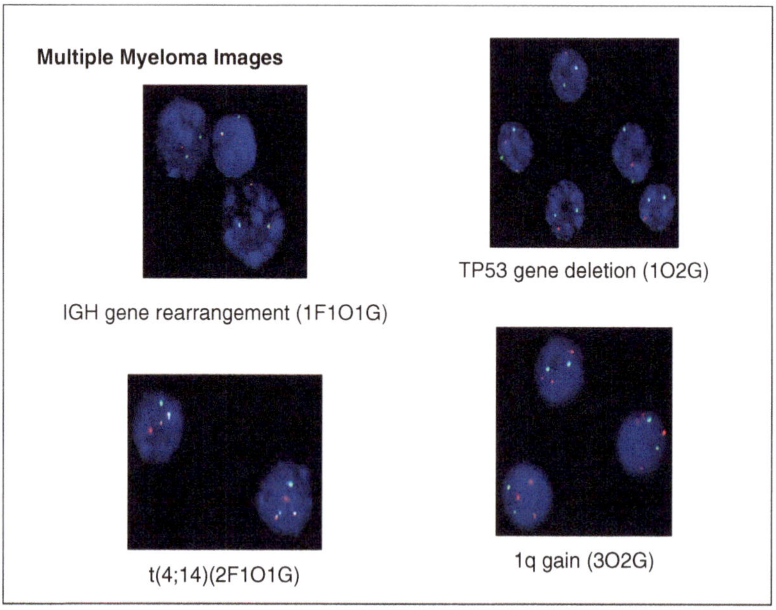

Fig. 1.34 Similar depiction through images is seen in multiple myeloma showing some of the most important gene rearrangements, gene deletion, translocation, and gains. MM images

List of Probes available in Cytogenetics

Name of Probe	Probe design
RUNX1/RUNX1T1, t(8;21)	Dual color dual fusion
BCR/ABL1, t(9;22)	Dual color dual fusion
PML/RARA, t(15;17)	Dual color dual fusion
ETV6/RUNX1, t(12;21)	Dual color dual fusion
CBFB	Dual Color breakapart
CEP17/TP53	Dual color
Inv(16)	Dual Color breakapart
IgH(14q32) GENE REARRANGEMENT	Dual Color breakapart
MYEOV/IGH, t(11;14)	Dual color dual fusion
FGFR3/IGH, t(4;14)	Dual color dual fusion
IGH/MAF, t(14;16)	Dual color dual fusion
CDKN2C/CKS1B, 1p/1q	Dual color
DLEU/LAMP/12cen	Tri color
ATM/TP53	Dual color
MYC	Dual Color breakapart
BCL2	Dual Color breakapart
BCL6	Dual Color breakapart

Fig. 1.35 Shows the list of probes that are used in most laboratories for FISH in a spectrum of hematologic malignancies. For case studies presented here most of the probes were from Zytovision, GMBH, Germany. For Multiple Myeloma, the probes used were from Metasystem, GMBH, Germany

2 Techniques in Molecular Hematology

Molecular diagnostic testing has virtually become a routine part of characterization of hematologic malignancies. In addition to confirming or refining or establishing new diagnoses, it is used extensively to stage patients, to follow them for response to therapy and disease recurrence as well as for predicting prognosis.

The mainstay of a molecular hematology lab is the Real Time PCR (RT-PCR). This chapter describes the principles of PCR, the procedural steps performed sequentially for generating PCR data. Disease directed panels are used to identify the alterations in the genes specific to the disease. In addition, information on equipment, etc. is provided towards infrastructural requirements.

RT-PCR has been used extensively in the analysis of the case studies in this book.

2.1 Polymerase Chain Reaction

Principles: The Polymerase chain reaction is a temperature-dependent process of DNA amplification using a Thermocycler.

The requirements for the procedures are

- A DNA template containing the target region to amplify
- DNA polymerase (Taq polymerase) for polymerizing DNA strands
- Specific primers complementary to the DNA target region
- Deoxynucleotide triphosphates (dNTPs) building blocks used by DNA polymerase to synthesize new strands
- Buffer solution

Procedures

- The reaction is carried out in a volume of 10–200 ml in small reaction tubes (0.2–0.5 ml volume) in the PCR machine (Thermocycler).

The steps are:

- Denaturation: Denaturation of double stranded DNA to two single stranded DNA molecules at 94 °C–98 °C.
- Annealing: Annealing of primers to two single stranded complements containing the target region at 50 °C–65 °C.
- Extension/elongation.
- Synthesis of a new DNA strand by Taq polymerase at 75 °C–80 °C. Complementary to DNA template strand by adding free dNTPs.

This chapter is presented in a format that facilitates the establishment of a molecular hematology laboratory.

The next step is to plan the layout of the plates, panel wise for analysis in the PCR machine which provides copy numbers of the genes and the amplification plots. The following figures provide this information on individual panels.

The next most important and widely used investigation is BCR-ABL International scale.

M. Bhargava, *Hematologic Malignancies*, https://doi.org/10.1007/978-981-33-4799-1_2

List of Equipments

- Real time PCR Machine: Quantsudio 5Dx (Applied Biosystems)
- PCR machine: Veritti (Applied Biosystems)
- Refrigerated Centrifuge (Eppendorf)
- Rotospin (Tarsons)
- Gel Electrophoresis system (Biorad)
- Pipettes (Eppendorf)
- Laminar Flow Hood (Thermo)
- Spinwin (Tarsons)
- Vortex (Tarsons)
- Thermomixer (Eppendorf)
- Deep freezer (Vestfrost)

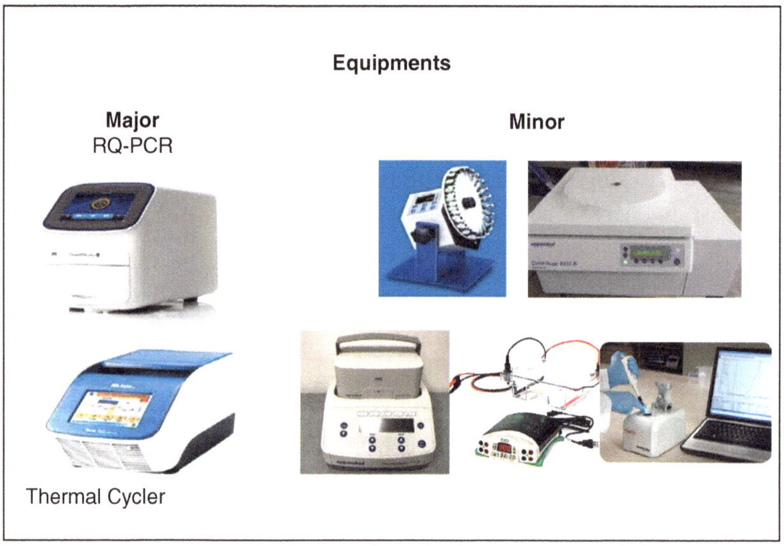

Fig. 2.1 and 2.2 Show the list of equipment, the major equipment of RQPCR together with other smaller but essential instruments

RNA Extraction (Qiagen Kit)

- Take 5ml peripheral blood in EDTA
- Add erythrocyte Lysis buffer and incubate at ice for 10 min and repeat once more
- Pellet down the cells and add lysis buffer to lyse the WBCs
- Pass the lysate through spin columns and wash with 70% ethanol
- Wash the columns with RW1 and RPE wash buffers and finally elute the RNA with nuclease free water
- Quantify the RNA with Nanodrop. A260/A280 should be betwen 1.7-2.0. The required RNA for further Application should be 1000-1500ng/5μL

DNA – Extraction Kit (Qiagen Kit)

- Take 200ul peripheral blood in EDTA
- Add Lysis buffer and proteinase K and incubate at 70°C for 30 min in a dry bath
- Add ethanol and pass the lysate through spin columns
- Wash the columns with wash buffers and finally elute the DNA with elution buffer pre warmed at 70°C
- Quantify the DNA with Nanodrop. A260/A280 should be betwen 1.8-2.0.
- The required DNA for further Application should be 1000-1500ng/μL

Fig. 2.3 and 2.4 Molecular procedures begin with RNA and DNA extraction from a blood/BM sample. It show stepwise procedural details including the kits for the same. All the kits used are TRU-PCR kits

Fig. 2.5 Shows how cDNA is synthesized, the reverse transcription product for further analysis in PCR

cDNA Synthesis

- Take 1ul of RNA in a sterile nuclease free PCR tube.
- Add oligo dT primers followed by reaction buffer, 20U Ribolock RNAse inhibitor, 10mM dNTP and 200U Revert Aid M-MuLV RT.
- Incubate the reaction for 60 min at 42°C.
- Terminate the rection by heating at 70°C for 5 min.
- The reverse transcription reaction product can be directly used in PCR applications or stored at -20°C for less than one week. For longer storage, -70°C is recommended

Fig. 2.6 For the case studies presented in this book test panels were used for analyzing BCR-ABL (qualitative and quantitative), JAK2 mutation, MPN mutations and AML Panel and ALL comprehensive panels. Figure 2.5 gives the details of the genes that these panels are designed to detect

Molecular Investigations

S.NO	TEST / PANEL	DESCRIPTION
1	**BCR-ABL Qualitative**	Major, Minor, Micro
2	**BCR-ABL (IS Scale)**	Major, Minor, Micro
3	**JAK2 Mutation**	V617F
4	**MPN Mutation panel**	BCR-ABL1, JAK2, CALR and MPL
5	**AML Comprehensive panel**	BCR-ABL1, PML-RaRa, AML ETO gene (RUNX1;RUNX1T1), Inv16, FLT3 (ITD/TKD), NPM1, c-KIT, CEBPA
6	**ALL Comprehensive panel**	BCR-ABL1, ETV6-RUNX1(TEL-AML1), MLL-AF4 t(4;11), MLL-AF9 t(9;11), MLL-ENL t(11; 19), TCF3-PBX1 (E2A-PBX1)

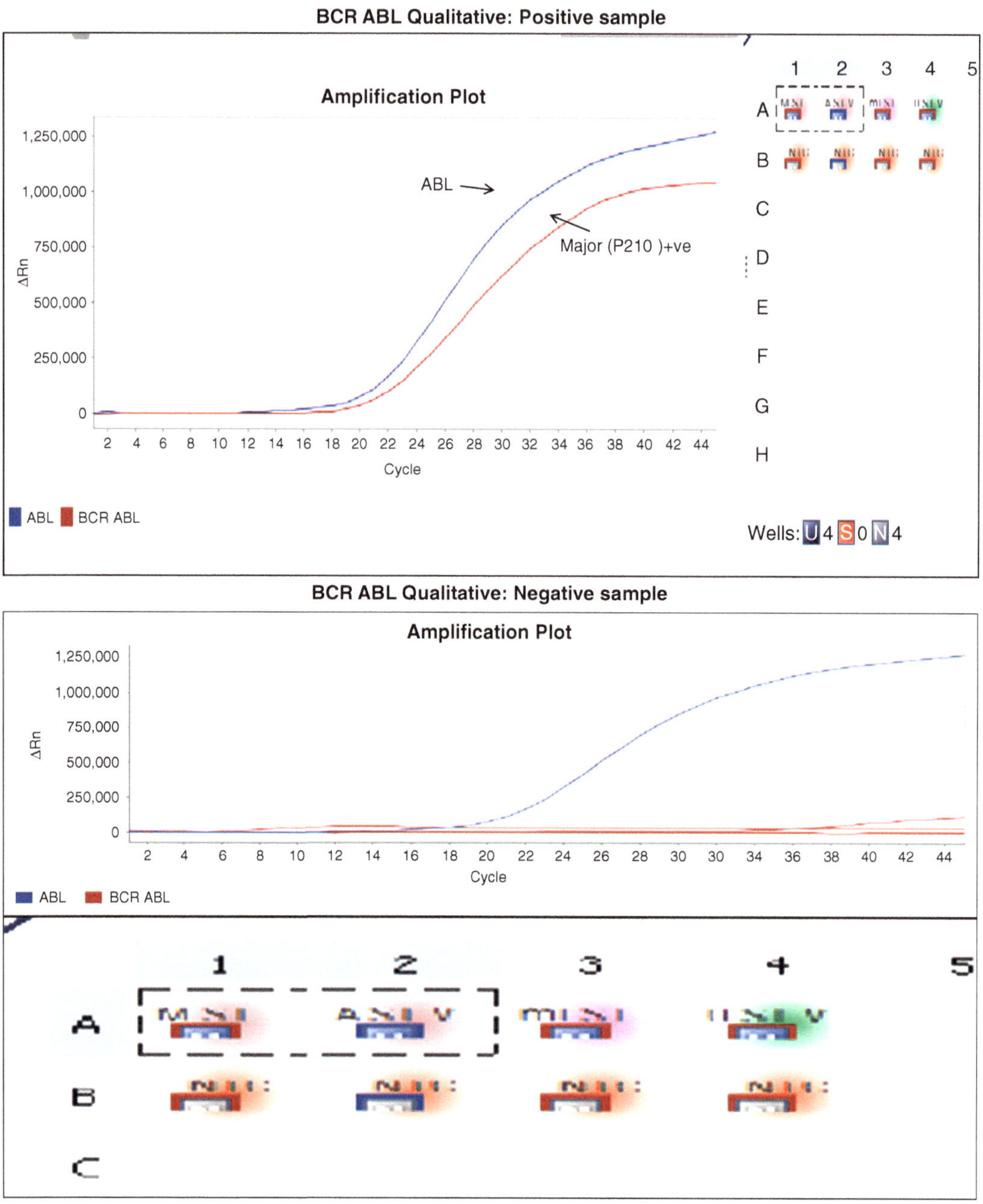

Fig. 2.7 and 2.8 Show the plate layout and amplification plot for BCR-ABL positive and negative samples (qualitative)

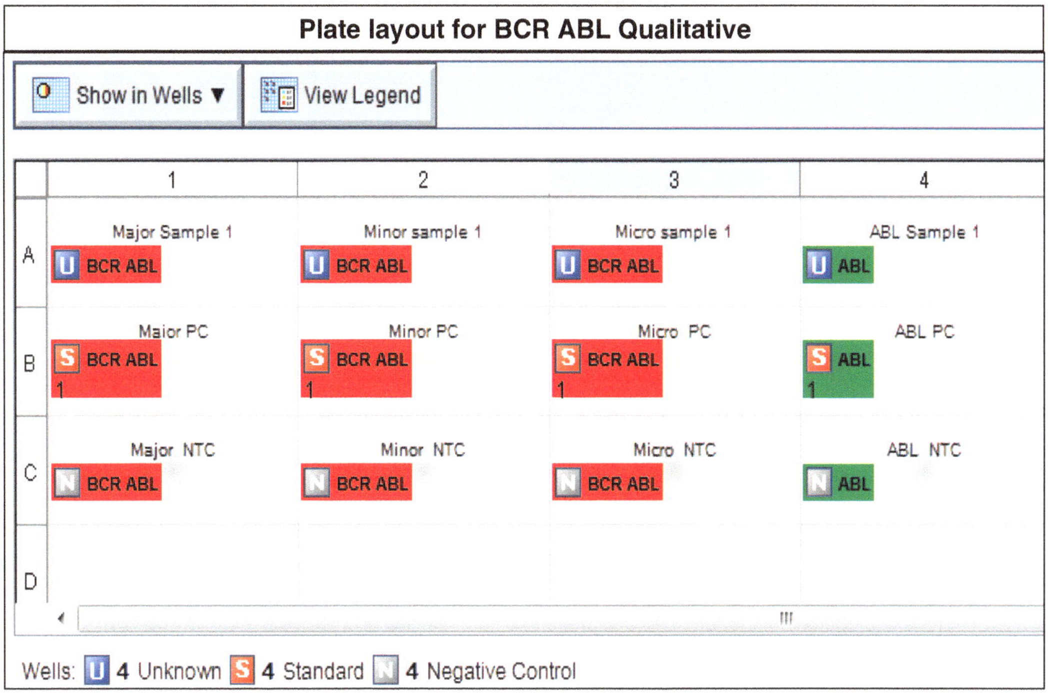

Fig. 2.9 Since all the 3 BCR-ABL forms major, minor and micro are looked for, this provides a plate layout for the same

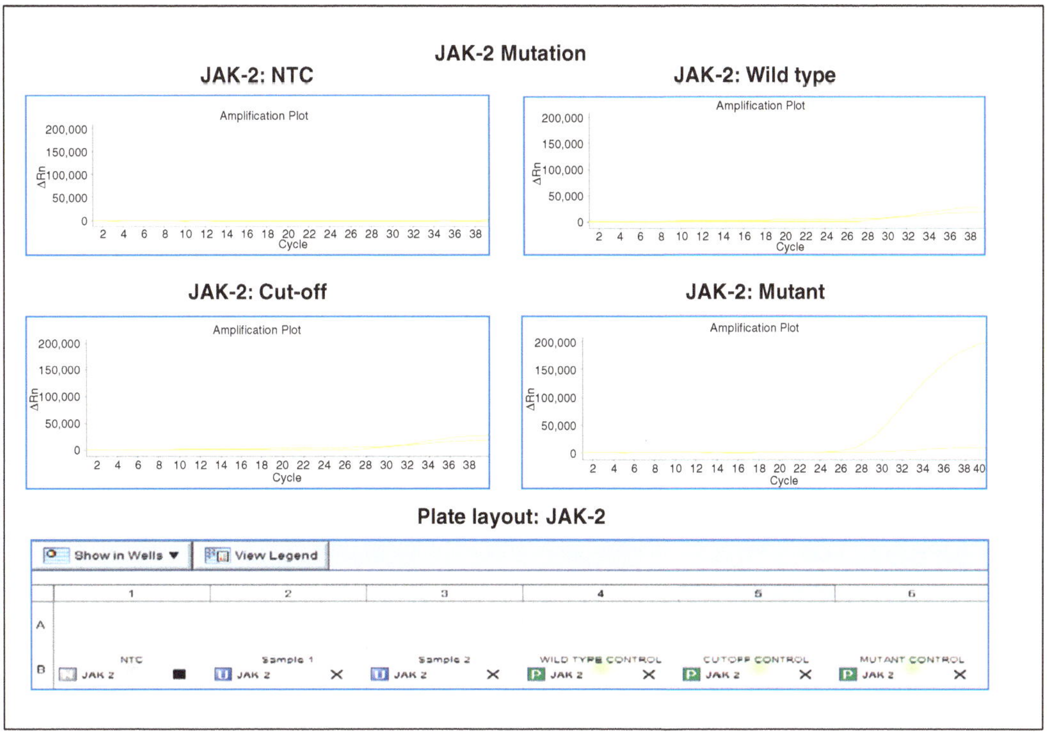

Fig. 2.10 Shows the standardization of JAK2 mutation giving the plate layout and amplification plots for JAK2 wild type and mutant JAK2

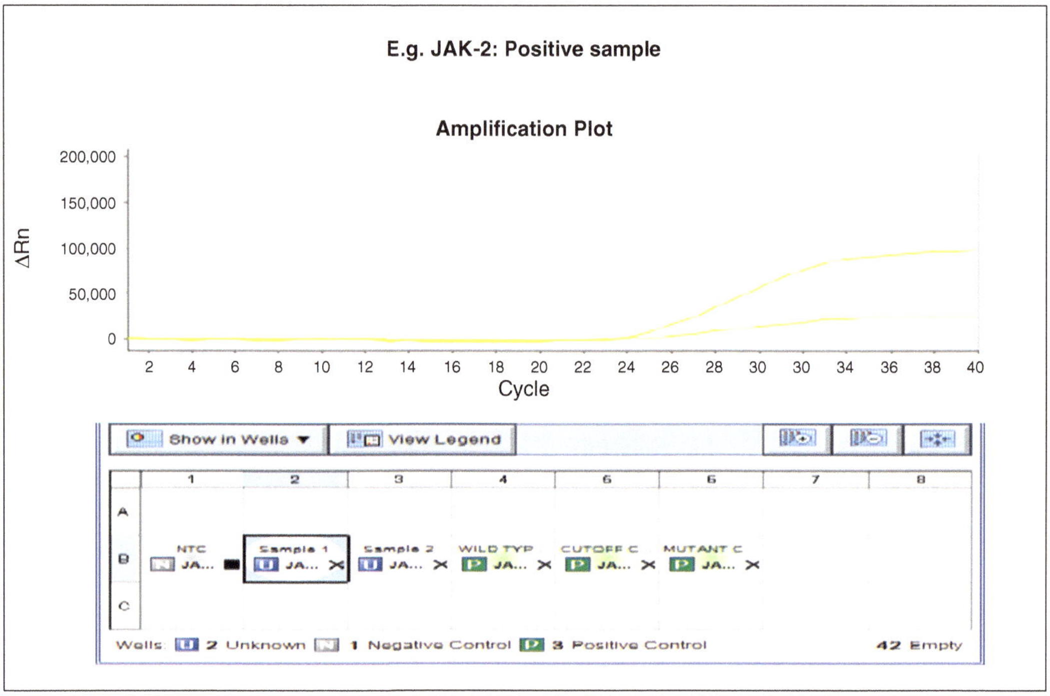

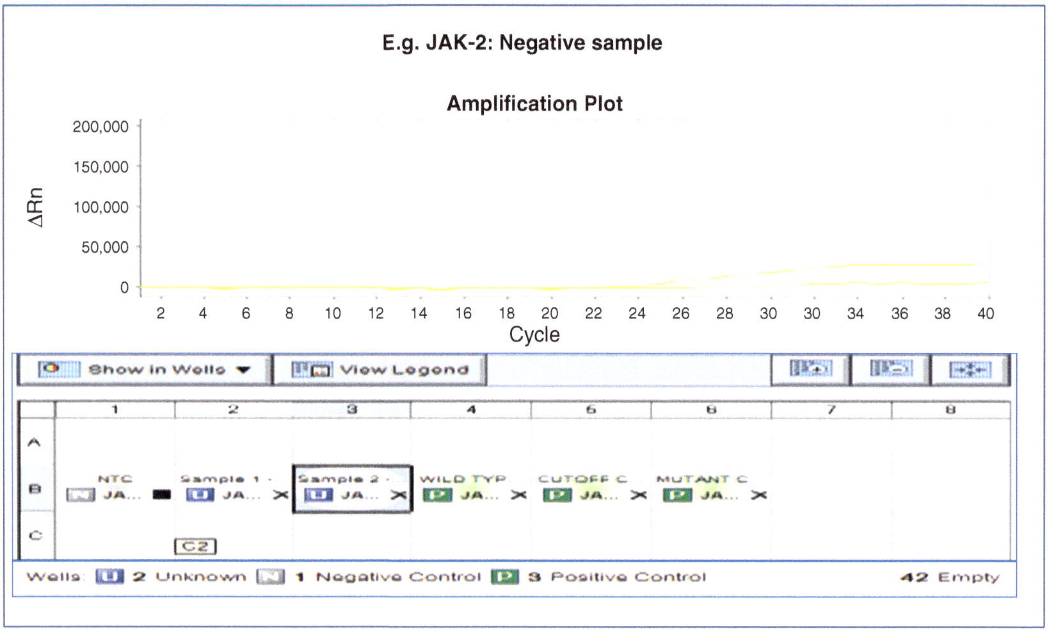

Figs. 2.11 and 2.12 These figures show the amplification plots and plate layouts for JAK2 positive and JAK2 negative samples

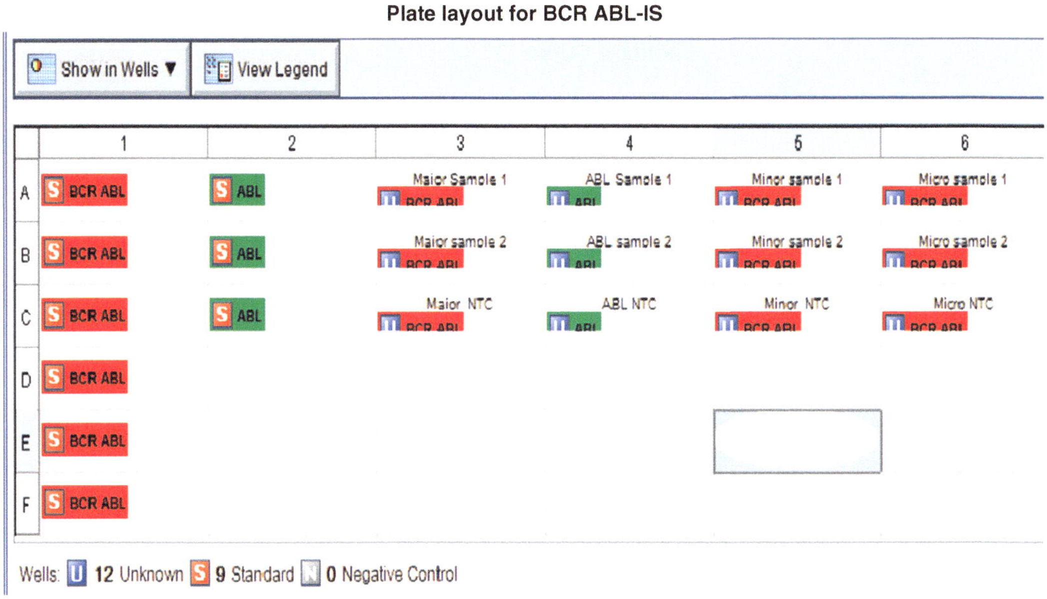

Fig. 2.13 Gives the platelet layout for standards and for testing major, minor and micro BCR-ABL

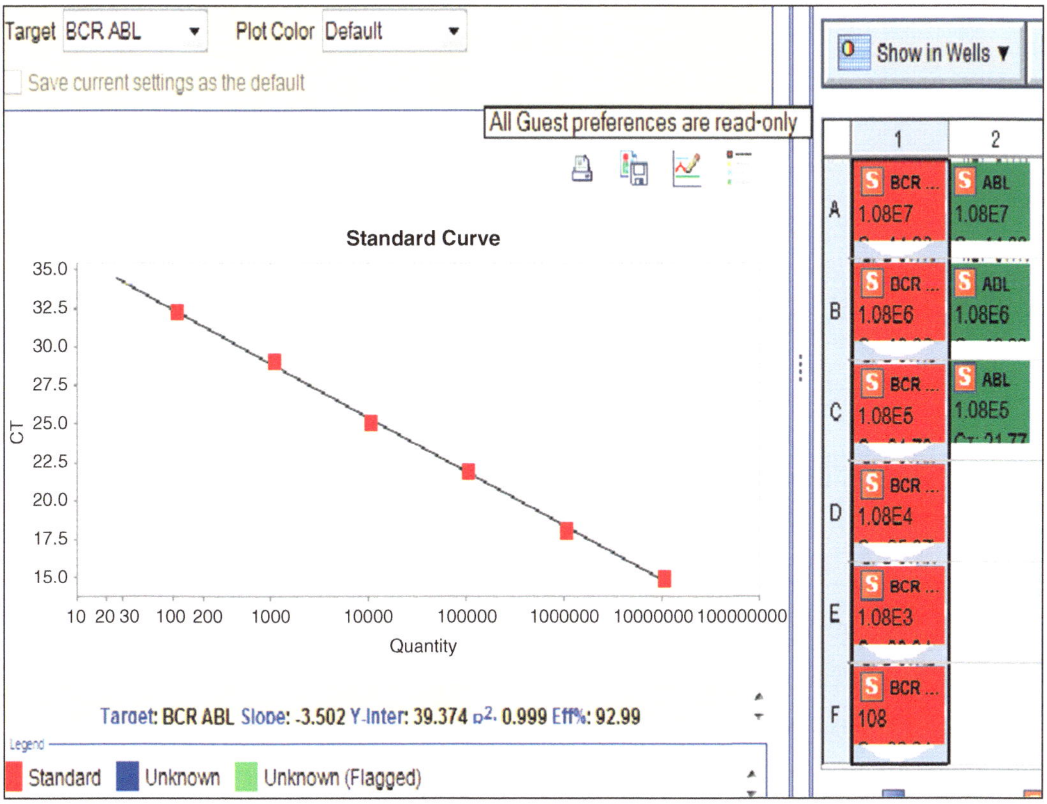

Fig. 2.14 and 2.15 Provide the standard curves for BCR-ABL and ABL

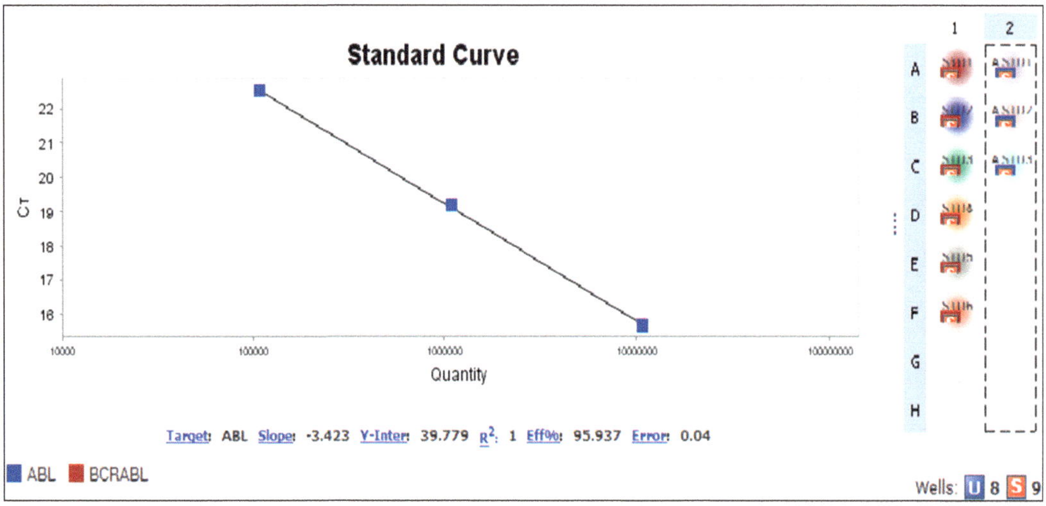

Fig. 2.14 and 2.15 (continued)

Fig. 2.16 and 2.17 Show the amplification plots for the standards of BCR-ABL and ABL and their plate layouts

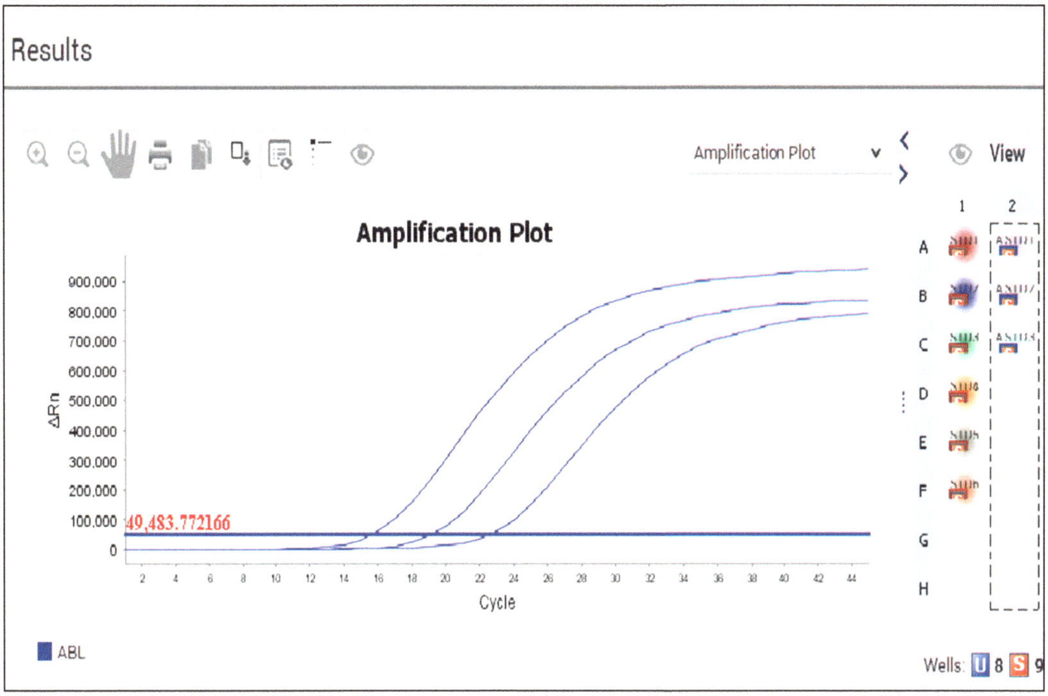

Fig. 2.16 and 2.17 (continued)

Fig. 2.18 and 2.19 Show the results in a BCR-ABL negative and BCR-ABL positive samples

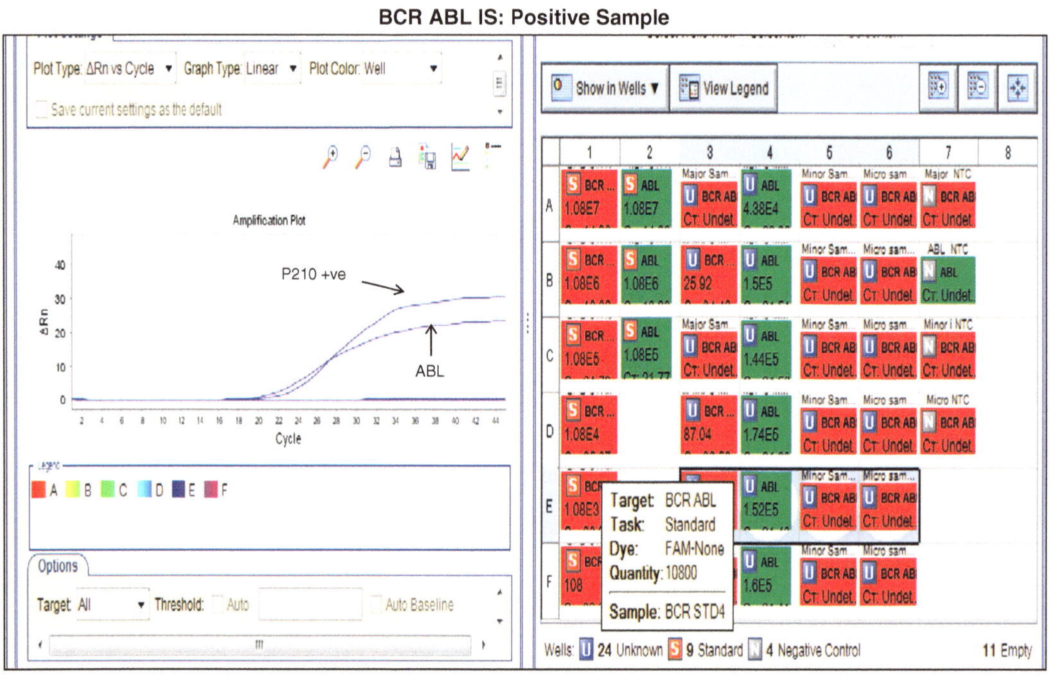

Fig. 2.18 and 2.19 (continued)

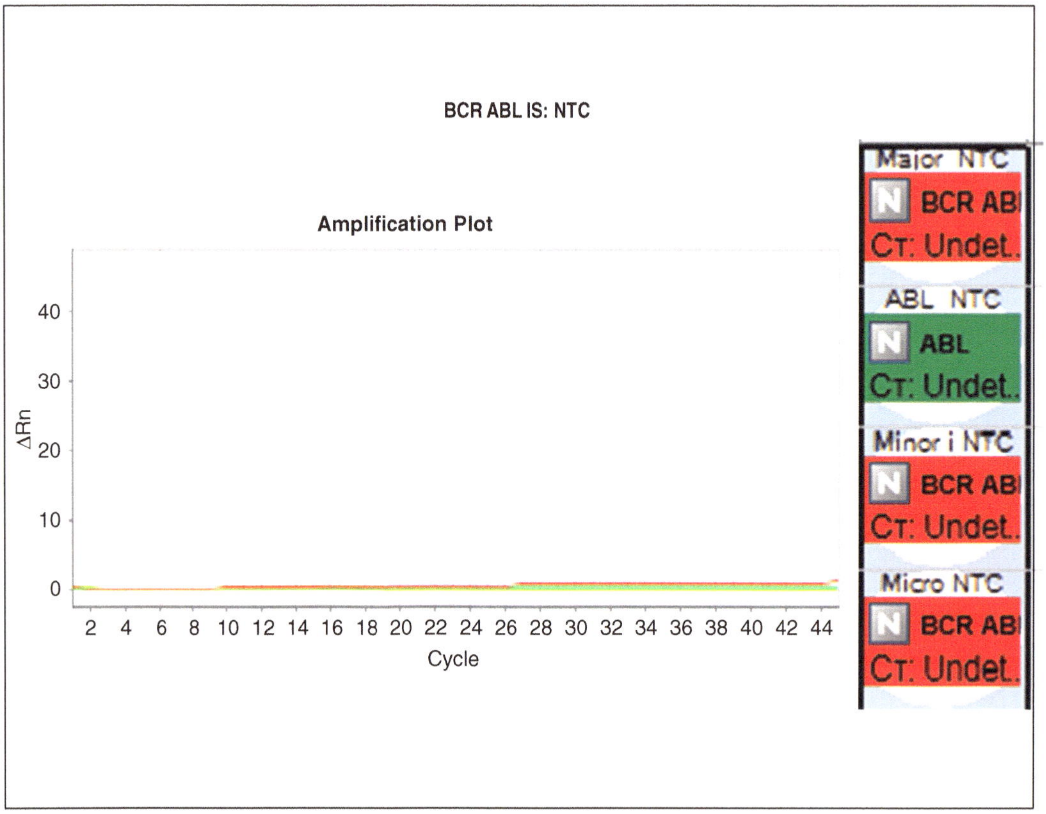

Fig. 2.20 Shows the NTC (controls) for BCR-ABL major, minor, and microforms

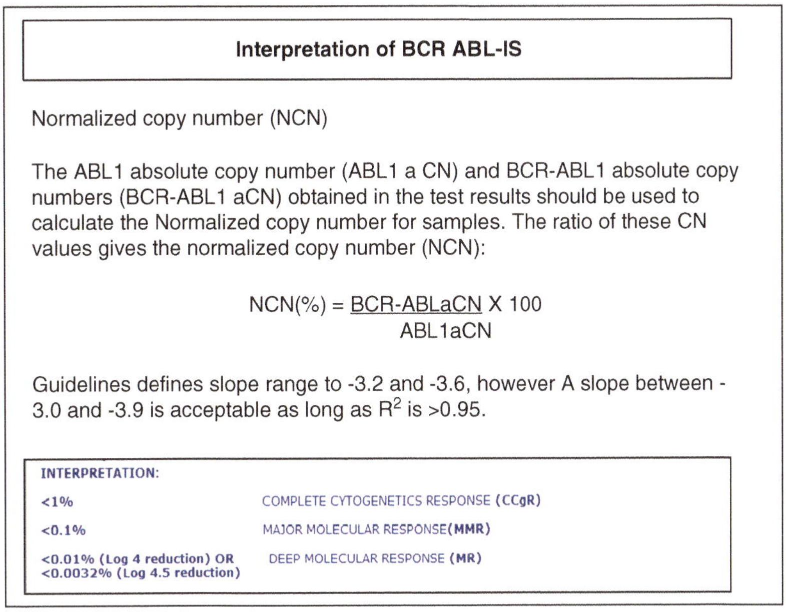

Fig. 2.21 Interpretation of BCR-ABL-IS is an extremely important step based on the absolute copy numbers (aCN) obtained in the test. Calculation of the normalized copy number (NCN) is shown in this figure. A clinical response to treatment is based on these values. For example, in CML, to determine complete cytogenetic response (CGR), Major Molecular Response (MMR), and Deep Molecular Response (MR) based on the percent values obtained for BCR-ABL

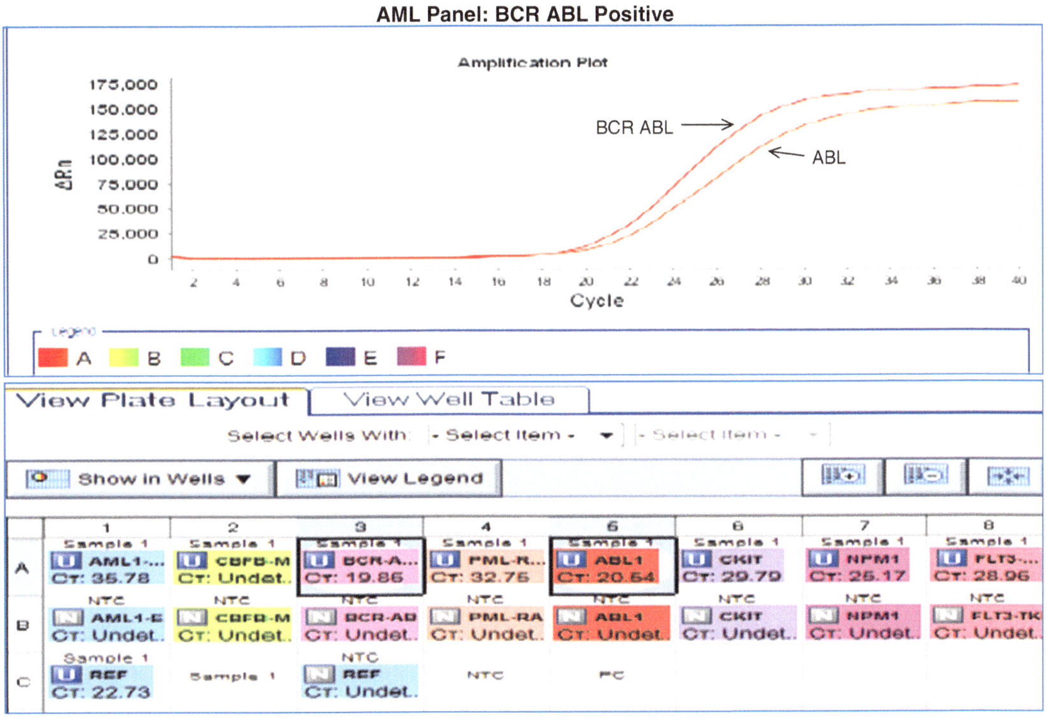

Fig. 2.22 An example of a BCR-ABL positive AML is shown in this figure. It shows again the plate layout and amplification plot in AML

AML Panel: PML RARA Positive

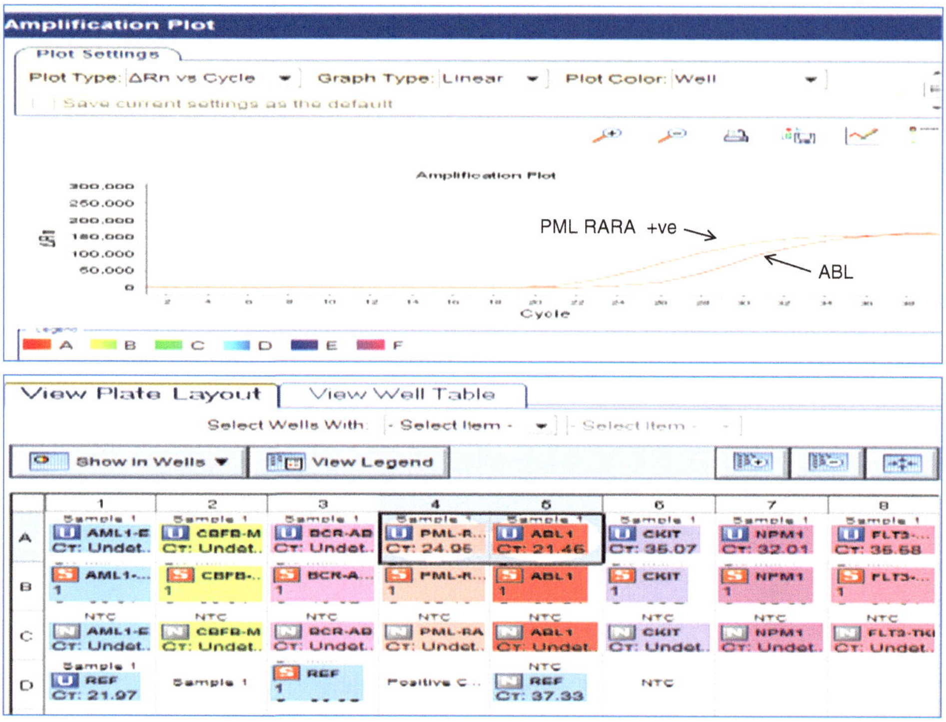

Fig. 2.23 Shows an example of results obtained for PML-RARA in a case of Acute promyelocytic leukemia

AML Panel: Negative Sample

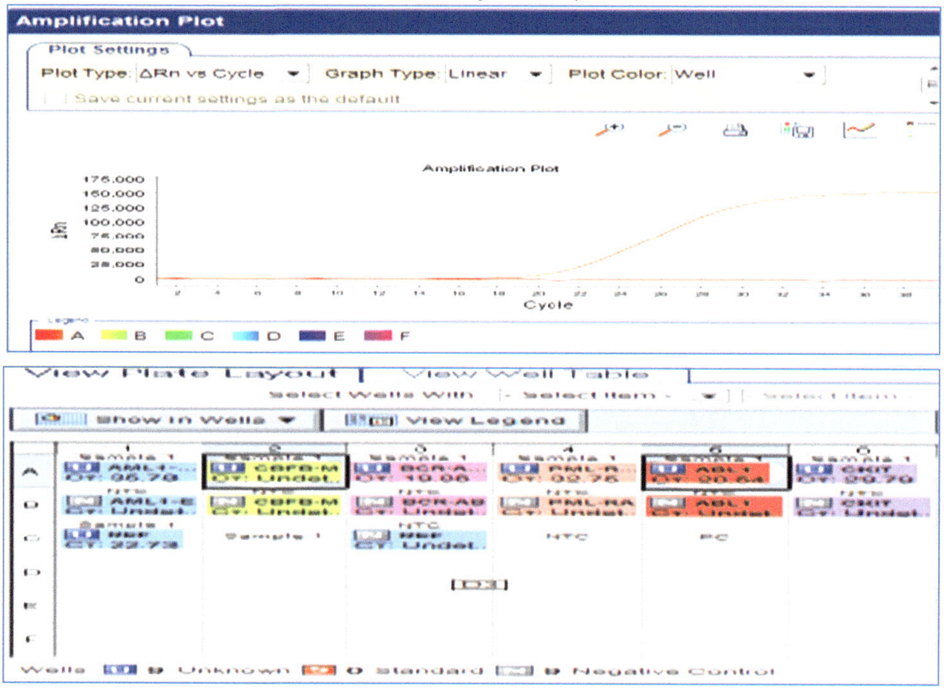

Fig. 2.24 Shows negative result obtained in a case of AML showing the AML panel plate layout and amplification plot

Plate layout for AML Panel

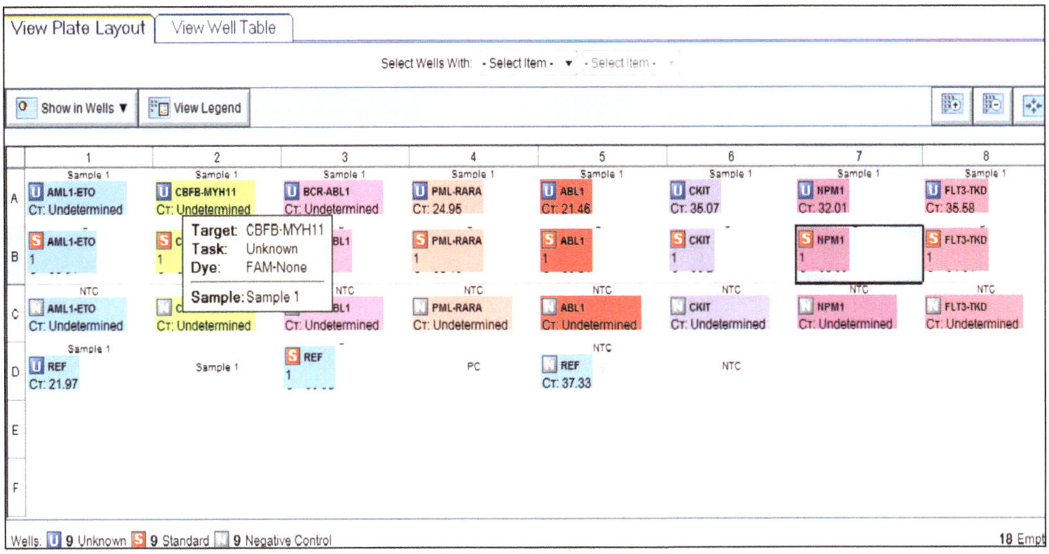

Fig. 2.25 AML panel plate layout using standards and controls

Interpretation of AML Panel

Case	Amplification signals in ABL1	Amplification Signals				Interpretation
		AML-ETO	CBFB-MYH11	BCR-ABL	PML-RARA	
1	Present	Present	Absent	Absent	Absent	Sample is positive for AML-ETO translocation
2	Present	Absent	Present	Absent	Absent	Sample is positive for CBFB-MYH11 translocation
3	Present	Absent	Absent	Present	Absent	Sample is positive for BCR-ABL translocation
4	Present	Absent	Absent	Absent	Present	Sample is positive for PML-RARA translocation
5	Present	Absent	Absent	Absent	Absent	Sample is Negative for all fusion genes
6	Absent	Absent	Absent	Absent	Absent	PCR inhibition, retest the sample

To avoid false negativity on or before the ABL1 should be detected on or before 26th cycle
Any amplification after 34th cycle should not be consider positive
* Other than above mentioned cases, there could be rare possibilities of detecting more than one fusion transcript in the sample.

Fig. 2.26 Shows the interpretation of the AML panel results based on amplification signals seen in the tests

Fig. 2.27 FLT 3-ITD mutation in AML is a clinically important mutation carried out on gel electrophoresis and not on PCR. This shows the gel image of an FLT-3 mutation-positive AML

FLT3 Mutation

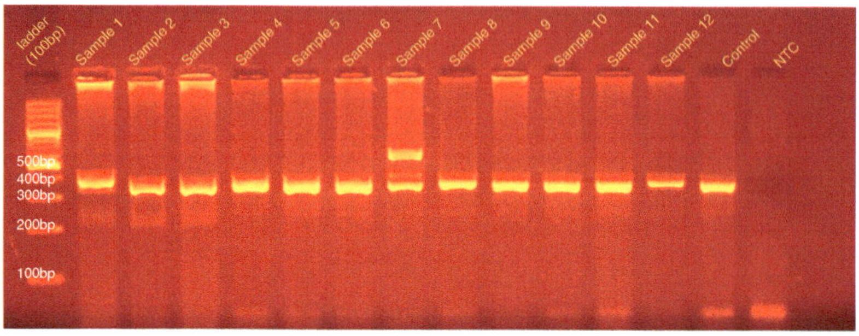

Gel image: Sample 7 is positive for FLT3-ITD mutation

Fig. 2.28 Shows an interpretation of the AML panel for FLT-3ITD

Interpretation of AML Panel : FLT3-ITD

Well	Band Size	Result
Negative control	No Band	Ok
Control	330bp	OK
Unknown Samples	330bp	Sample is Negative for FLT3-ITD Mutation
Unknown Samples	More than 330bp	Sample is positive for FLT3-ITD Mutation
Unknown Samples	More than 330bp & 330bp	Sample is positive for FLT3-ITD Mutation

- The analysis of FLT3-ITD products is performed by horizontal electrophoresis in low EEO agarose gels.
- Add 3g of agarose to 120ml of 1X TAE or 0.5X TBE Buffer. Microwave it at interval swirling till it completely dissolves in the Buffer. Cool it for about 5-8 minutes or 60°C. Add 4μl of EtBr solution (10mg/ml). Mix properly and pour in gel casting tray. Allow it solidify for about 30 min.
- Electrophoresis the gel at 100V in TAE 1X or TBE 0.5X tank buffer or 30min. or till the dye front reaches 3/4th of the gel

Fig. 2.29 Shows interpretation for c-KIT in the AML Panel

Interpretation of AML Panel: C-KIT

Control/Sample	ΔCt	Interpretation
Negative Control	No amplification	OK
Positive control	$\Delta Ct \le 8$	OK
Unknown samples	$\Delta Ct \le 8$	Sample is positive for C-KIT D816V Mutation
Unknown samples	$\Delta Ct \ge 8$	Sample is Negative for C-KIT D816V Mutation

Fig. 2.30 shows interpretation for NPM1 mutation in thc AML Panel

Interpretation of AML Panel: NPM1

Control/Sample	ΔCt	Interpretation
Negative Control	No amplification	OK
Positive control	ΔCt ≤ 6.2	OK
Unknown samples	ΔCt ≤ 6.2	Sample is positive for NPM1 Mutation
Unknown samples	ΔCt ≥ 6.2	Sample is Negative for NPM1 Mutation

Fig. 2.31 Shows interpretation for FLT-3-D 385 (TKD) in the AML panel

Interpretation of AML Panel: FLT3-D835(TKD)

Control/Sample	ΔCt	Interpretation
Negative Control	No amplification	OK
Positive control	ΔCt ≤ 8.5	OK
Unknown samples	ΔCt ≤ 8.5	Sample is positive for FLT3-D835 Mutation
Unknown samples	ΔCt ≥ 8.5	Sample is Negative for sting Mutation

Fig. 2.32 Gives the actual calculation for c-KIT, NPM1, and FLT3-D835 based on CT method (threshold cycle CT)

Result Analysis for C-KIT, NPM1 and FLT3-D835

In TRUEPCR AML Panel kit, the analysis of C-kit, NPM1 and FLT3-D835 is based on ΔCt method. Analyze the samples and note down the threshold cycle (Ct) for REF (tube 9) and mutation tubes C-kit, NPM1 and FLT3-D835

$$\Delta Ct = Ct\ Mut\text{-}Ct\ REF$$

ALL Panel: Positive Sample

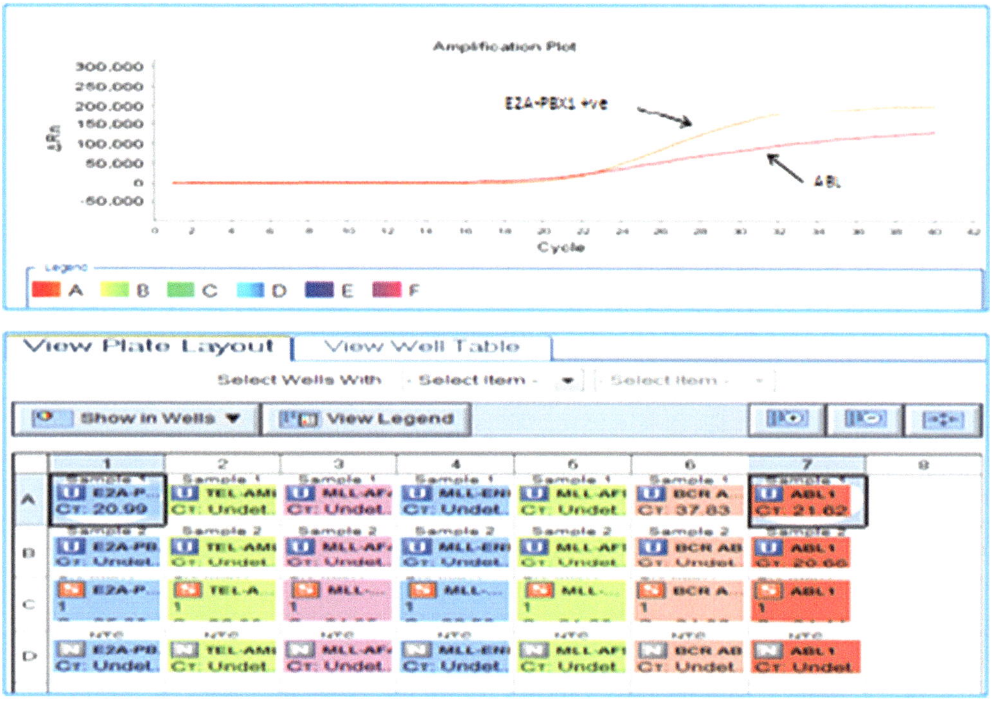

ALL Panel: Negative Sample

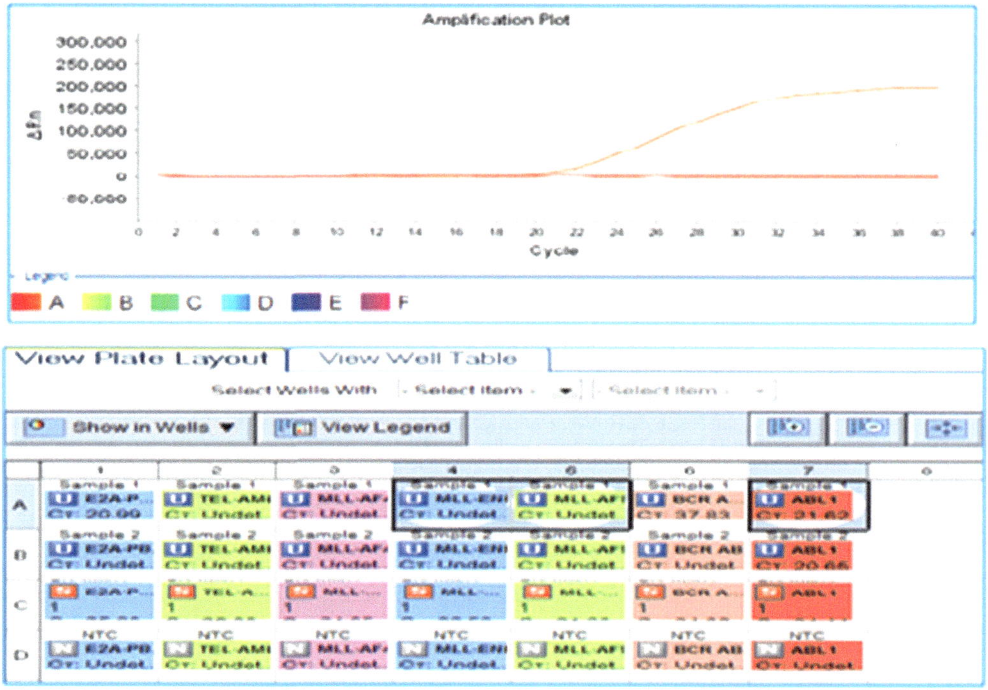

Figs. 2.33 and 2.34 The following figures show a similar approach to analyze ALL using ALL comprehensive panel outlined before. This show a positive and a negative sample for E2A-PBX1

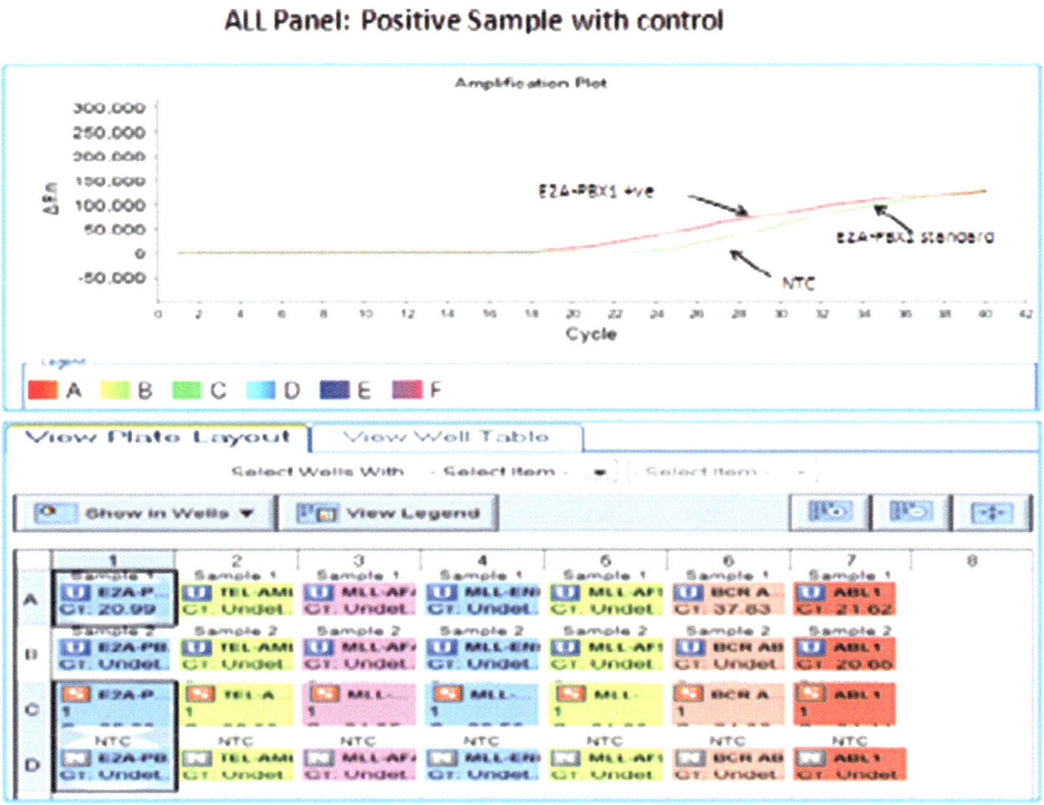

Fig. 2.35 Shows E2A-PBX1 sample run with a control in ALL Panel

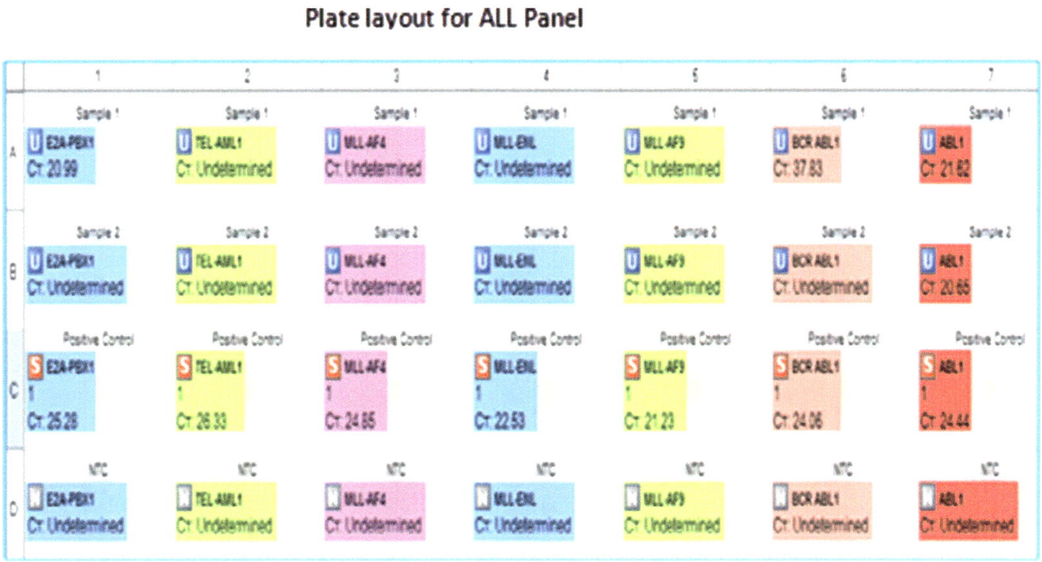

Fig. 2.36 Shows the plate layout using standards, control, and samples

Interpretation of ALL Panel

		Amplification Signals						
No	Amplification signals in ABL1 (Tube 7)	E2A-PBX1 (Tube 1)	TEL-AML1 (Tube 2)	MLL-AF4 (Tube 3)	MLL-ENL (Tube 4)	MLL-AF9 (Tube 5)	BCR-ABL1 (Tube 6)	Interpretation
1	Present#	Present	Absent	Absent	Absent	Absent	Absent	Positive for E2A-PBX1
2	Present#	Absent	Present	Absent	Absent	Absent	Absent	Positive for TEL-AML1
3	Present#	Absent	Absent	Present	Absent	Absent	Absent	Positive for MLL-AF4
4	Present#	Absent	Absent	Absent	Present	Absent	Absent	Positive for MLL-ENL
5	Present#	Absent	Absent	Absent	Absent	Present	Absent	Positive for MLL-AF9
6	Present#	Absent	Absent	Absent	Absent	Absent	Present	Positive for BCR-ABL1
7	Present#	Absent	Absent	Absent	Absent	Absent	Absent	Negative for all fusion genes
8	Absent	Absent	Absent	Absent	Absent	Absent	Absent	PCR inhibition, retest the sample

To avoid false negativity the ABL1 should be detected on or before 26th cycle.
Any amplification after 34th cycle should not be consider positive
* Other than above mentioned cases, there could be rare possibilities of detecting more than one fusion transcripts in the sample

Fig. 2.37 Shows the interpretation for the genes detected by the ALL comprehensive panel

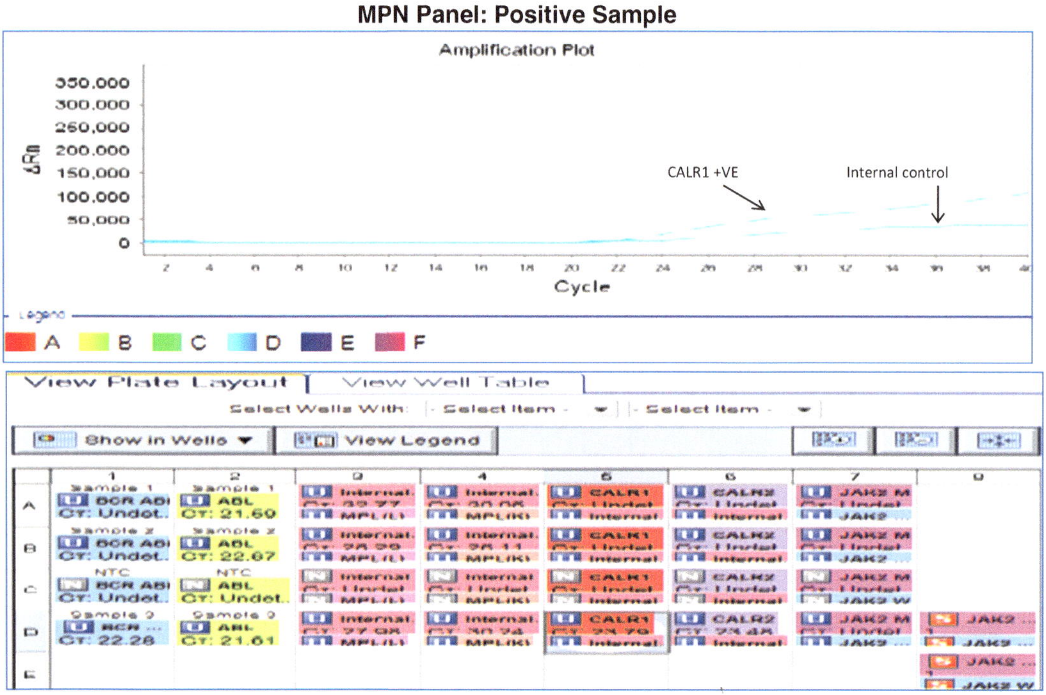

Fig. 2.38 Shows results obtained using an MPN panel (BCR-ABL, JAK2, CALR, and MPL mutations). This shows a case of MPN (PMF) found to be CALR positive

Part I

Case Studies in Myeloid Malignancies

3 Acute Myeloid Leukemia

3.1 Case Study 3.1: Acute Myeloid Leukemia with t(8;21)(q22;q22.1);RUNX1-RUNX1T1 and KIT Gene Mutation

Brief Clinical History
A 34-year-old female presented with complaints of headache for 1 week. There were no comorbidities, fever, bleeding manifestations, cough, weight loss, or icterus. No organomegaly, lymphadenopathy, or any other physical signs of clinical importance were found. There was no history of any blood transfusions.

CBC: *Hb:* 6.6 g/dl, TLC: 72,100/cu mm and Platelets: 60,000/cu mm

DLC: Blasts: 27.0%: Neutrophils: 31%, Lymphocytes: 30%, Monocytes: 3.0%, Myelocyte: 1.0%

Peripheral Blood Smear and Bone Marrow
Blasts constituted 27% of the DLC, large in size with a high N:C ratio, moderate granular cytoplasm, and 2–3 prominent nucleoli. Occasional blasts showed Auer rods. Many dyspoietic neutrophils were seen. Sixty percent of these blasts infiltrated the bone marrow aspirate and trephine biopsy.

Flow Cytometric Immunophenotyping
A large population of myeloid blasts (50% of acquired events), in CD45 moderate blast region with SSC low with cMPO+, CD34+, CD117+, CD13+, CD33+, HLA-DR+, *CD56+*, CD38+.

Cytogenetics
Conventional Karyotype

45,X,-X,t(8;21)(q22;q22.1)[18]/46,XX[2]

Cytogenetic analysis of 20 metaphases showed a neoplastic clone characterized by a balanced translocation between the long arms of chromosomes 8 and 21 and loss of sex chromosome (loss of X) as a secondary change in 18 metaphases analyzed.

M. Bhargava, *Hematologic Malignancies*, https://doi.org/10.1007/978-981-33-4799-1_3

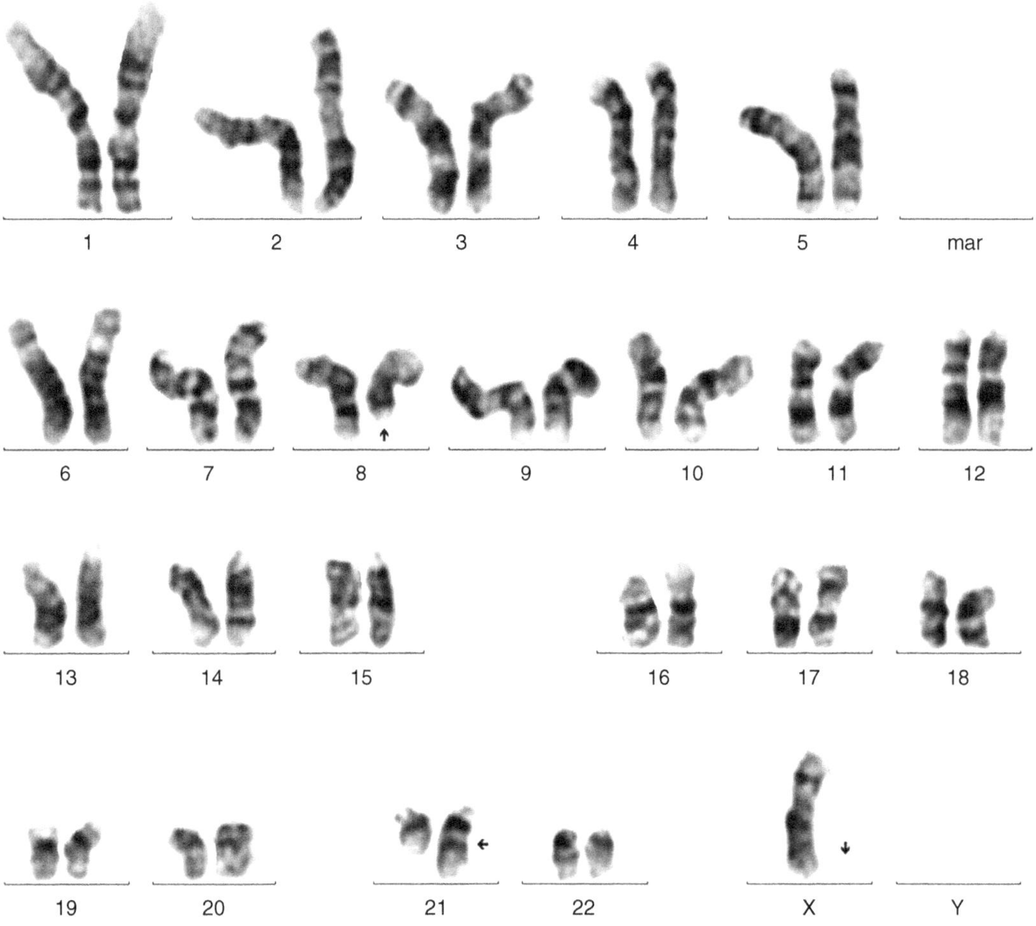

Case Study 3.1 Karyotype showing t(8;21) along with loss of chromosome X as secondary abnormality

Molecular Findings

The Real-Time Quantitative PCR was positive for *RUNX1-RUNX1T1 gene rearrangement t(8;21) (q22;q22.1), and KIT gene mutation.*

Discussion

As per the 2017 WHO classification [1], the presence of t(8;21) is diagnostic of acute myeloid leukemia irrespective of the percentage of blasts in the marrow. It is the commonest recurrent cytogenetic abnormality occurring in acute myeloid leukemia and is associated with a good prognosis.

Additional cytogenetic abnormalities are seen in nearly 70% of cases such as loss of a sex chromosome or del (9q). The presence of additional cytogenetic abnormalities does not alter the good prognosis [2]. The loss of the sex chromosome is the most frequent secondary cytogenetic abnormality in t(8;21) AML. Secondary cooperating mutations such as *KIT*, KRAS, or NRAS occur in 20%–30% of cases. The presence of *KIT mutation* and *CD56 expression* as seen in this case adversely affects prognosis. RUNX1:RUNX1T1 is an informative molecular marker that can be tracked by RNA-based RT-qPCR assays for molecular MRD monitoring in these patients at various time points during therapy [3].

3.2 Case Study 3.2: Acute Myeloid Leukemia with a Variant Three-Way Translocation t(8;17;21)(q22;q23;22.1);RUNX1-RUNX1T1

Brief Clinical History
A 12-year-old male presented with complaints of fever and bleeding gums for 1 week. No organomegaly, lymphadenopathy, or any other physical signs of clinical importance were found. There was no history of any blood transfusions.

CBC: *Hb:* 4.4 g/dl, TLC: 91,800/cu mm and Platelets: 34,000/cu mm.

DLC: Blasts: 80%: Neutrophils: 69: 4%, Lymphocytes: 16%

Peripheral Blood Smear and Bone Marrow
Blasts constituted 80% of the DLC, large in size with a high N:C ratio, convoluted nucleus with 1–2 prominent nucleoli, and moderate amount of granular cytoplasm. Few blasts showed Auer rods. They almost completely replaced the bone marrow aspirate and trephine biopsy and were positive for myeloperoxidase.

Flow Cytometric Immunophenotyping
A large population of myeloid blasts (76% of acquired events), in CD45 moderate blast region with SSC moderate with cMPO+, CD34+, CD117+, CD13+, CD33dim, HLA-DR+, *CD56+, CD19+*, CD38+.

Cytogenetics
Conventional Karyotype
45,X,-Y,t(8;17;21)(q22;q23;22.1)[20]

Cytogenetic analysis of 20 metaphases showed a neoplastic clone characterized by a balanced three-way translocation between the long arms of chromosomes 8, 17, and 21 also showing loss of sex chromosome (loss of Y) as a secondary change in 20 metaphases analyzed.

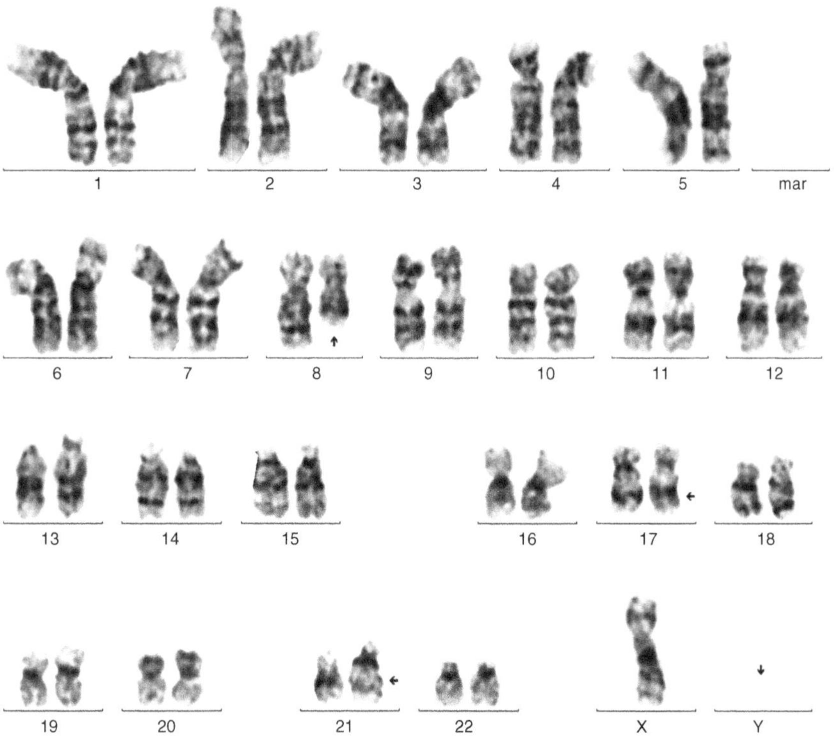

Case Study 3.2 Karyotype showing three-way rearrangement showing t(8;17;21) along with loss of chromosome Y as a secondary abnormality

Molecular Findings
The Real-Time Quantitative PCR was positive for *RUNX1-RUNX1T1 gene rearrangement in concordance with the karyotyping findings.*

Discussion
As per the 2017 WHO classification [1], the presence of t(8;21) is diagnostic of acute myeloid leukemia irrespective of the percentage of blasts in the marrow. It is the commonest recurrent cytogenetic abnormality occurring in acute myeloid leukemia and is associated with a good prognosis.

Complex three-way translocations, t(8;21;Var) involving a (variable) third AML chromosome have been described in 3% of cases. Many third chromosome partners have been reported in the literature including chromosome 12p13, 13q14, Xp22 [4]. Such findings are shown to be associated with poor response to therapy in some studies. A rare four-way variant t(8;17;15;21) (q22;q23;q15;q22) has also been reported in the literature. The loss of the sex chromosome is the most frequent secondary cytogenetic abnormality in t(8;21) AML, like the loss of Y chromosome as seen in this case.

AML with t(8;21) shows strong expression of CD34, HLA-DR, MPO, and CD13 and a weak expression of CD33. The aberrant expression of lymphoid markers *CD19* (as seen in this case) and *cCD79a, CD56* has been frequently observed. CD56 expression adversely affects prognosis.

3.3 Case Study 3.3: Acute Myeloid Leukemia, t(8;21) (q22;q22) with Rare Duplication of der (21)

Brief Clinical History
A 55-year-old male presented with complaints of fever for 10 days. No organomegaly, lymphadenopathy bleeding manifestations, or any other physical signs of clinical importance were found. There was no history of any blood transfusions.

CBC: *Hb*: 10.3 g/dl, TLC: 11,360/cu mm and Platelet count: 1,10,000/cu mm

DLC: Blasts: 54%, Neutrophils: 14%, Lymphocytes: 28%, Monocytes: 04%

Peripheral Smear and Bone Marrow
Blasts constituted 54% of the DLC, large in size with a high N:C ratio, convoluted nucleus with 1–2 prominent nucleoli, and scant cytoplasm. Few blasts showed the presence of thin Auer rids. Bone marrow showed similar blasts 48% with dysmyelopoiesis and dysmegakaryopoiesis.

Flow Cytometric Immunophenotyping
This confirmed a population of myeloid blasts (~40% of acquired events), in the CD45 moderate blast region with SSC moderate with cMPO+, CD34+, CD117+, CD13+, *CD33dim,* HLA-DR+, *CD56+,* CD38+.

Cytogenetics
Conventional Karyotype and FISH
46,X,-Y,t(8;21)(q22;q22),+der(21)t(8;21)(q22;q22)[15]/45,X,-Y,t(8;21) (q22;q22)[1]/46,XY[4]

Cytogenetic analysis showed a neoplastic clone characterized by a balanced reciprocal translocation between the long arms of chromosomes 8 and 21. The clone also showed gain of an extra derivative chromosome 21, i.e., der(21)t(8;21)(q22;q22) and loss of sex chromosome (loss of Y) as a secondary change in 15 of the 20 metaphases analyzed.

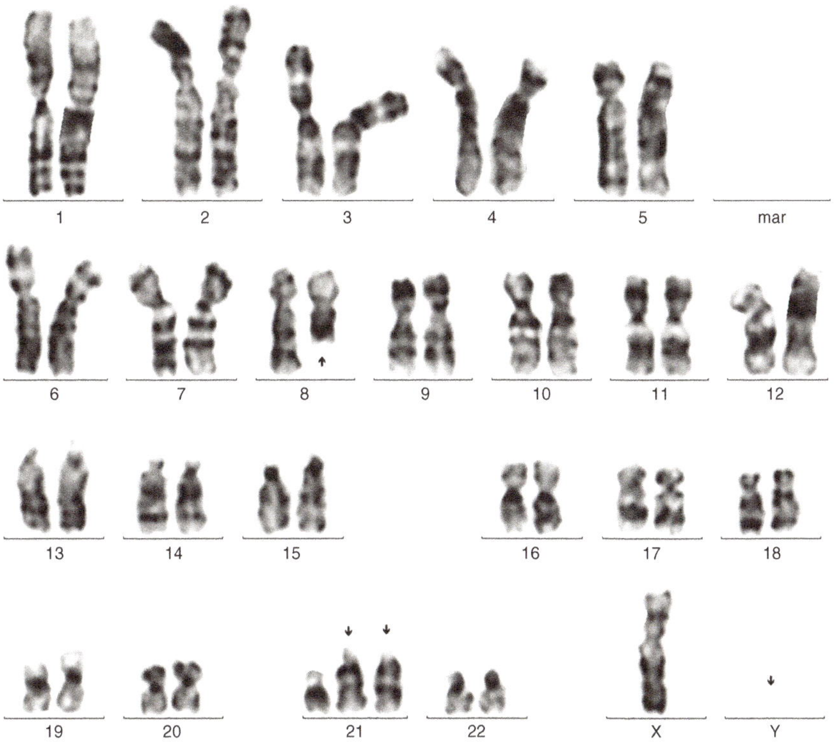

Case Study 3.3 (1) Karyotype showing t(8;21) along with loss of chromosome Y and gain of derivative 21 formed by t(8;21), as secondary abnormality

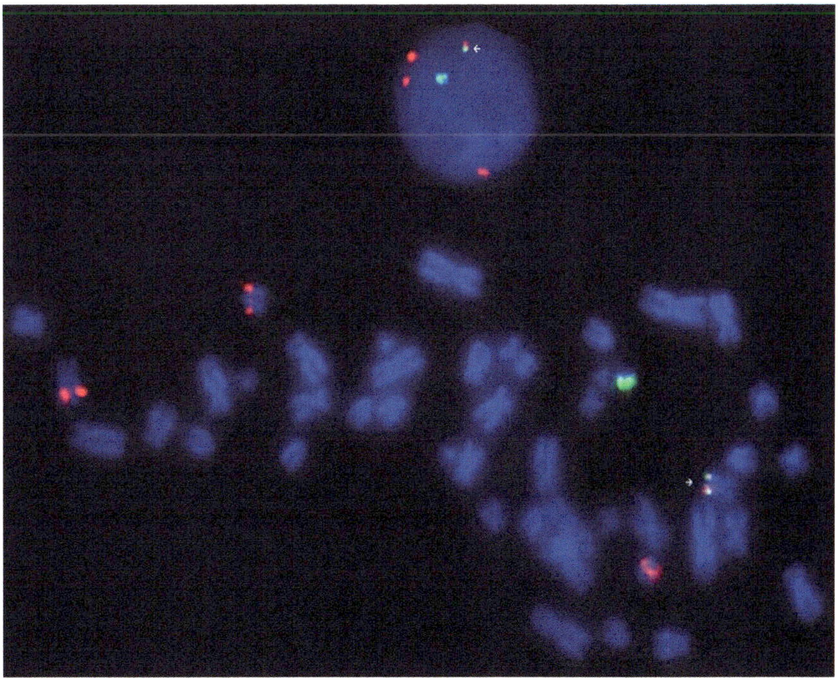

Case Study 3.3 (2) FISH showing 1F3R1G aberrant signal pattern using *RUNX1/RUNX1T1(=AML1/ETO)* probe indicating *RUNX1/RUNX1T1*, t(8;21) fusion

Molecular Findings
The Real-Time Quantitative PCR was *positive for RUNXI-RUNX1T1, gene rearrangement in concordance with the karyotyping findings.*

Discussion
The case described above is morphologically and immunophenotypically AML with recurrent cytogenetic abnormality t(8;21) with a rare additional finding of duplication of der (21).

Duplication of derivative (21)t(8;21) is a rare phenomenon and very few cases are reported in the literature. One case of triplication of derivative (21)t(8;21) has been described.

The duplication of der (21) can be present at diagnosis and also during relapse. Due to the rarity of its occurrence, the clinical implication of duplication or triplication of der(21) is unknown [5].

Complicons imply that two different genomic events, translocations, and duplications, act simultaneously on the same allele. Duplication of der(21) is the result of t(8;21) and the formation of the hybrid ETO-AML1 gene with subsequent duplication of der(21) resulting in two copies of the hybrid gene, as in this case.

Complicons are well described in other leukemias and lymphomas. Some of the other examples of complicons occurring in hematological malignancies include duplication of PML-RARA resulting in inder(17q), double Philadelphia chromosome in BCR-ABL-positive ALL and CML, Duplication of der(18)t(14;18) IGH-BCL2 fusion in follicular lymphomas, dup der (14) t(8:14)(q24;q32) in Burkitt's lymphoma and duplication of der(11)t(11;14) in mantle cell lymphoma.

3.4 Case Study 3.4: AML with inv(16)(p13;q22)/t(16;16)p(13;q22);CBFB-MYH11

Brief Clinical History
A 4-year-old female presented with complaints of fever, fatigue, and pallor for 2 weeks, along with bleeding gums for the past 4 days. No other physical signs of clinical importance were found. There was no history of any blood transfusions.

CBC: *Hb:* 5.6 g/dl, TLC: 1,43,190/cu mm and Platelets: 10,000/cu mm.

DLC: Blasts + Promonocytes: 62%, Neutrophils: 3%, Lymphocytes: 15%, Monocytes: 20 %

Peripheral Blood Smear and Bone Marrow
Smear showed hyperleukocytosis. Blasts + promonocyts constituted 62%. Few blasts showed azurophilic granules and Auer rods. Sheets of these cells along with the prominence of eosinophilic precursors infiltrated the bone marrow aspirate and trephine biopsy.

Flow Cytometric Immunophenotyping
The immunophenotype confirmed two populations:

One population of myeloblasts (~37%) with a phenotype of CD34+, CD117+, MPO+, CD13+,CD33+, HLA-DR+, CD38+ and another population of blasts and promonocytes (32%) which had a phenotype of CD34+, CD64+, CD14+, CD13+, CD33+, CD4dim.

Overall the findings were suggestive of AML with monocytic differentiation.

Cytogenetics
Conventional Karyotype
45,XX,inv(16)(p13;q22)[20]

Cytogenetic analysis of 20 metaphases showed the presence of neoplastic clone characterized by a pericentric inversion of the chromosome 16 in 20 metaphases.

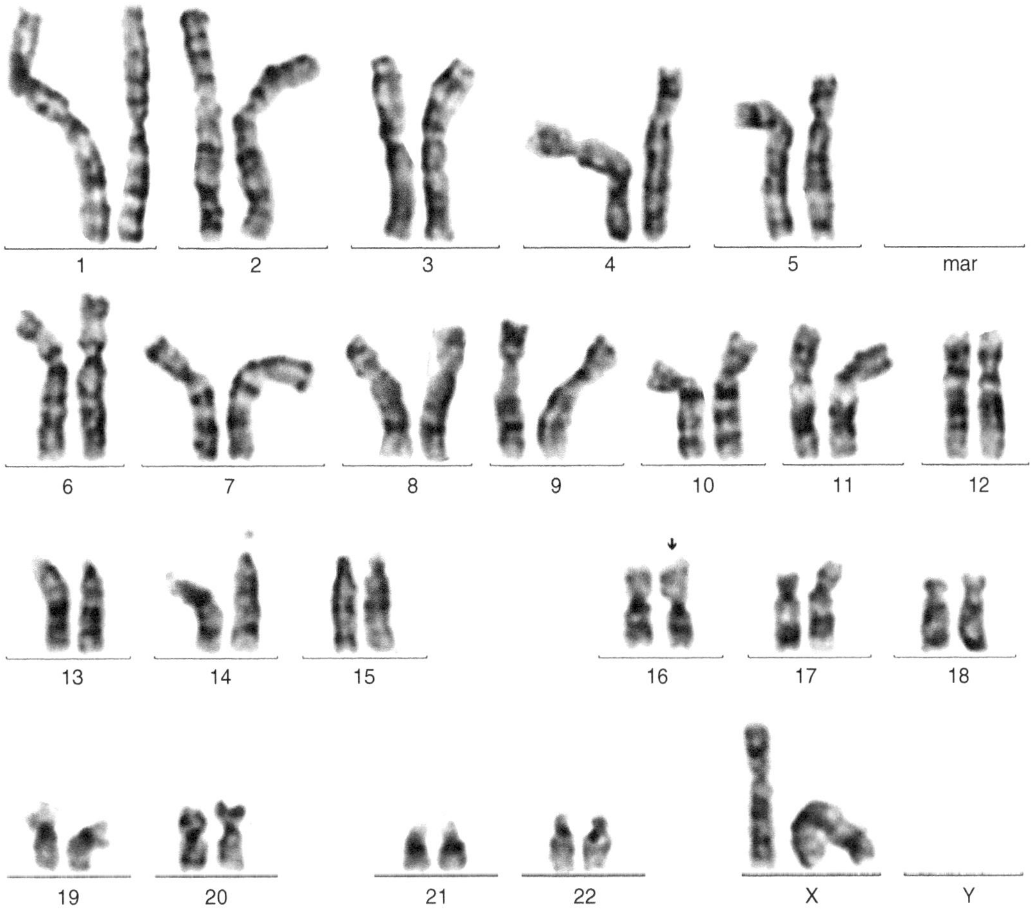

Case Study 3.4 Karyotype showing inversion of chromosome 16

Molecular Findings

Real-Time Quantitative PCR was *positive for CBFB–MYH11 gene rearrangement inv(16) (p13;q22)/t(16;16)p(13;q22) in concordance with karyotypic findings.*

Discussion

Core-binding factor (CBFB) AML includes AML with t(8;21)(q22;q22) and AML with inv(16) (p13q22) or t(8;21)(p13;q22) chromosomal rearrangements, respectively. AML with inversion of chromosome 16 is more often diagnosed with FAB m4Eo or acute myelomonocytic leukemia with abnormal marrow eosinophils, as in this case [1].

The fusion gene CBFB-MYH11 is more often seen in younger patients (5%–8% of all AML). The fusion transcripts are well-established markers for measurable residual disease (MRD) monitoring by real-time quantitative PCR [6].

CBFB AML are associated with a high rate of complete remission and favorable overall survival with intensive consolidation therapy.

3.5 Case Study 3.5: Acute Myeloid Leukemia with t(9;11)(p21;q23.3) (KMT2A-MLL3) with Trisomies of Chromosomes 8 and 21

Brief Clinical History

A 57-year-old male presented to the medicine department with complaints of fever, fatigue for 2 weeks. No history of organomegaly, lymphadenopathy, bleeding from any sites, or any other physical signs of clinical importance were found.

CBC: *Hb*: 7.5 g/dl, TLC: 6000/cu mm and Platelet count: 1,52,000/cu mm

DLC: Neutrophils: 5%, Monocytes: 2%, Lymphocytes: 92%

Peripheral Blood Smear and Bone Marrow

Smear showed severe leukopenia and 2nRBCs/100 WBCs. No blasts were seen. Bone marrow showed blasts (~90%) which were large with high N:C ratio, abundant finely granular basophilic cytoplasm, round nucleus with delicate lacy chromatin, and 2–3 nucleoli, morphologically suggestive of monoblasts.

Flow Cytometric Immunophenotyping

The leukemic cells (~55% of acquired events) in the blast region on CD45 vs SSC dot plot were CD34 negative, cMPO negative, CD13 negative, CD33 bright, *HLA-DR*+, CD117 dim, *CD11c*+, *CD64*+, CD4+, *CD56*+, CD38+.

Cytogenetics

Conventional Karyotype

48,XY,t(9;11)(p22;q23),+8,+21[10]

Cytogenetic analysis of 10 metaphases showed the presence of a neoplastic clone characterized by hyperdiploidy (modal number 48) with trisomy 8 and 21 and a structural abnormality showing balanced translocation between the short arm of chromosome 9 and long arm of chromosome 11.

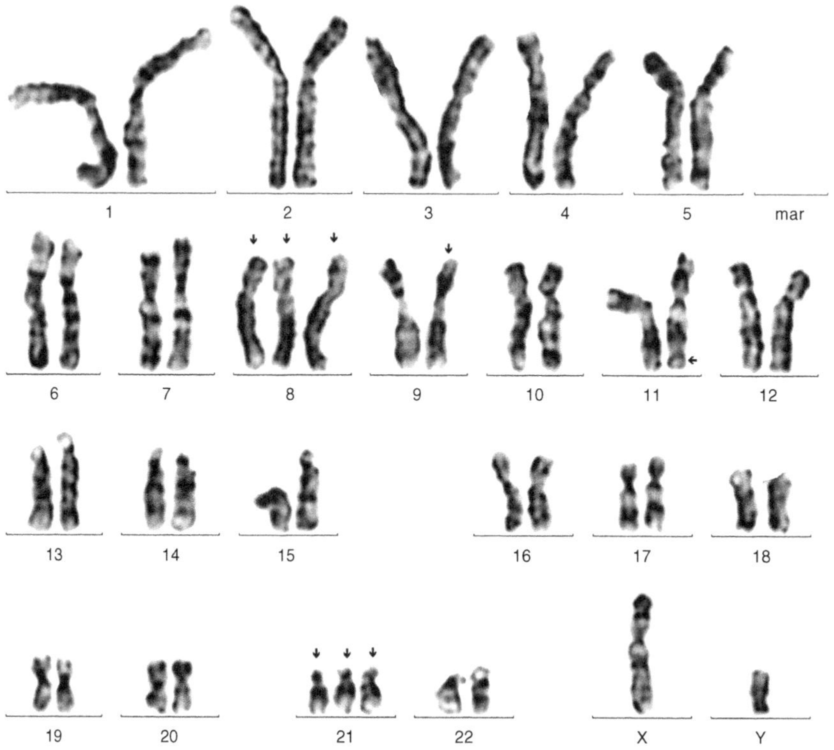

Case Study 3.5 Karyotype showing t(9;11) along with trisomy 8 and 21 as secondary abnormality

Molecular Findings
The Real-Time Quantitative PCR for *AML panel was negative.*

Discussion
AML with t(9;11) can occur at any age but it accounts for 9%–12% of all pediatric AML cases and only ~2% of adult AML. In this case t(9;11) was found in a 57-year-old male. The common presenting manifestations are coagulopathy due to DIC, myeloid sarcomas, and gingival infiltrations.

It is usually associated with monocytic features in morphology and immunophenotyping, as was the case in this patient [1].

More than 120 different translocations have been identified involving KMT2A in acute leukemias. t(9;11)(p21.3;q23.3) involving MLLT3(AF9) is the most common in AML. Secondary cytogenetic abnormalities such as trisomy 8 are common; however, they do not have an influence on survival in these patients.

Other common partner genes in KMT2A translocations resulting in AML are MLLTI(ENL), MLLT10(AF10), AFDN(MLLT4, AF6), or ELL.

According to the 2017 ELN AML risk stratification, AML with t(9:11)(p21.3;q23.3) has an intermediate survival, whereas the other 11 q23.3 translocations are included in the adverse risk category [7].

3.6 Case Study 3.6: Acute Promyelocytic Leukemia with PML-RARA, t(15;17) with Mutated FLT3-TKD (D835)

Brief Clinical History
A 36-year-old male presented to the gastroenterology department with complaints of abdominal pain, hematochezia, and hematuria for the past 2 days. No other physical signs of clinical importance were found, including bleeding from any other site.

CBC: *Hb*: 9.7 g/dl, TLC: 52.500/cu mm and Platelet count: 10,000/cu mm

DLC: Abnormal promyelocytes: 98%, Lymphocytes: 2%

Peripheral Blood and Bone Marrow
Smear showed leukocytosis with the presence of 98% abnormal promyelocytes which were large with moderate to abundant granular cytoplasm, bilobed reniform nuclei, and fine chromatin. Many showed Auer rods. Few faggot cells were also seen. Similar cells almost totally replaced the bone marrow aspirate and trephine biopsy.

Flow Cytometric Immunophenotyping
The leukemic cells (~93% of acquired events) had a characteristic high SSC extending from the edge of the blast region and merging with the granulocytic region on CD45 vs SSC dot plot. The neoplastic cells were CD34 negative, HLA-DR-negative and cMPO+, CD33 bright, CD13 moderate, CD64+, CD117+.

Morphology and Immunophenotyping features were consistent with Classical Acute promyelocytic leukemia (Classical APML).

Cytogenetics
Conventional Karyotype and FISH

46,XY,t(15:17)(q24;q21)[18]/46,XY[2]

Cytogenetic analysis of unstimulated bone marrow aspirate culture showed a neoplastic clone characterized by a balanced reciprocal translocation between the long arm of chromosomes 15 and 17 in 18 of the 20 metaphases analyzed.

FISH analysis of 200 interphase cells showed PML/RARA fusion positivity in 98% of the cells confirming the presence of t(15;17).

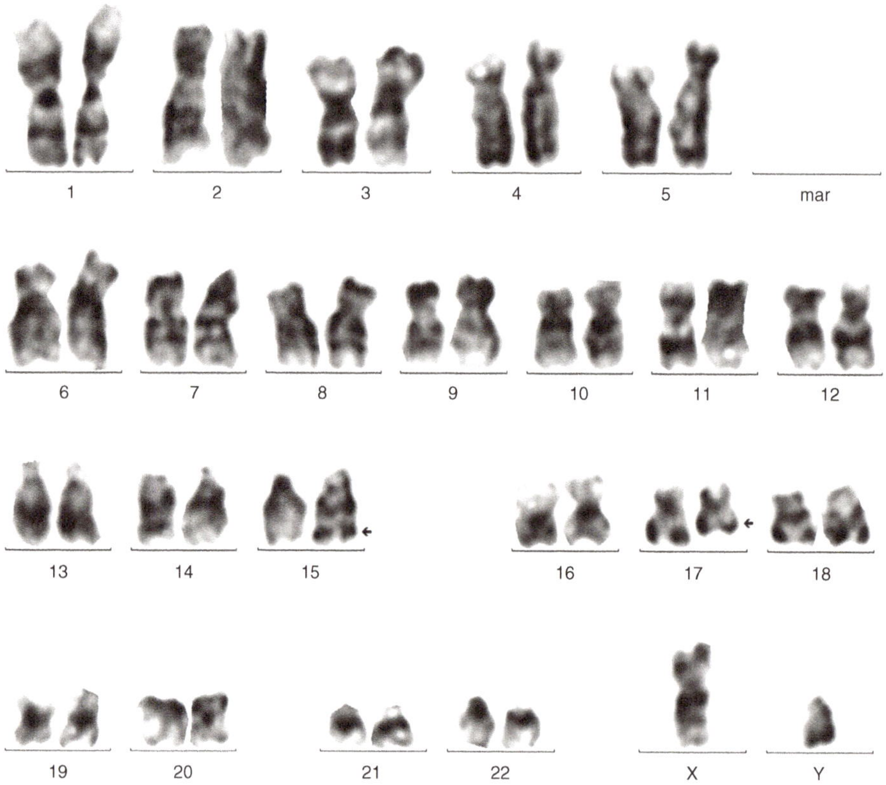

Case Study 3.6 (1) Karyotype showing t(15;17)(q24;q21)

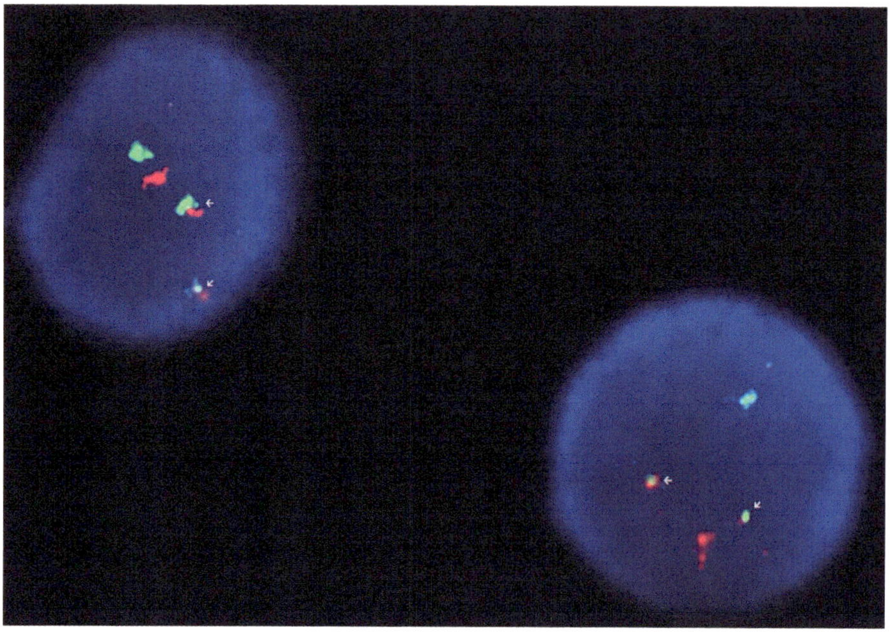

Case Study 3.6 (2) FISH image showing 2F1G1R signal pattern using DC DF *PML/RARA* probe indicating *PML/RARA* fusion, t(15;17)

Molecular Findings

The Real-Time Quantitative PCR was positive for *PML-RARA gene rearrangement* in concordance with the karyotyping findings and additional *FLT3 gene mutation FLT3-TKD (D835).*

Discussion

APML can occur at any age but in younger patients, it accounts for 5%–8% of AML cases. It has a relatively lower frequency in elderly patients. It is frequently associated with disseminated intravascular coagulation which is responsible for significant early deaths [1].

Classical APML has a reciprocal balanced translocation t(15;17)(q24.1;q21.2) giving rise to PM-RARA fusion gene product accounting for approximately 95% of all APML cases. The remaining 5% of APML cases are associated with many variant translocations. Secondary cytogenetic abnormalities are seen approximately in 40% of cases with trisomy 8 being the most common [2].

Karyotypically cryptic variants lacking classic t(15;17)(q24.1;q21.2) are rare resulting from submicroscopic insertion of RARA into PML, thus necessitating conventional karyotyping with *FISH* studies and RT-PCR.

FLT3-ITD mutations (21%–32%) are the most frequent coexisting mutations occurring in APML and are associated with worse DFS and OS. FLT3-TKD mutations are seen in approximately 16% of cases and have been associated with lower EFS in some studies [7].

3.7 Case Study 3.7: Acute Promyelocytic Leukemia with Variant (11;17)(q23;q21.1)(PLZF/RARA)

Brief Clinical History

A 42-year-old male with a known case of hypertension had complaints of low backache for 1 month. He was diagnosed outside as L5S1 prolapsed intervertebral disc (PIVD) with sacroiliitis. An MRI done showed L5S1 PIVD left-sided with nerve root compression and marrow signal alterations in sacrum, ilium, and lower lumbar vertebra suggestive of marrow infiltrative disorder.

No fever, loss of weight or organomegaly or any other physical signs of clinical importance were found.

CBC: *Hb:* 11.8 g/dl, TLC: 18.190/cu mm, Platelet count: 81,000/cu mm

DLC: Abnormal promyelocytes + blasts: 57%, Neutrophils: 26%, Monocytes: 1%, Eosinophils: 2%, Lymphocytes: 14%.

Peripheral Blood Smear and Bone Marrow

Smear showed leukocytosis with the presence of 57% abnormal promyelocytes and granular blasts which were large with abundant granular cytoplasm, round nuclei, fine chromatin, and inconspicuous to single nucleolus. No Auer rods were seen. Similar cells infiltrated the bone marrow aspirate and trephine biopsy.

Flow Cytometric Immunophenotyping

The leukemic cells (~70% of acquired events) had a characteristic high SSC extending from the edge of the blast region and merging with the granulocytic region on CD45 vs SSC dot plot.

The neoplastic cells were CD34-negative, HLA-DR-negative and cMPO+, CD33 bright, CD13 moderate, CD117+, *CD56*+. The picture was suggestive of Acute Promyelocytic Leukemia.

Cytogenetics

Conventional Karyotype and FISH

46,XY, t(11;17)(q23;q21)[4]/46,XY[16]

Cytogenetic analysis of 20 metaphases showed an abnormal cell clone characterized by a balanced translocation involving the long arm of chromosome 11 (11q) and chromosome 17 (17q) in 4 metaphases analyzed.

FISH analysis of 200 interphase cells showed two copies of PML, 15q24 and three copies of RARA, 17p21 in 85% of cells. These three copies of RARA indicate a variant translocation between RARA and an unknown partner chromosome.

Based on karyotyping the unknown partner chromosome was identified as 11q23.

The findings were consistent with Acute Promyelocytic leukemia with variant RARA translocation involving the PLZF gene at 11q23 and RARA gene at 17q21.

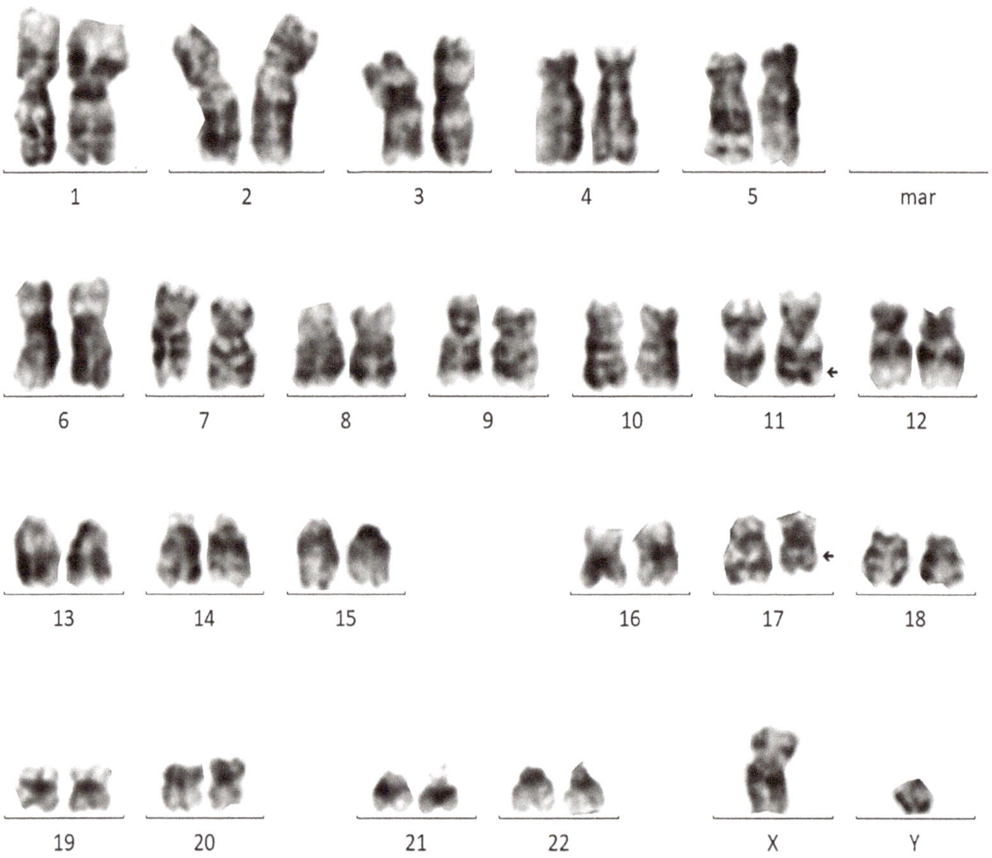

Case Study 3.7 (1) Karyotype showing t(11;17)(q23;q21)

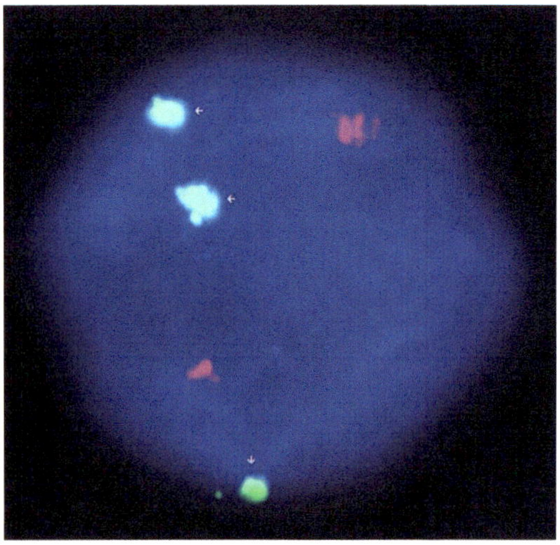

Case Study 3.7 (2) FISH image showing three copies of *RARA* gene and two copies of *PML* gene (3G2R) using *PML/RARA* DC DF probe indicating *RARA* gene rearrangement

Molecular Findings

The Real-Time Quantitative PCR was *negative for PML-RARA t(15;17)(q24.1;q21.2) gene rearrangement.*

Discussion

Classical APML has a reciprocal balanced translocation t(15;17)(q24.1;q21.2) giving rise to PML-RARA fusion gene product [1]. This abnormality accounts for approximately 95% of all APML cases. The remaining 5% of APML cases are associated with many variant translocations.

In a vast majority of patients with APL, the t(15;17) is detected. However, in a minority (5%) classic t(15;17) is lacking. Variant translocations have been noted. Some of which are t(11;17), t (5;17), t(9;17), del(17), der (17), and t(2;17). The sensitivity of (PLZF-RARA) to ATRA is not known. Sensitivity to ATO has not been documented outside PML-RARA-positive APL, except for PLZF-RARA-positive APL, which has been shown to be resistant [7].

3.8 Case Study 3.8: Acute Promyelocytic Leukemia with t(15;17) and t(9;22) Translocations, Trisomy 8 and Multiple Molecular Abnormalities

Brief Clinical History

A 65-year-old female admitted for high-grade fever, cough, and breathlessness for 15 days. Found to have anemia, *thrombocytopenia,* and *hyperleukocytosis*. Investigated further for evaluation of the disease.

CBC: *Hb*: 4.9 g/dl, TLC: 88,750/cu mm and Platelets: 13,000/cu mm.

DLC: Abnormal promyelocytes and blasts: 96%,neutrophils 4%.

Peripheral Blood Smear and Bone Marrow

Smear showed hyperleukocytosis with ~96% Blasts large in size, with fine homogenous chromatin, 1–2 nucleoli, and moderate amount of cytoplasm. No Auer rods were seen. Picture was consistent with acute leukemia.

Flow Cytometric Immunophenotyping

Blasts and abnormal promyelocytes (~77.8% of acquired events; extending from the edge of the blast region and merging with the maturing myeloid region) CD117+, CD13+, CD33+, MPO+, ~36% of blasts and abnormal promyelocytes), CD38+, CD123+, CD64+, CD11c+ , CD11b (dim)+,CD34+,HLA-DR-. The picture was suggestive of Acute Promyelocytic Leukemia.

Cytogenetics

Conventional Karyotype and FISH

47,XX,+8,t(9;22)(q34;q11.2)[20]

Cytogenetic analysis of unstimulated peripheral blood culture showed the presence of an abnormal clone showing translocation between the long arms of chromosome 9 (9q) and chromosome 22 (22q) forming Philadelphia chromosome (Ph) and gain of chromosome 8 as trisomy in all the analyzed 20 metaphases. In addition, FISH analysis using PML/RARA probe showed fusion positivity in 40% of the cells indicating that the cryptic t(15;17) was present.

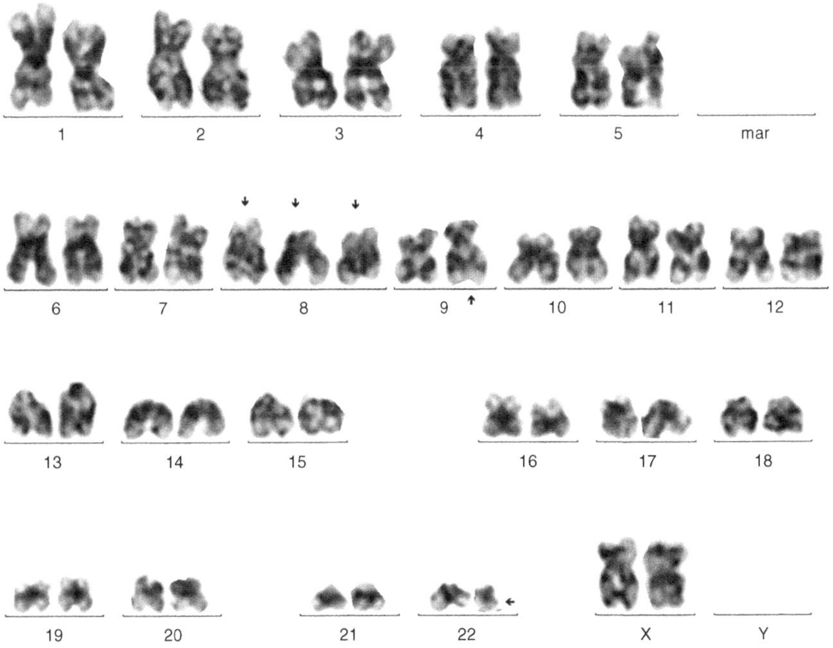

Case Study 3.8 (1) Karyotype showing Philadelphia chromosome formed by t(9;22)(q34;q11.2) along with trisomy 8

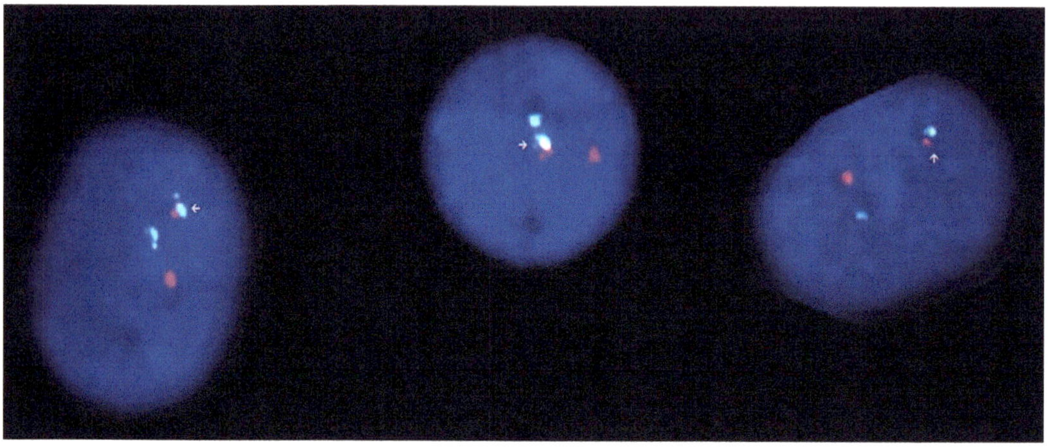

Case Study 3.8 (2) FISH image showing 1F1G1R aberrant signal pattern using *PML/RARA* DC DF probe indication *PML/RARA* fusion by FISH

Molecular Findings
The Real-Time Quantitative PCR on the AML panel showed *BCR-ABL, PML-RARA, c-KIT, NPM, and FLT-3TKD positive.*

Discussion
This patient presented morphologically and cytogenetically as APL with both Ph 1 chromosome t(9;22) and t(15;17) together with multiple other molecular abnormalities. Such a combination is rare. Mao et al. [8] described a patient with fever, pancytopenia, and promyelocytes on PBS with cytogenetic analysis showing a complex karyotype with 46XYt(9;22)(q34;q11) in addition to t(15;17) (q22;q21) as well as trisomy 8 and t(9:14). There are two other previously reported cases of coexistence of t(15;17) and t(9;22) in APL [9, 10]. Whether patients with concurrent t(9;22) and t(15;17) represent primary APL versus blast crisis of previously undiagnosed CML can be difficult to differentiate Takahashi et al. [11]. Evidence of preexisting splenomegaly or other findings can be helpful toward distinguishing the two. The appropriate treatment and prognosis is yet to be determined. The patient presented here had no organomegaly, showed severe anemia, leukocytosis, and thrombocytopenia with APL features on PBS. Karyotyping showed t(9;22) and trisomy 8. FISH showed cryptic t(15;17). Molecular analysis using AML panel (RQ-PCR) showed multiple abnormalities of BCR-ABL, PML-RARA, c-KIT, NPM, and FLT-3TKD indicating multiple clones in the evolution of this disease.

3.9 Case Study 3.9: Acute Promyelocytic Leukemia with PML-RARA t(15;17) with Mutated FLT3-ITD

Brief Clinical History
A 27-year-old male presented to the emergency department with complaints of two episodes of hemoptysis and fever for 1 week. No organomegaly, lymphadenopathy, or any other physical signs of clinical importance were found, including bleeding from any other site.

CBC: *Hb:* 7.7 g/dl, TLC: 49.930/cu mm and Platelet count: 13,000/cu mm

DLC: Abnormal promyelocytes + Blasts: 92%, Neutrophils: 6%, Lymphocytes: 2%

Peripheral Blood Smear and Bone Marrow
Smear showed leukocytosis with the presence of 92% abnormal promyelocytes + blasts which

were large with moderate to abundant granular cytoplasm, bilobed reinform nuclei, and fine chromatin. Few faggot cells containing bundles of Auer rods were also seen. Bone marrow showed near-total replacement by abnormal promyelocytes (~90%) many with azurophilic granules and Auer rods.

Flow Cytometric Immunophenotyping

The leukemic cells (~90% of acquired events) had a characteristic high SSC extending from the edge of the blast region and merging with the granulocytic region of CD45 vs SSC dot plot. The neoplastic cells were CD34-negative, HLA-DR-negative, CD56-negative and cMPO+, CD33 bright, CD13 moderate, CD64+, CD117+, CD11c+.

Morphology and Immunophenotyping features were consistent with Classical Acute promyelocytic leukemia (Classical APML).

Cytogenetics

Conventional Karyotype and FISH

46,XY,t(15;17)(q24;q21)[19]/46,XY[1]

Cytogenetic analysis of unstimulated bone marrow aspirate culture showed a neoplastic clone characterized by a balanced reciprocal translocation between the long arm of chromosomes 15 and 17 in 19 of the 20 metaphases analyzed.

FISH analyses of 200 interphase cells showed PML/RARA fusion positively in 90% of the cells confirming the presence of t(15;17).

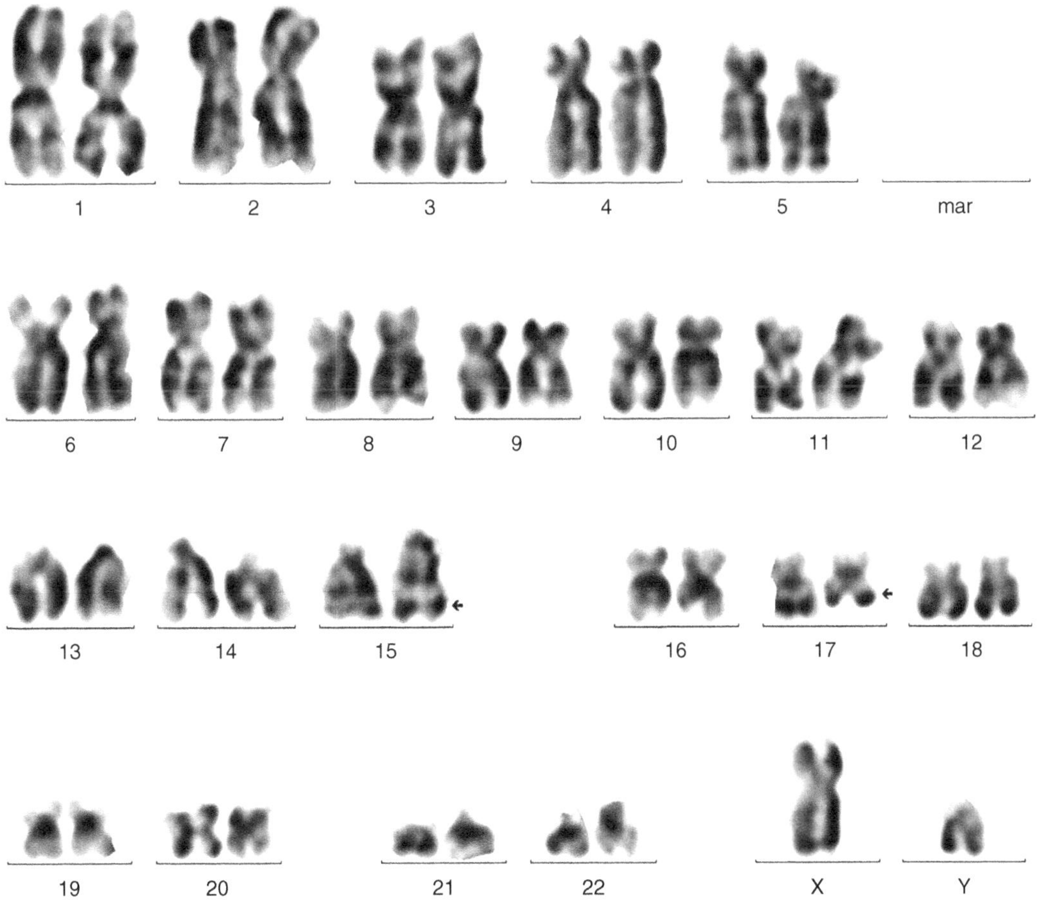

Case Study 3.9 (1) Karyotype showing t(15;17)(q24;q21)

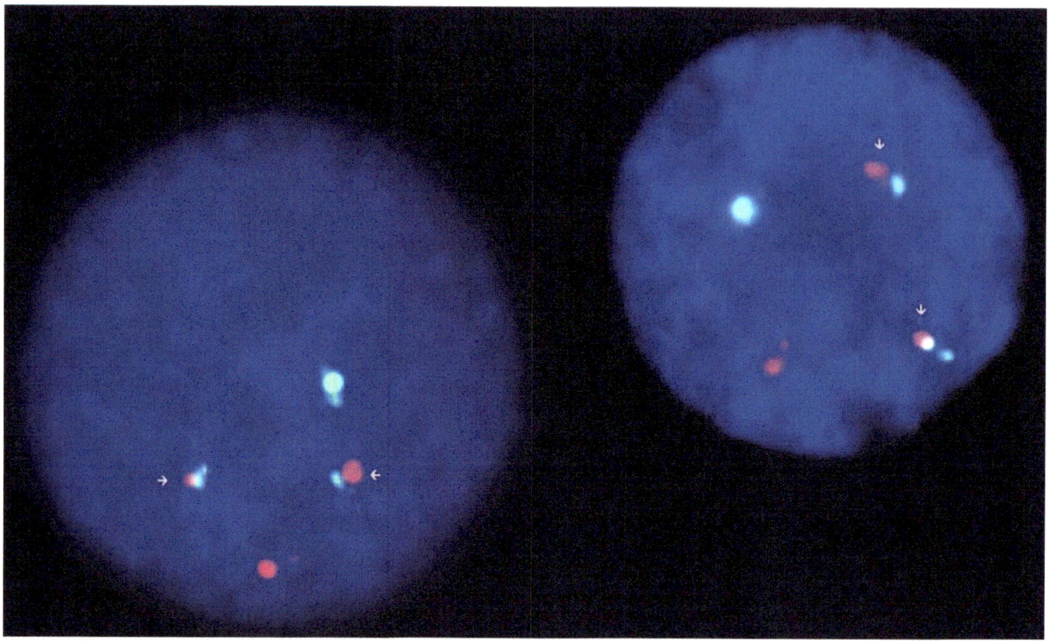

Case Study 3.9 (2) FISH image showing 2F1G1R signal pattern using DC DF *PML/RARA* probe indicating *PML/RARA* fusion, t(15;17)

Molecular Findings

The Real-Time Quantitative PCR was positive for *PML-RARA gene rearrangement* in concordance with the karyotyping findings and additional *FLT3 gene mutation FLT3-ITD* (Internal Tandem Duplications).

Discussion

FLT3 (FMS-like tyrosine kinase 3) is a transmembrane tyrosine kinase receptor that stimulates cell proliferation upon activation. FLT3 mutations (21%–32%) are associated with a higher white blood cell (WBC) count at presentation and are the most frequent coexisting mutations in APML [1]. Their impact on prognosis is controversial. Recently, the expert panel on behalf of the European Leukemia Net has recommended against the inclusion of the analysis of FLT3 mutations in the routine work of APML [12].

3.10 Case Study 3.10: Acute Myeloid Leukemia with Normal Karyotype, Mutated NPM1 and FLT3-ITD

Brief Clinical History

A 32-year-old male presented with complaints of fever on and off for 2 weeks. No comorbidities or history of regular medications. No organomegaly, lymphadenopathy, bleeding from any sites were found. General physical examination was unremarkable.

CBC: *Hb*: 5.7 g/dl, TLC: 1,64,970/cu mm and Platelet count: 23,000/cu mm

DLC: Blasts: 94%, Neutrophils: 1%, L: 3%, Myelocyte: 1%, Monocyte: 1%. Occasional nRBC was seen.

Peripheral Blood Smear and Bone Marrow
Smear showed hyperleukocytosis. Blasts constituted 94% of the DLC, large in size with a high N:C ratio, round nucleus with 1–2 prominent nucleoli, and scant cytoplasm. Few blasts showed Auer rods. Bone marrow showed similar blasts almost replacing it.

Flow Cytometric Immunophenotyping
A large population of moderate blasts (~90% of acquired events) were seen in CD45-moderate blast region with SSC moderate with *CD34-negative, HLA-DR-negative*, cMPO+, CD117+, CD13+, CD33+, *CD56+*, CD38+, CD123+.

Cytogenetics
Conventional Karyotype

46 XY [20]

Cytogenetic analysis showed normal karyotype in all the analyzed 20 metaphases.

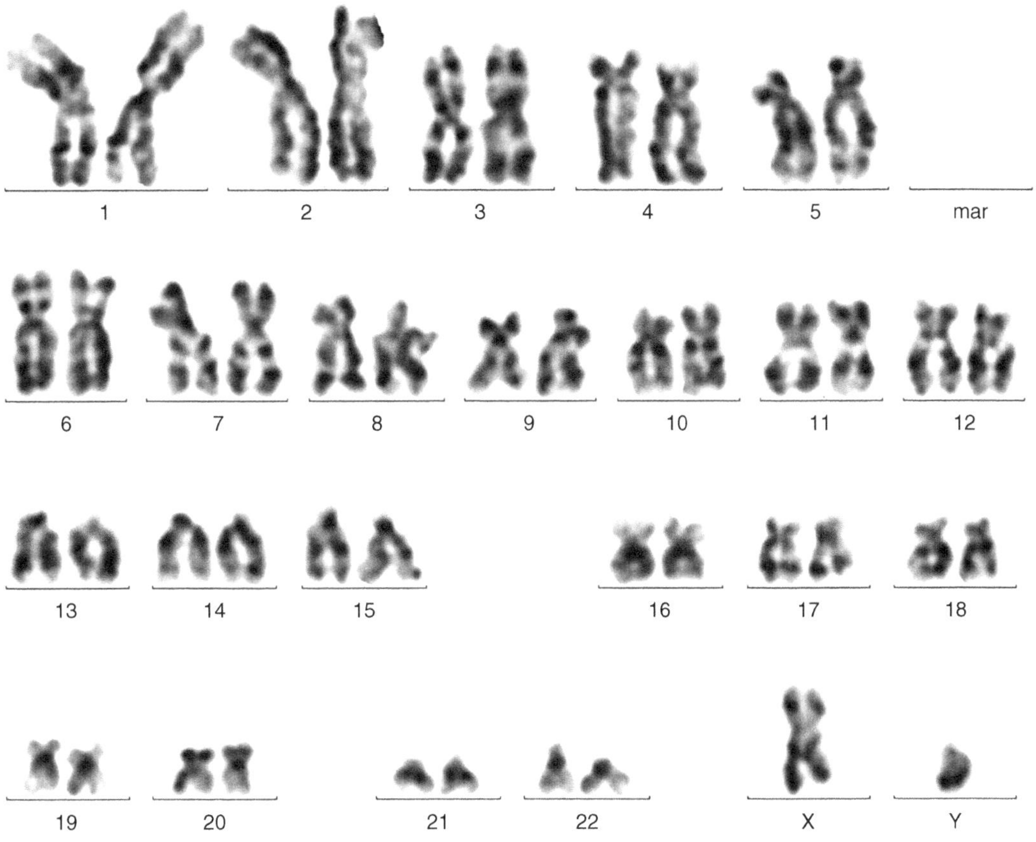

Case Study 3.10 Karyotype showing normal male karyotype

Molecular Findings
Real-Time Quantitative PCR was *positive for NPM1 and FLT3-ITD mutations*.

Discussion
The case described above is morphologically AML without maturation, cytogenetically normal with absent CD34 and HLA-DR, an immunophenotype characteristically seen in NPM1 mutated AML [1]. Mutations in exon 12 of the gene encoding nucleophosmin NPM1 is seen in approximately 25%–50% of cases of de novo AML. NPM1 mutation has a strong association with both acute myelomonocytic and acute monocytic leukemia. However, some cases with morphology of AML without maturation and pure erythroid leukemia have also been shown to expresses this mutation. NPM1 mutation is also associated with multilineage dysplasia in 25% of de novo AML cases.

A unique feature on immunophenotyping of NPM1 mutated AML is the absence of CD34 and HLA-DR in most of the cases. In this case, CD56 positivity may indicate a poor prognosis.

Secondary mutations that are common in NPM1 mutated AML involve FLT3 (as seen in this case), DNMT3A, IDHI, KRAS, and NRAS genes.

The NPM1 mutation can be detected by molecular techniques and/or immunohistochemistry. On IHC, the NPM1 is seen aberrantly expressed in the cytoplasm of the leukemic cells.

NPM1 mutation is associated with improved outcomes in both young and older adults, as well as in children with AML who have normal karyotype and do not have concomitant FLT3-ITD. The ELN risk stratification stratifies patients with AML with NPM1 mutation and normal karyotype and absence of FLT3-ITD (or with FLT3-ITD with low allelic ratio) as favorable and those with NPM1 mutated and FLT3-ITD high allelic ratio as intermediate-risk category [7].

3.11 Case Study 3.11: Acute Myeloid Leukemia with normal Karyotype and NPM 1 Gene Mutation

Brief Clinical History
A 51-year-old Nigerian patient was admitted for investigating acute leukemia.

CBC: *Hb*: 5.1 g/dl, TLC: 18,090/cu mm and Platelets: 50,000/cu mm.

DLC: Neutrophils: 3.0%, Lymphocytes: 6.0%, Eosinophils: 0.0%, Monocytes: 0.0%, Basophils: 0.0%, Blasts: 89.0%; Myelocytes: 2.0%.

Peripheral Smear and Bone Marrow
Smear showed Leukocytosis with ~89% blasts. The blasts were medium in size with a high N:C ratio, scant to moderate amount of cytoplasm, few showing prominent granules, and large immature nuclei with few showing prominent nucleoli. Similar blasts replaced almost the entire bone marrow (77%). Peripheral blood and the bone marrow aspirate showed features consistent with acute leukemia.

Flow Cytometric Immunophenotyping
Blasts (~73.0% of acquired events; present in the blast region), MPO+, (~38 blasts), CD13+, CD33+, CD38+, CD56+, CD123+, CD64+, CD11c+, CD34−, HLA-DR−, CD117− (moderate to heterogeneous). The picture showed Acute Myeloid Leukemia.

Cytogenetics
Conventional Karyotype
46,XX[20]

Cytogenetic analysis of unstimulated bone marrow aspirate culture showed normal karyotype in all the analyzed 20 metaphases.

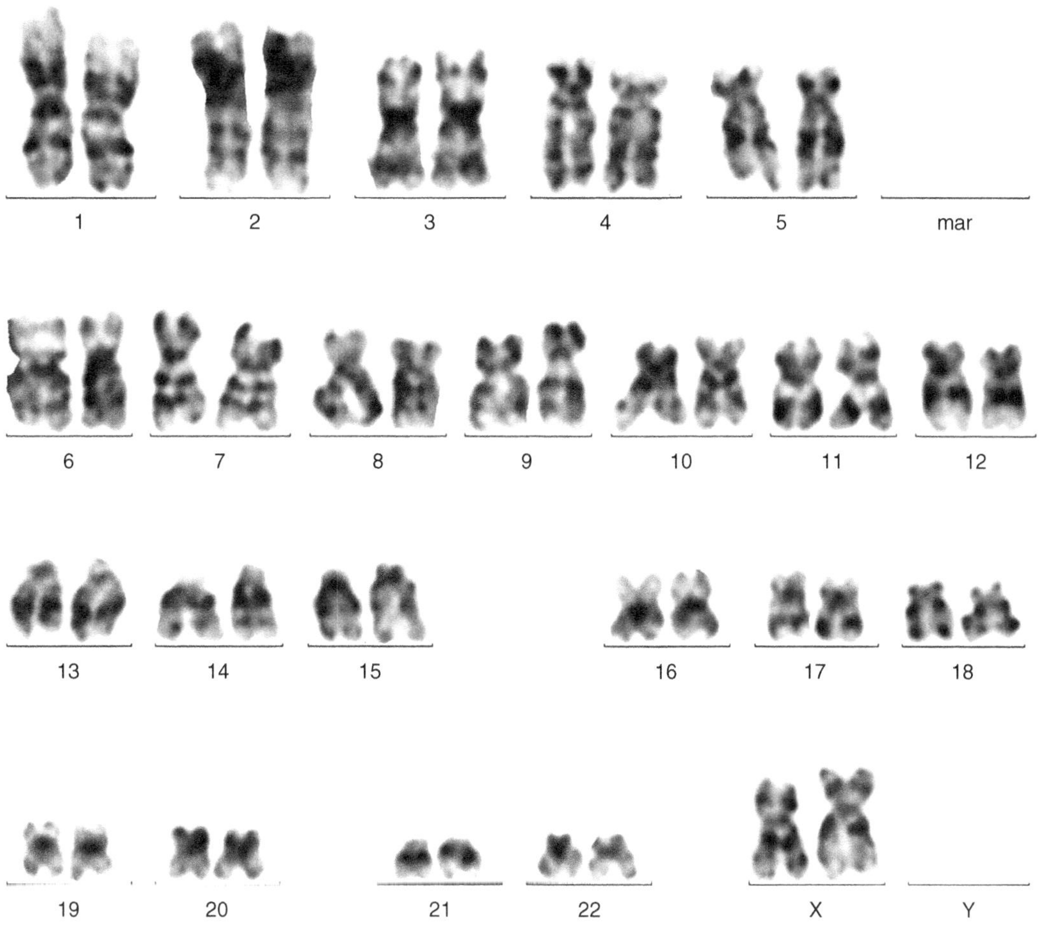

Case Study 3.11 Karyotype showing normal female karyotype

Molecular Studies

The Real-Time Quantitative PCR showed *AML comprehensive panel for NPM1 gene mutation positive.*

Discussion

This patient was morphologically AML without maturation, cytogenetically normal with absent CD34 and HLA-DR, an immunophenotype characteristically associated with NPM1 mutated AML. Mutations of the nucleophosmin gene have been reported as the most frequent mutation in AML, especially in the presence of a normal karyotype [7]. It accounts for nearly one-third of AML patients and shows mutual exclusion, of recurrent genetic abnormalities, high frequency of FLT3-ITD mutations, frequent M4 or M5 morphology, multilineage involvement, and distinctive gene expression signature.

In this patient, there were no secondary mutations, such as FLT3, DNMT3A, IDH1, KRAS, or NRAS genes. NPM1 mutation is associated with good prognosis in both young and older adults as well as in children with normal karyotype and no associated FLT3 mutation or FLT3–ITD allele ratio low. As per ELN risk stratification, this patient would fall in the favorable prognosis group [13].

3.12 Case Study 3.12: Acute Myeloid Leukemia with Normal Karyotype and FL3-ITD Gene Mutation

Brief Clinical History

A 67-year-old male was admitted for marked pallor and body aches. Evaluated for Acute Leukemia.

CBC: *Hb:* 6.7 g/dl, TLC: 1,17,020/cu mm and Platelets: 50,000/cu mm.

DLC: Blasts and Promonocytes: 82.0%, Neutrophils: 2.0%, Lymphocytes: 8.0%, Eosinophils: 1.0%, Monocytes: 6.0%, Metamyelocytes: 1.0%, Basophils: 0.0%.

Peripheral Smear and Bone Marrow

Smear showed Hyperleukocytosis with blasts and promonocytes constituting ~82%. They were large cells with a high N:C ratio, fine nuclear chromatin with inconspicuous to prominent nucleoli, and a moderate amount of cytoplasm with vacuolation. No Auer rods were seen.

Similar blasts (89.0%) infiltrated the bone marrow. The picture in peripheral blood and BM was morphologically consistent with AML with monocytic differentiation.

Flow Cytometric Immunotyping

Blasts constituted ~83% of acquired events expressing moderate CD45 and having moderate SSC. They were CD117+, HLA-DR+, CD13+, CD33+, CD64+, CD11c+, CD11b+, CD123+, CD38+, CD14+ , CD 34+. The picture was compatible with Acute Myeloid Leukemia with monocytic differentiation.

Cytogenetics

Conventional Karyotype

46,XY[20]

Cytogenetic analysis of unstimulated bone marrow aspirate culture showed normal karyotype in all the analyzed 20 metaphases.

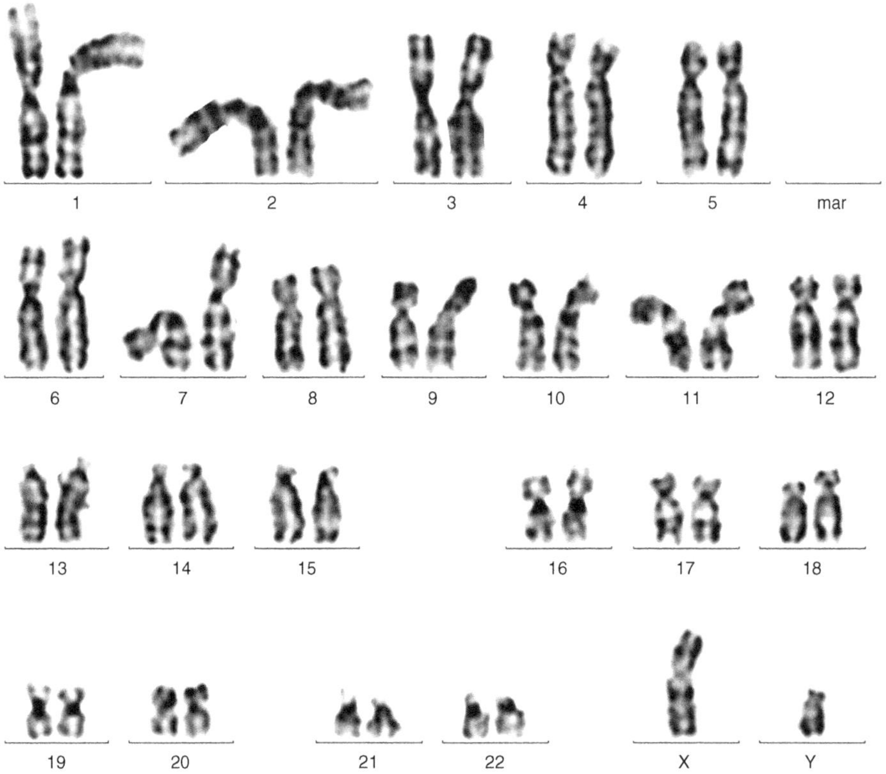

Case Study 3.12 Karyotype showing normal male karyotype

Molecular Studies

The Real-Time Quantitative PCR AML comprehensive panel *showed FLT3-ITD positive.*

Discussion

This patient with normal karyotype in cytogenetic analysis showed FLT3-ITD mutation molecularly. FMS-like tyrosine kinase 3(FLT3) is preferentially expressed on hematopoietic progenitor cells and mediates proliferation and differentiation of these cells [7]. Two types of activating FLT3 mutations have been described in AML. An internal tandem duplication (ITD) of the FLT3 gene, seen in 20%–30% of young adult patients and FLT3-TKD (FLT310835) in ~7% of patients. The presence of FLT3-ITDs is reported in most studies as a marker of poor prognosis associated with an increased risk of relapse and decreased DFS and OS. As per ELN classification also, this patient would fall in the adverse risk category [14].

3.13 Case Study 3.13: Acute Myeloid Leukemia with t(11;19) and FLT3-ITD Mutation

Brief Clinical History

A 30-year-old female was admitted with high-grade fever (on and off) for 1 month with nausea and vomiting, painful gums, and breathlessness on exertion. Peripheral smear showed ~81% blasts. Bone marrow was done for evaluation.

CBC: *Hb*: 4.9 g/dl, TLC: 52,700/cu mm and Platelets: 50,000/cu mm

DLC: Neutrophils: 1.0%, Blasts: 80.0%, Lymphocytes: 11.0%, Eosinophils: 0.0%, Monocytes: 8.0%, Basophils: 0.0%.

Peripheral Blood Smear and Bone Marrow

Smear showed leukocytosis with the presence of ~80.0% blasts and promonocytes. Monocytes were ~8.0%. The bone marrow was replaced by variable-sized blasts and promonocytes constituting ~87.0% of the total nucleated cells. The blasts were large in size with a high N:C ratio, moderate amount of granular cytoplasm, and immature large convoluted nuclei, few showing prominent nucleoli.

Flow Cytometric Immunophenotyping

Blasts (~79.4% of acquired events) expressing moderate CD45, moderate SSC.

CD34+, CD117+, HLA-DR+, CD13+, CD33+, CD38+, CD64+, CD19+, CD7+, CD123+, CD11c+, CD11b +, CD4+, MPO+, cCD3.

The picture was consistent with Acute Myeloid Leukemia with monocytic differentiation with aberrant expression of CD19 and CD7.

Cytogenetics

Conventional Karyotype and FISH

46,XX,t(11;19)(q23;p13.1)[18]/46,XX[2]

Cytogenetic analysis of 20 metaphases showed an abnormal cell clone showing translocation between the long arm of chromosome 11 (11q) and the short arm of chromosome 19 (19p) in 18 metaphases analyzed, which was also confirmed on FISH.

This translocation results in the fusion of the MLL gene at 11q23 with the ELL gene at 19p13.1 and is consistent with a diagnosis of acute myeloid leukemia (AML). It is associated with a poor prognosis.

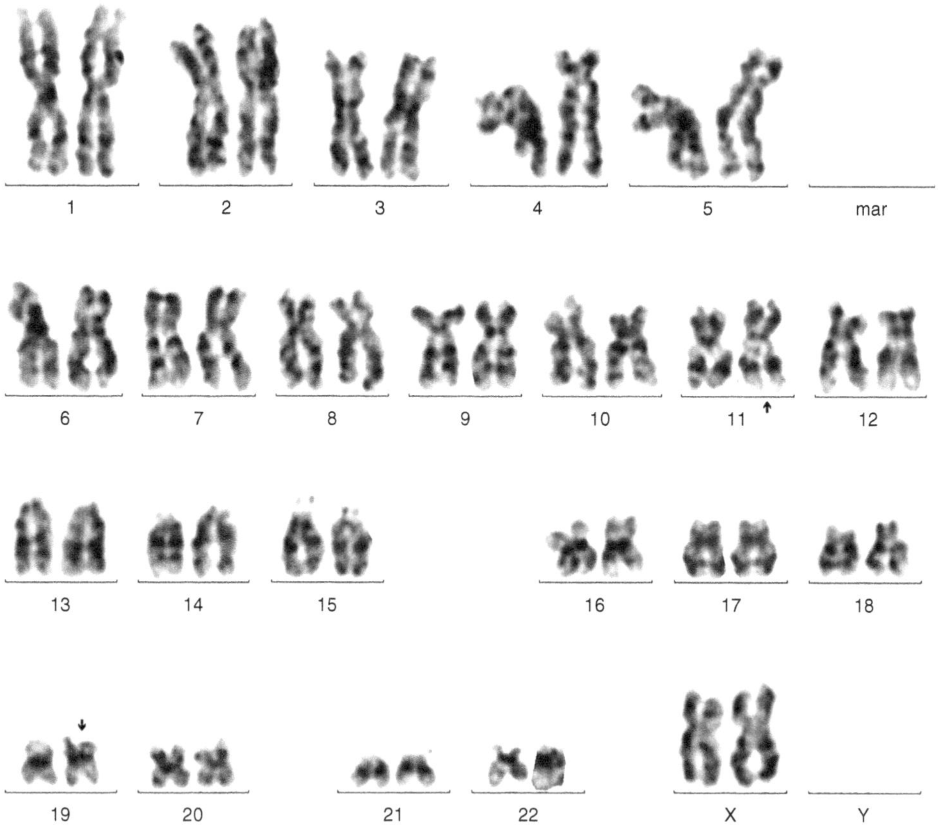

Case Study 3.13 Karyotype showing t(11;19)

Molecular Studies

Since the karyotyping and FISH showed t(11;19) (q23p13.1) MLL-ENL from ALL panel was performed and found to be negative because Molecularly MLL-ENL picks up the Breakpoint at (q23p13.3) and not (q23p13.1). *The AML comprehensive panel was positive for FLT3-ITD gene mutation.*

Discussion

This patient of AML showed a rare and unusual combination of t(11;19) with FLT-3 gene mutation. t(11;19) is seen in 2% of adults and is usually associated with monocytic features as in this case [1]. Translocations involving KMT2A including MLLT3 (AFa) are most common in AML [15]. Other common partners are MLLT1 (ENL) which was investigated in this patient but was found to be negative because the panel used picks up the breakpoint at (q23;p13.3) and not (q23; p13.1), the latter being the case in this patient.

FLT-3 gene mutation in this patent showed the heterogeneity of this mutation in de novo AML and may have developed as a secondary event with a leukemogenic effect in a subset of patients with AML and indicates a poor prognosis.

3.14 Case Study 3.14: t-Acute Myeloid Leukemia with Myelodysplasia-Related Changes (AML-MRC) with Monosomy 7 Transformed from Aplastic Anemia

Brief Clinical History

A 56-year-old male diagnosed with aplastic anemia a year back treated with cyclosporine with partial response has now presented with recurrent skin infections for 1 month. Suspected AML transformation (PBS showed blasts). Bone marrow was done for evaluation.

CBC: *Hb:* 6.3 g/dl; TLC: 1380/cu mm and Platelet count: 60,000/cu mm

DLC: Neutrophils: 35%, Lymphocytes: 32%; Myelocyte: 0.5%; Monocytes: 9.5%; Blasts: 23%; 12nRBC/100 WBC.

Peripheral Blood Smear and Bone Marrow

Smear showed marked pancytopenia, marked leukopenia with neutropenia (ANC ~ 483 cells/cu mm). Approximately 23% of blasts were seen which were medium to large in size with high N:C ratio, opened up chromatin, 1–2 nucleoli, and scant basophilic cytoplasm showing granules in some of the blasts. No Auer rods were seen. Bone marrow showed trilineage dysplasia with blasts (21%). The findings were consistent with *Acute Myeloid Leukemia with myelodysplasia-related changes (AML-MRC)*

Flow Cytometric Immunophenotyping

Blasts constituted 9% of acquired events expressing moderate CD45 and low SSC. They were CD34+, CD117+, cMPO+, CD13+, CD33+, HLA-DR+, CD38+

The flow cytometric immunophenotyping was consistent with acute myeloid leukemia.

Cytogenetics

Conventional Karyotype

45,XY,-7[16]/46,XY[4]

Cytogenetic analysis of unstimulated bone marrow aspirate culture showed neoplastic clone characterized by monosomy 7 in 16 of the 20 analyzed metaphases.

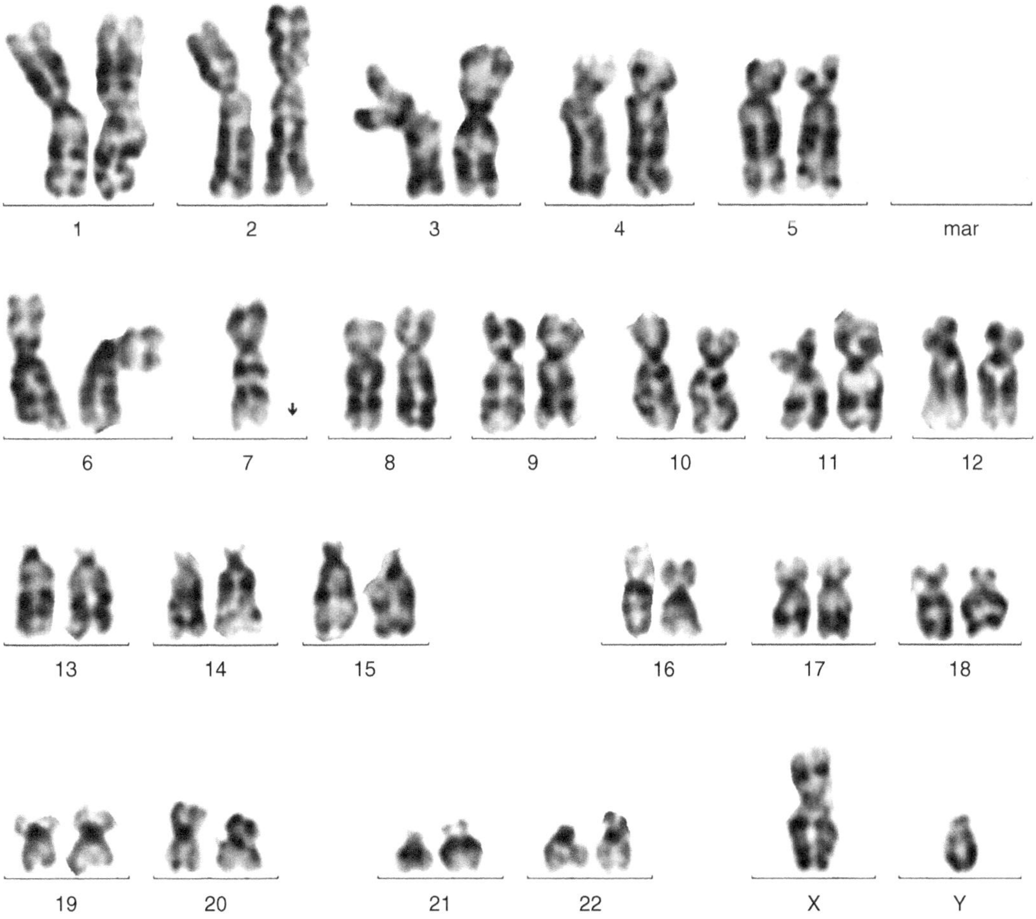

Case Study 3.14 Karyotype showing loss of chromosome 7, i.e., monosomy 7

Discussion

This patient of aplastic anemia transformed into myelodysplasia to acute myeloid leukemia. Almost 10%–15% of aplastic anemia patients typically progress to MDS and/or AML. In a multicenter study by Maciejewski of Cleveland clinic [16], they have identified acquired mutations normally found in leukemia in a large proposition of patients with aplastic anemia which puts them at a higher risk for leukemia. The clonal evolution involved genes such as DNMT 3A, ASXL1, BCoR, and BCORL1, the last two being favorable mutations for immunosuppressive therapy.

In a case report, Takai et al. [17] described a patient of severe aplastic anemia progressing to MDS with monosomy 7 and becoming transfusion independent after immunosuppressive therapy as in this case.

Monosomy 7 does not seem to be a specific feature in MDS, probably a secondary event contributing to lencogenesis. Monosomy 7 is a poor prognostic factor in aplastic anemia [18].

3.15 Case Study 3.15: Acute Myeloid Leukemia with Myelodysplasia-Related Changes with Monosomy 7

Brief Clinical History

A 63-year-old male presented with persistently low hemoglobin and colonic diverticulitis. Bone marrow was done for evaluation of suspected MDS.

CBC: *Hb:* 7.8 g/dl; TLC: 50,200/cu mm and Platelets: 3,58,000/cu mm

DLC: Neutrophils: 20.0%, Lymphocytes: 72.0%, Eosinophils: 2.0%, Monocytes: 3.0%, Basophils: 2.0%

Peripheral Smear and Bone Marrow

Smear showed neutropenia (ANC~1004 cells/cu mm) with an occasional blast and significant dysgranulopoiesis and pseudo Pelger–Huet cells. Bone marrow showed trilineage dysplasia with 24% blast cells.

Cytogenetics

Conventional Karyotype

45,XY,-7,[17]/46,XY[3]

Cytogenetic analysis of unstimulated bone marrow aspirate culture showed a neoplastic clone characterized by monosomy 7 in 17 of the 20 analyzed metaphases.

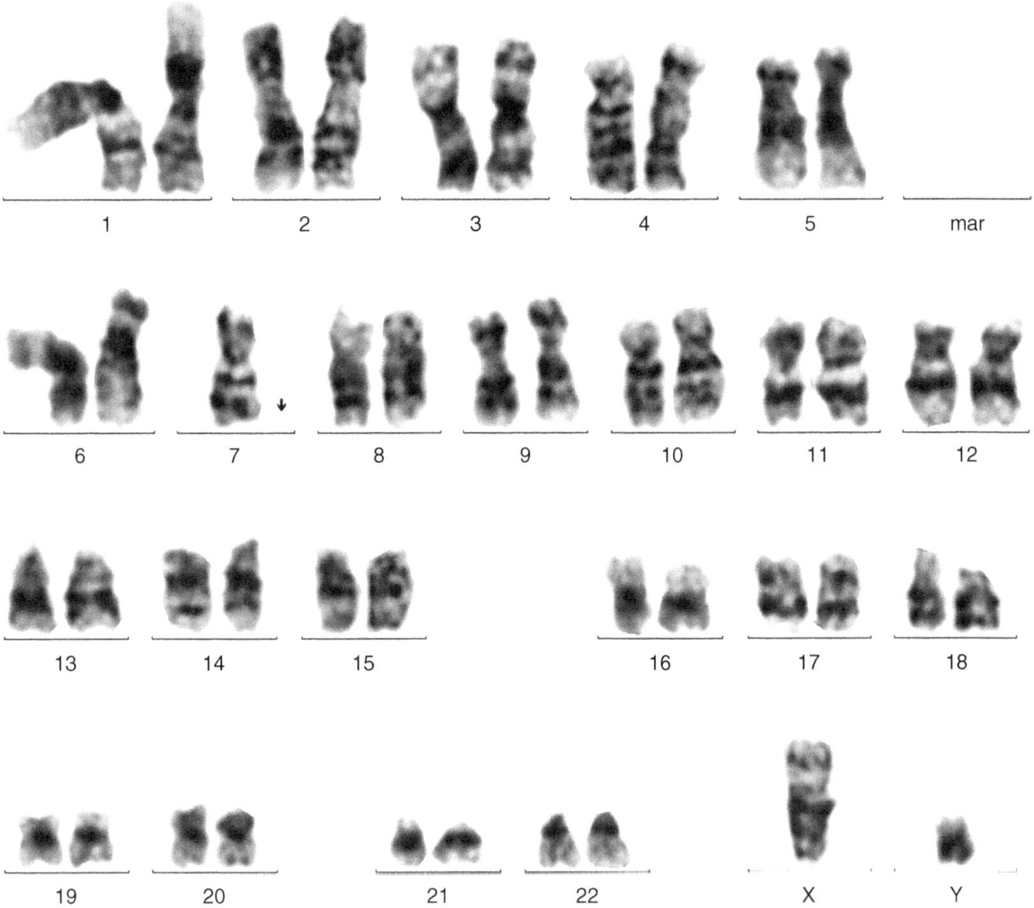

Case Study 3.15 Karyotype showing loss of chromosome 7, i.e., monosomy 7

Discussion

AML-MRC is a heterogeneous disorder defined by morphologic, genetic, or clinical features. Genetic abnormalities are often associated with adverse prognostic features and many are preceded by MDS or MDS/MPN [19]. This patient of AML-MRC on karyotyping showed an isolated monosomy 7. In a study of 415 patients of AML-MRC reported by Montalban Bravo et al. [20], the following cytogenetic changes were found: complex karyotype (76%), monosomy 5 or del(5q) (7%) monosomy 7 or del(7q), in (16%), concurrent chromosome 5 or 7 abnormalities in (1%), del(13q) in (0.5%), i (17) in (0.5%). Most cases of AML-MRC are usually associated with adverse genetic abnormalities, particularly 5/del(5q)or-7/del(7q). It is important to note that in (IPSS-R) del(7q) is in intermediate but monosomy 7 in poor prognosis category [21].

3.16 Case Study 3.16: Acute Myeloid Leukemia with Myelodysplasia-Related Changes with Monosomal Karyotype

Brief Clinical History

A 56-year-old male presented with fever for 2 months and Pancytopenia. Bone marrow was done for evaluation for MDS.

CBC: *Hb:* 9.5 g/dl, TLC: 1690/cu mm and Platelets: 18,000/cu mm.

DLC: Neutrophils: 30.0%, Lymphocytes: 57.0%, Eosinophils: 1.0%, Monocytes: 8.0%,

Basophils: 0.0%, Myelocytes: 1.0%, Blasts: 3.0%

Peripheral Smear and Bone Marrow
Peripheral smear showed leukopenia with neutropenia and 3% blasts as well as dysgranulopoiesis. Bone marrow showed marked trilineage dysplasia with blasts 21% of total nucleated cells. An occasional Auer rod was seen. Iron stores were increased.

Flow Cytometric Immunophenotyping
Blast (~13% of acquired events) was present in CD45 moderate blast region. They were CD34+, CD117+, HLA-DR+, CD13+, CD33+ (moderate to bright), CD38+. All the gated myeloid cells showed evidence of dysplasia on (CD13-CD16) and (CD11b-CD16) dot plots.

Cytogenetics
Conventional Karyotype
43,XY,t(3;17)(q12;p13),-5,-7,-20[18]/46,XY[2]

Cytogenetic analysis of 20 metaphases showed the presence of a neoplastic clone characterized by hypodiploidy with structural and numerical abnormalities. Structural abnormalities showed balanced translocation between the long arms of chromosome 3 (3q) and the short arm of chromosome 17 (17p). The numerical abnormalities included monosomies of chromosomes 5, 7, and 20. These abnormalities indicate a poor prognosis.

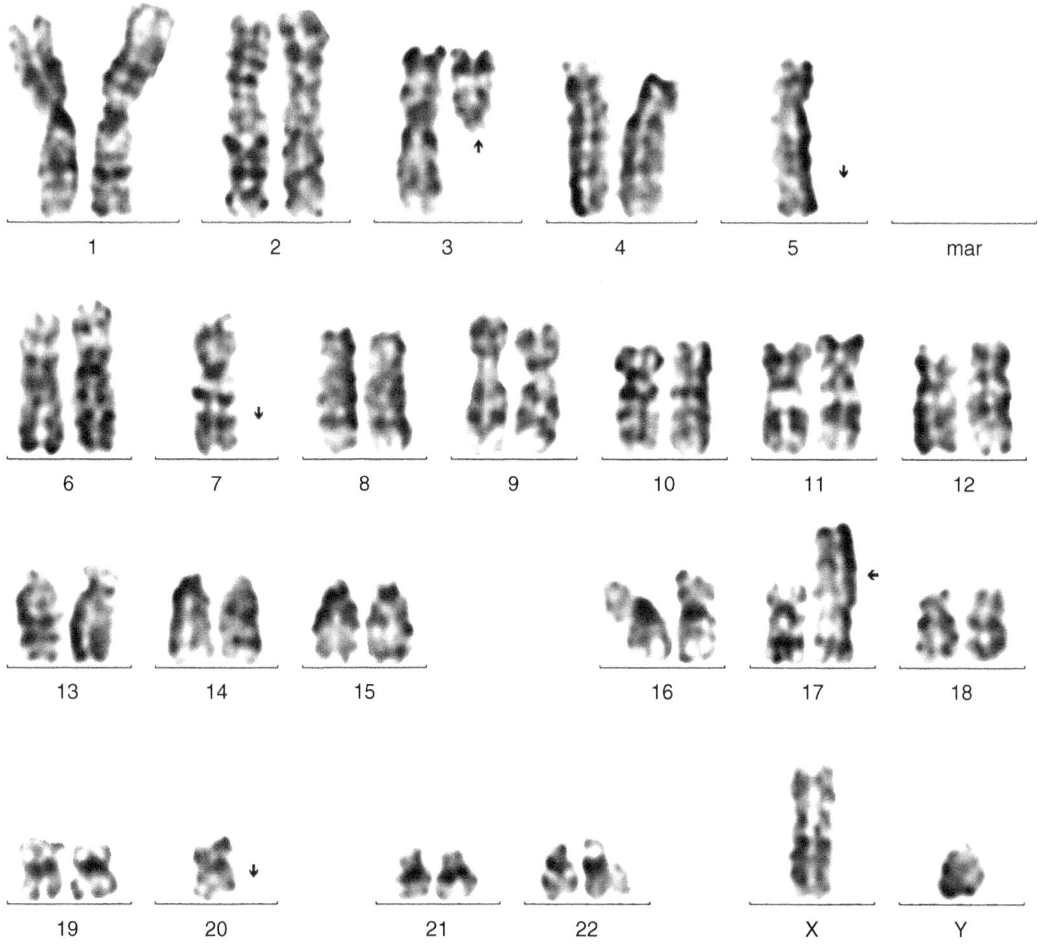

Case Study 3.16 Complex karyotype showing abnormality of 3q, i.e., t(3;17)(q12;p13), -5, -7, and -20

Discussion

This patient with AML-MRC showed morphologic and immunophenotypic features of AML-MRC. Cytogenetics revealed a monosomal karyotype (MK). Monosomal karyotype is defined as the presence of at least two autosomal monosomies or a single autosomal monosomy with at least one structural abnormality (as seen in this case) [22]. In IPSS-R for MDS, the poor and the very poor prognostic groups include patients with MK [21]. Reports on MK as an independent predictor of survival in MDS are conflicting depending on whether MK is present with or without complex karyotype. Patnaik et al. [23] showed that monosomal karyotype in MDS with or without monosomy 7 or 5 is prognostically worse than otherwise a complex karyotype.

3.17 Case Study 3.17: Acute Myeloid Leukemia (s-AML) from (MDS-EB2) with Myelodysplasia-Related Changes with Complex Karyotype

Brief Clinical History

A 62-year-old male was investigated in September 2017 for anemia and thrombocytopenia. Bone marrow reported as hypoplastic marrow. In June 2018, bone marrow repeated, found to have MDS-EB2. In July 2018, bone marrow showed secondary AML. Follow-up case of MDS(EB-2) diagnosed outside, progressed to acute leukemia.

CBC: *Hb*: 5.2 g/dl; Platelets: 11,000/cu mm and TLC: 3660/cu mm.

DLC: Blasts: 34%, Neutrophils: 6.0%, Lymphocytes: 57.0%, Eosinophils: 0.0%, Monocytes: 2.0%

Peripheral Blood Smear and Bone Marrow

Smear showed leukopenia with 34% blasts and few dyspoietic neutrophils. Bone marrow showed erythroid hyperplasia and dyserythropoiesis with blasts constituting 25% of the total nucleated cells. Myeloid and megakaryocytic cells were reduced but showed features of dysplasia.

Flow Cytometric Immunophenotyping

CD45+ blasts in the blast region, CD34+, HLA-DR+,CD117+,CD13+, CD33+,CD38+.

A CD45-negative cluster comprised 55% erythroid cells positive for glycophorin. Picture was consistent with acute myeloid leukemia.

Cytogenetics

Conventional Karyotype

46,XY,del(5)(q14q31),del(7)(q21q32)[3]/44~45,idem,-7,-13,-16, del(20) (q12q13.1)[cp17]

Cytogenetic analysis of unstimulated bone marrow aspirate culture showed a myeloid neoplastic clone characterized by complex karyotype with multiple structural and numerical abnormalities in all the analyzed 20 metaphases. Numerical abnormalities included monosomy 7 (in 1 metaphase), monosomy 13 (in 15 metaphases), and monosomy16 (in 9 metaphases). The structural abnormalities included interstitial deletion in the long arm of chromosome 5 (in 20 metaphases), chromosome 7 (in 18 metaphases), and chromosome 20 (in 15 metaphases).

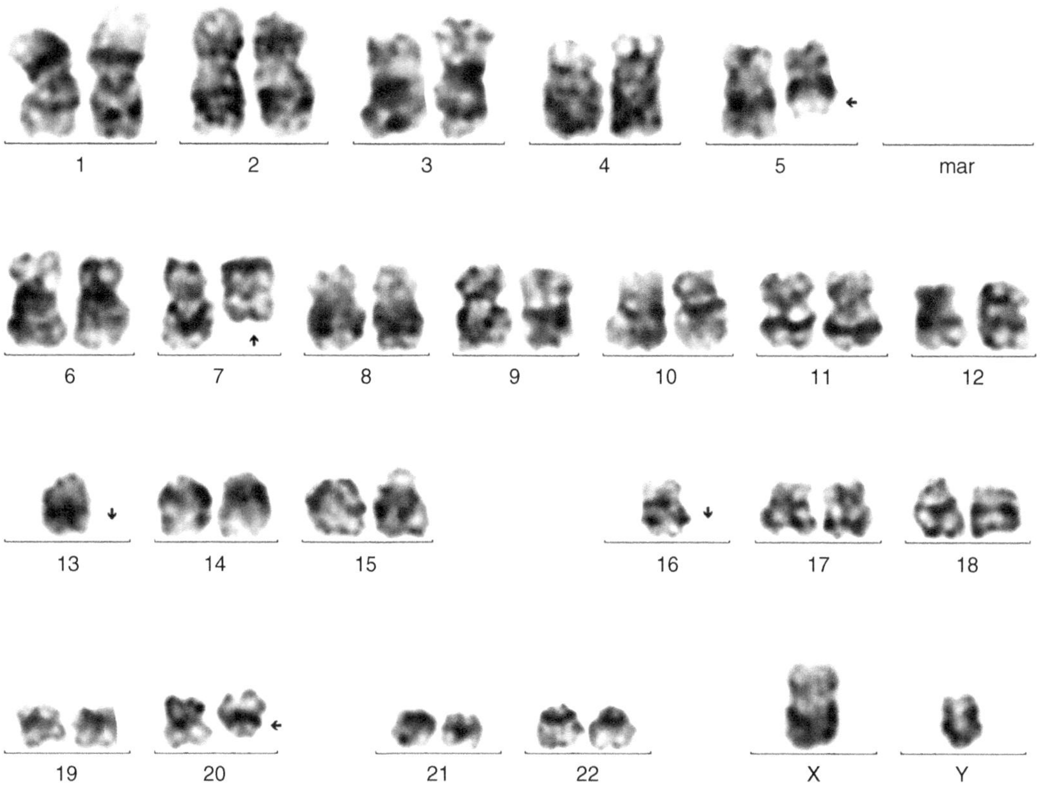

Case Study 3.17 (1) Complex karyotype showing deletion 5q, deletion 7q, and deletion 20q along with loss of chromosomes 13 and 16

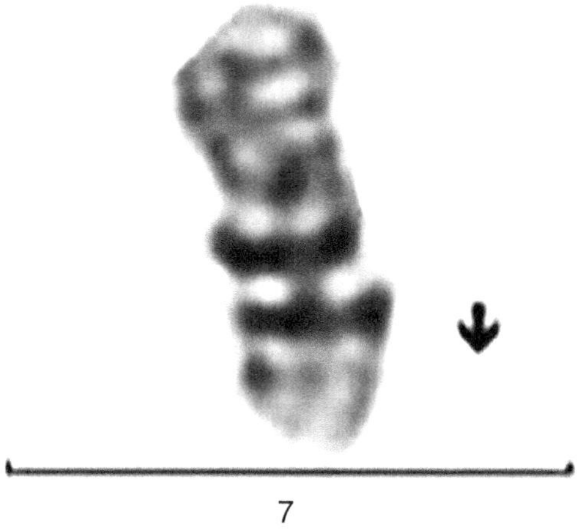

Case Study 3.17 (2) Partial karyotype showing monosomy 7

Discussion

This patient was first found to have a hypocellular marrow with anemia and thrombocytopenia. A year later, BM revealed MDS-EB2 and a month later had evolved into AML, qualifying for s-AML [19]. At this time, he showed a complex karyotype. Acute Myeloid Leukemia (s-AML) from (MDS-EB2) with Myelodysplasia-Related Changes with Complex Karyotype

Secondary acute myeloid leukemia refers to a condition in patients who had prior MDS, MPN, or aplastic anemia with or without treatment that converts to AML (s-AML) or as a product or previous exposure to a previous leukemogenic chemotherapeutic agent (t-AML).

Evolution to s-AML occurs as the acquisition of genetic changes such as activating mutations in signaling molecules like FLT-3 and NRAS as well as inactivating mutations in CEBPA.

Alkylating agents increase the risk of s-AML usually preceded by an MDS phase. Typical cytogenetic abnormalities include aberrations in the long arm of chromosomes 5, 7 or both and the latency period is typically 3.5–5.5 years with a cumulative risk of 10% at 20 years. The prognosis is poor. The treatment of choice is Allogeneic hematopoietic stem cell transplantation [24].

In the WHO classification [1], the criteria for AML-MRC were defined as 20% blasts in blood or BM, history of MDS or MDS/MPN, MDS-related cytogenetics, multilineage dysplasia (MLD), no recurring cytogenetic abnormalities, and no prior cytotoxic therapy. As per these criteria, this patient had s-AML with complex karyotype, del(5q), del(7q), del(20q), and monosomy 7, 13, and 16. The prognosis in these patients is poor [25].

3.18 Case Study 3.18: Acute Myeloid Leukemia (t-AML) with Myelodysplasia-Related Changes with Monosomy 7

Brief Clinical History

A 31-year-old male presented with a history of fever and left lower limb Cellulitis. His peripheral smear showed ~5.0% blasts with suspected acute leukemia. He is a follow-up case of Aplastic Anemia diagnosed and treated in 2015.

CBC: *Hb:* 8.1 g/dl; Platelets: 40,000/cu mm and TLC: 1860/cu mm.

DLC: Neutrophils: 25.0%, Lymphocytes: 25.0%, Eosinophils: 0.0%, Monocytes: 40.0%,
Basophils: 0.0%. Blasts: 10.0%

Peripheral Smear and Bone Marrow

Smear showed leukopenia with neutropenia (ANC~480 cells/cu mm). There were ~10.0% blasts. Bone marrow showed marked trilineage dysplasia with blasts constituting 24% of the total nucleated cells.

Flow Cytometric Immunophenotyping

The immunophenotype showed two populations of cells: one of blasts (~5.0%) with positivity for CD117/CD34/HLA-DR/CD38 and the other of abnormal monocytic cells (~11.0) with positivity for CD64/CD14/CD13/CD33/CD4/CD11c/CD11b. The picture was consistent with AML with monocytic differentiation.

Cytogenetics

Conventional Karyotype

45,XY,-7[11]/46,XY[9]

Cytogenetic analysis of unstimulated bone marrow aspirate culture showed a neoplastic clone characterized by monosomy 7 in 11 of the 20 metaphases analyzed. Monosomy 7 is associated with a poor prognosis.

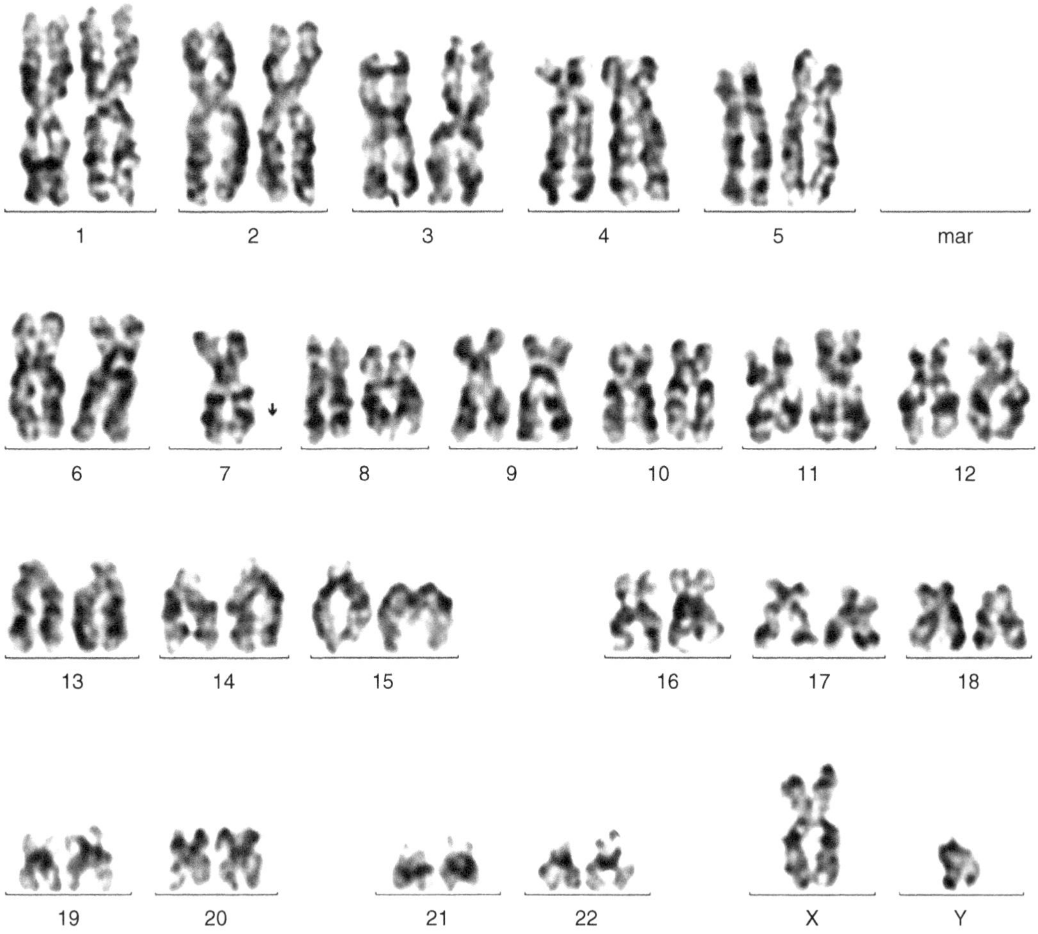

Case Study 3.18 Karyotype showing loss of chromosome 7, i.e., monosomy 7

Discussion

This patient of aplastic anemia treated on immunosuppressive therapy in 2015 developed AML detected in 2018 and found to have AML-MRC with isolated monosomy 7. Inherited bone marrow failure syndromes (IBMFs) have a high risk of transformation to MDS and/or AML. This patient could well have gone through a phase of MDS before developing full-fledged AML-MRC. In a retrospective study of 738 MDS/AML cases of 4293 individuals, monosomy 7 or del (7q) occurred in 17% [26]. This cytogenetic abnormality carries a poorer prognosis as compared to other cytogenetic abnormalities.

In a study of 200 patients of t-AML, Kayser et al. [27] reported that patients of t-AML were older, had lower WBC counts, and abnormal cytogenetics including complex and monosomal karyotypes.

Of these, most frequent were t(9;11),-5/5q-,inv(16,-7/7q-,TP53, complex and monosomal karotype).

Approximately 10% of all AMLs arise after a patient's exposure to chemotherapy and/or radiotherapy for a primary malignancy or autoimmune disease. The prognosis is uniformly poor [28].

3.19 Case Study 3.19: Acute Myeloid Leukemia with Myelodysplasia-Related Changes with Complex Karyotype

Brief Clinical History

An 18-year-old male presented with pancytopenia with suspected aplastic anemia. Bone marrow was done for evaluation of the disease.

CBC: *Hb*: 9.3 g/dl, TLC: 1130/cu mm and Platelets: 40,000/cu mm.

DLC: Neutrophils: 11.0%, Lymphocytes: 79.0%, Eosinophils: 1.0%, Monocytes: 5.0%,
Basophils: 3.0%. Occasional blast seen.

Peripheral Blood Smear and Bone Marrow
Smear showed marked leukopenia and neutropenia (ANC = 124 cells/cu mm) with occasional blasts. The bone marrow aspirate was cellular and showed trilineage dysplasia with ~28% blasts.

The overall morphologic features were consistent with acute myeloid leukemia with myelodysplasia-related changes (AML-MRC)

Flow Cytometric Immunophenotyping
Total Blasts were 17% of acquired events expressing moderate CD45 and low SSC. Two distinct blast populations were seen on CD34 vs. SSC plot. First population constituted 9% blasts CD34+, CD117+, CMPO+, CD13+, and CD33+. Second population were 8% blasts, CD3+, CD14+, CD4+, CD64+, CD11c+, CD16+. Immunophenotype was consistent with AML with monocytic differentiation.

Cytogenetics
Conventional Karyotype

45,XY,dic(1;2)(q10;q10),del(7)(q11.2),+8, del(9)(p13),add(16)(q24),-18[18]/46,XY[2]

Cytogenetic analysis of 20 metaphases showed a complex karyotype with numerical and structural abnormalities. The former were trisomy of chromosome 8 and monosomy of chromosome 18. Structural abnormalities were the formation of the dicentric chromosome between long arms of chromosomes 1 and 2. In addition, there was a deletion of the long arm of chromosome 7, deletion of the short arm of chromosome 9, and addition of unknown material on the long arm of chromosome 16 in 18 of the 20 metaphases analyzed.

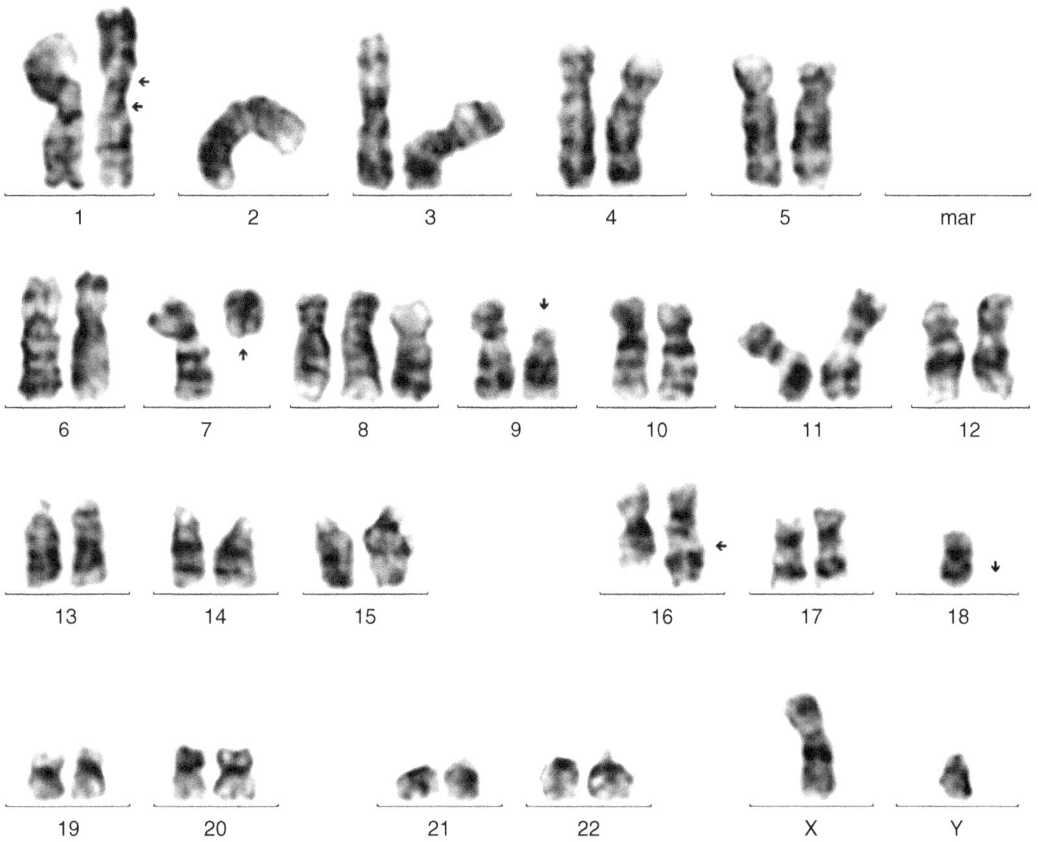

Case Study 3.19 Complex karyotype showing deletion 7q, deletion 9p, derivative 16, and monosomy 18

Discussion

This patient had AML-MRC with a complex karyotype. Since no mutations of NPM1 or biallelic mutations of CEBPA were known in this patient, multilineage dysplasia alone qualified him for AML-MRC. The presence of multilineage dysplasia alone will not classify a case as AML with myelodysplasia-related changes when a mutation of NPM1 or biallelic mutation of CEBPA is present. In cases lacking these mutations, the morphologic detection of multilineage dysplasia (presence of 50% or more dysplastic cells in at least two cell lines) is sufficient to make a diagnosis of AML-MRC. The presence of an MDS-related cytogenetic abnormality with one exception, del(9q) remains an inclusion criterion for AML-MRC [19]. There was formation of a dicentric chromosome (DC) between long arms of chromosomes 1 and 2 in addition to trisomy 8, deletion 7q, deletion 9p, and monosomy 18. Dicentric chromosomes are markers of cancer and are seen in many hematologic diseases including AML. FISH analysis often confirms the karyotypic findings of dicentric chromosomes. Karyotypes with DCs are often complex and indicate a poor prognosis [29].

3.20 Case Study 3.20: Acute Myeloid Leukemia with Myelodysplasia-Related Changes (MDS-EB1) with Complex Karyotype

Brief Clinical History

A 55-year-old female, a diagnosed case of MDS-EBI, presented with weakness and mouth ulcers for the past 2 months. Bone marrow was done for evaluation of disease status.

CBC: *Hb:* 6.8 g/dl, TLC: 2980/cu mm and Platelets: 12,000/cu mm

DLC: Neutrophils: 25.0%, Lymphocytes: 46.0%, Eosinophils: 8.0%, Monocytes: 10.0%, Basophils: 0.0%, Blasts: 11.0%:

Peripheral Smear and Bone Marrow

Peripheral smear showed leukopenia with severe neutropenia (ANC ~ 745 cells/cu mm and 11% blasts which were medium to large with high N:C ratio, scant to moderate amount of cytoplasm, with few showing prominent nucleoli. Bone marrow showed trilineage dysplasia with similar blasts (26%). The overall picture was consistent with AML—with myelodysplasia-related changes, in a known case of MDS-EB1.

Flow Cytometric Immunophenotyping

Total Blasts constituted ~6.04% of acquired events expressing CD45 moderate and SSC low, Blasts were CD34+, CD117+, CD13+, CD33+.

Cytogenetics

Cytogenetics Karyotype

42~44,X,-X,dup(1)(p36,3p22),del(1)(q21),-5,-7,-12,add(12)(p13),-14, del(17) (p13), add(18) (q23) [cp8]/45,X,-X[3]/46,XX[9]

Cytogenetic analysis of 20 metaphases showed a hypodiploid complex karyotype with numerical and structural abnormalities. Numerical abnormalities were loss of chromosomes 5, 7, 12, 14, and X. Structural abnormalities were duplication of the short arm of chromosome 1, i.e., dup (1p) along with deletion of its long arm. There was an addition of unknown material on the short arm of chromosome 12 and the long arm of chromosome 18. Also, there was a deletion of the short arm of chromosome 17, i.e., TP53 gene deletion which was also confirmed by FISH.

The above cytogenetics alterations are associated with poor prognosis.

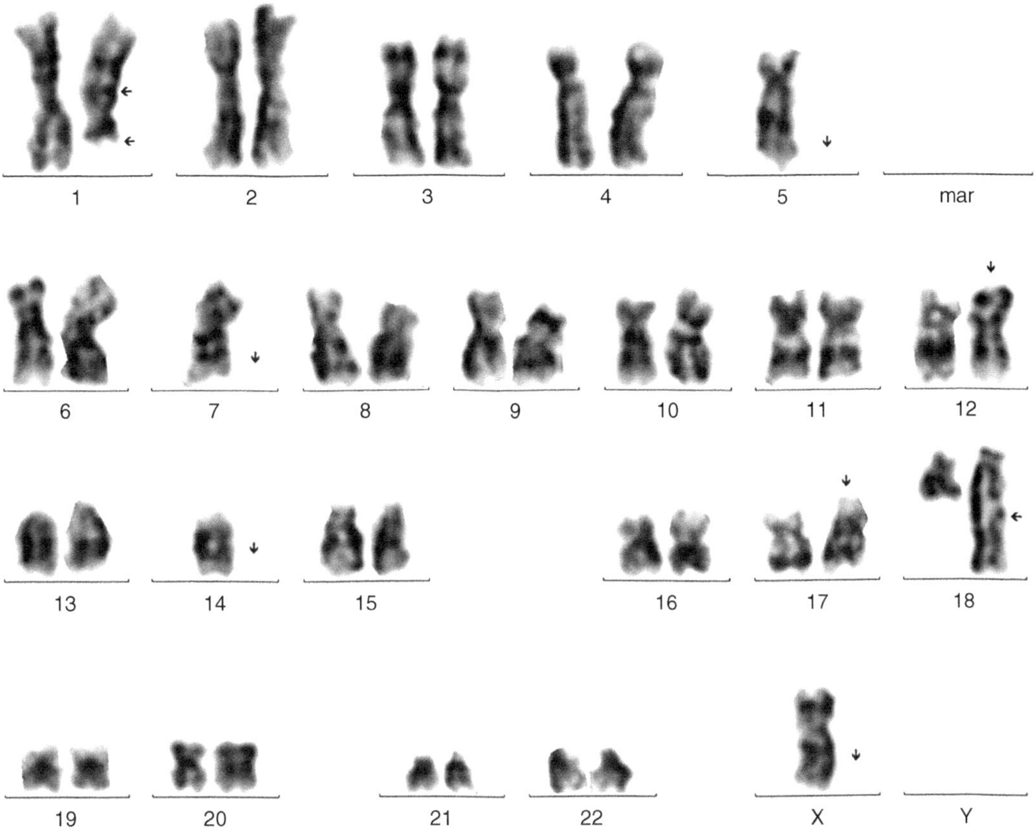

Case Study 3.20 (1) Complex karyotype showing monosomy 5, monosomy 7, monosomy 14, along with loss of X chromosome and other abnormalities

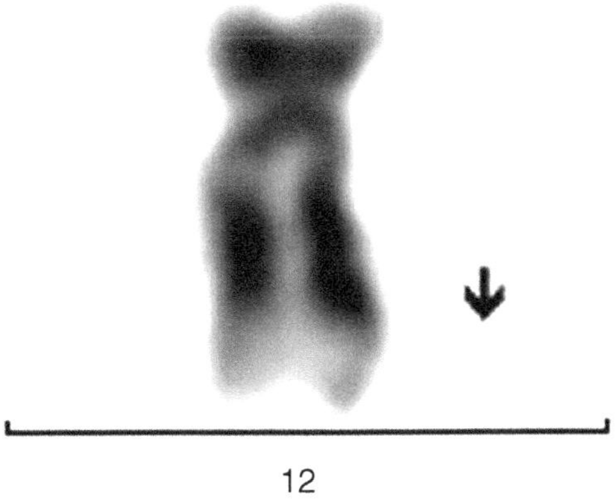

Case Study 3.20 (2) Partial karyotype showing monosomy 12

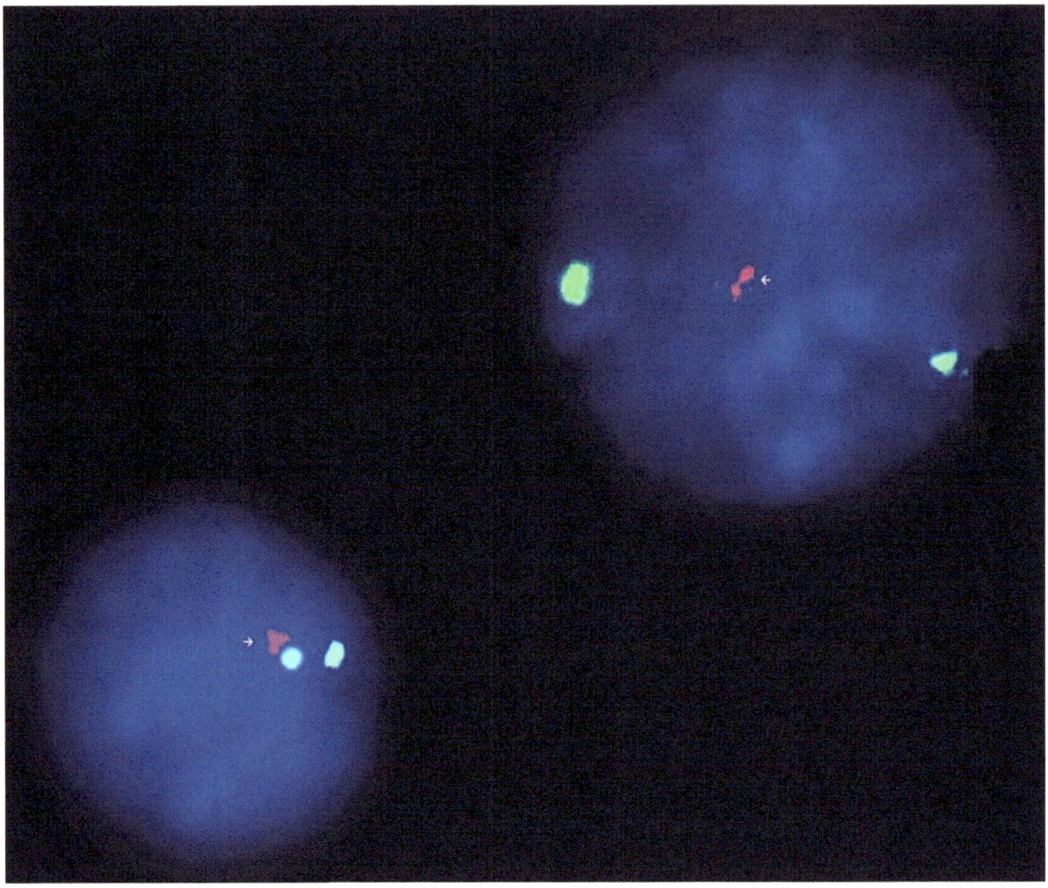

Case Study 3.20 (3) FISH image showing 2G1R signal pattern using *TP53* DC deletion probe indicating *TP53* deletion

Molecular Findings

The Real-Time Quantitative PCR showed *AML Comprehensive Panel: Negative.*

Discussion

This patient of AML had a previous diagnosis of MDS-EB1 with significant morphologic dysplasia. AML-MRC has no specific phenotype [30]. However, blasts show myeloid differentiation.

AML-MRC with NPM1 or CEBPA mutations are excluded from the AML-MRC category The presence of multilineage dysplasia alone will not classify a case as AML with myelodysplasia-related changes when a mutation of NPM1 or biallelic mutation of CEBPA is present. In cases lacking these mutations, the morphologic detection of multilineage dysplasia (presence of 50% or more dysplastic cells in at least 2 cell lines) is sufficient to make a diagnosis of AML-MRC. The presence of an MDS-related cytogenetic abnormality with one exception, del(9q) remains an inclusion criterion for AML-MRC [19]. In this case, cytogenetic abnormalities showed a complex karyotype, notably dup (1p) and TP53 gene deletion in addition to other numerical and structural abnormalities, also those seen in MDS such as loss of chromosomes 5, 7, 12, and 14. Complex karyotype in AML-MRC have very poor prognosis with an overall survival of 0.7 years [31].

Next-Generation Sequencing (NGS) on PGM (Ion Torrent)

Oncomine Myeloid Panel

This patient's sample was also subjected to NGS to see if any of the alterations seen on karyotyping would also be seen on NGS. No driver variant was identified.

3.21 Case Study 3.21: Acute Myeloid Leukemia (t-AML) with t(11;19)(q23;p13.1) (KMT2A-ELL)

Brief Clinical History

A 69-year-old female with a known case of carcinoma right breast post surgery and chemoradiotherapy 2 years ago presented with complaints of generalized weakness and breathlessness on exertion for 1 month. No history of fever, organomegaly, lymphadenopathy, bleeding from any sites, or any other physical signs of clinical importance were found.

CBC: *Hb:* 8.0 g/dl, TLC: 1,29,190/cu mm and Platelets: 60,000/cu mm.

DLC: Blasts: 93%, Neutrophils: 2%, Lymphocytes: 5%. Occasional nRBC was seen.

Peripheral Blood Smear and Bone Marrow

Smear showed hyperleukocytosis with the presence of ~93% blasts which were medium in size with high N:C ratio, moderate amount of cytoplasm, large nucleus with opened up chromatin, and prominent nucleoli.

Bone marrow smears showed blasts (~93%) with a morphology similar to that seen in the peripheral blood smear.

Flow Cytometric Immunophenotyping

The leukemic cells (~55% of acquired events) in the blast region on CD45 vs SSC dot plot had myeloid immunophenotype with CD34+, cMPO negative, CD123+, CD33 bright, CD117+, *CD11c+,* CD4 dim, *CD56+,* CD38+.

Cytogenetics

Conventional Karyotype and FISH

48,XX,t(11;19)(q23;p13.1)[11]

Cytogenetic analysis of 11 metaphases obtained showed the presence of an abnormal cell clone characterized by a translocation between the long arm of chromosome 11 (11q) and the short arm of chromosome 19 (19p). This finding was confirmed on FISH analysis using MLL breakapart probe.

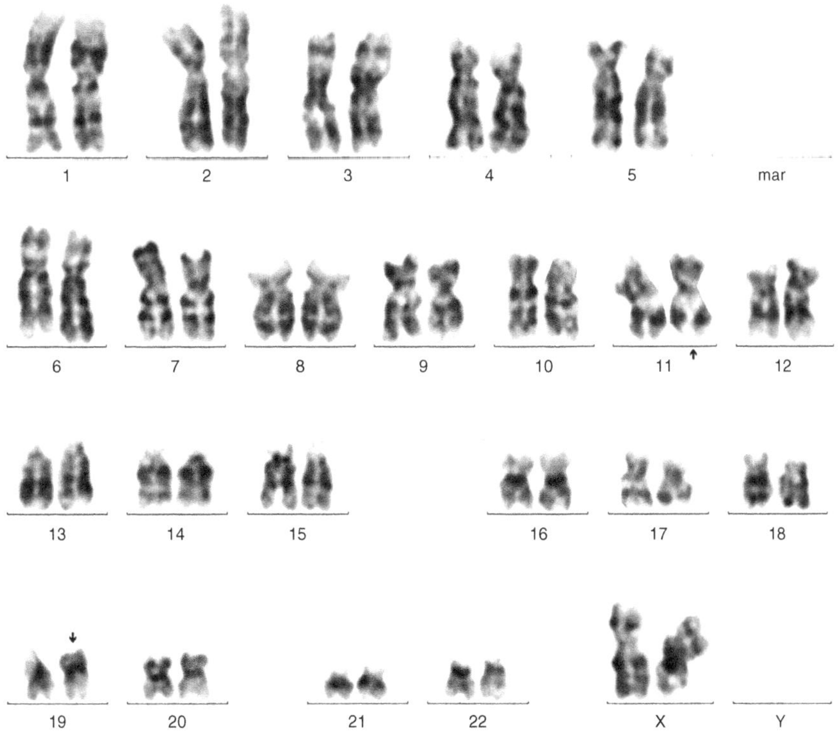

Case Study 3.21 (1) Karyotype showing t(11;19)

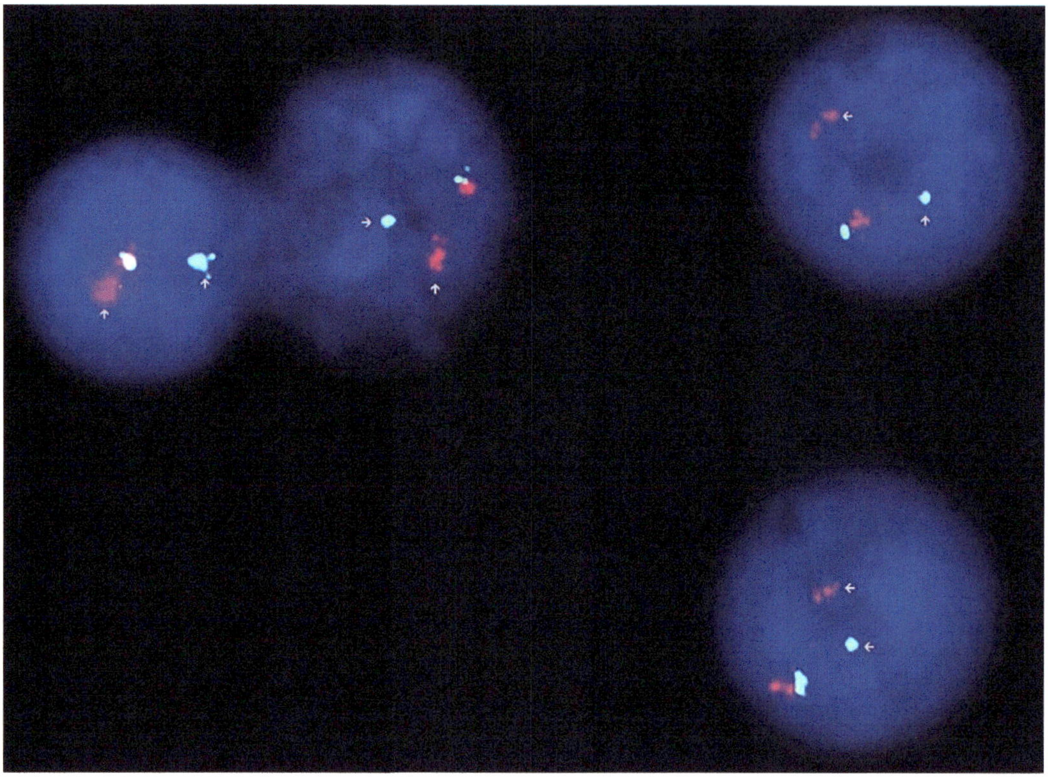

Case Study 3.21 (2) FISH image showing 1F1G1R signal pattern using *MLL(=KMT2A)* probe indicating *MLL(=KMT2A)* rearrangement

Molecular Findings

The Real-Time Quantitative PCR for the *AML panel was negative.*

Discussion

This patient of AML had a rare t(11;19)(q23;p13.1) (KMTZ1-ELL) translocation post chemotherapy and radiotherapy for breast cancer. Huret et al. [32] have described two different translocations (and two clinical entities) both involving 11q23 with a common breakpoint in MLL and 19p13 with different breakpoints also seen. In addition, the t(11;19)(q23;p13.1) was also seen as in this case. The phenotype was M4/M5 and most often therapy-related acute leukemia as in this case. Translocation 19p13.1 is found only in AML and involves the ELL gene. This rare cytogenetic abnormality is associated with a very poor prognosis.

In another study, Moorman et al. [33] reported that in 550 patients, 53 had a translocation involving chromosome 11, breakpoint q23 and chromosome 19, breakpoint (p13). Patients with t(11;19)(q23;p13.1) all had acute myeloid leukemia and predominantly in adults. The median survival in these patients was 2 years.

3.22 Case Study 3.22: Therapy-Related Acute Myeloid Leukemia (t-AML) with Hypodiploid Complex Karyotype

Brief Clinical History

A 76-year-old male diagnosed as Non-Hodgkin's Lymphoma (low grade) stage IV in February 2012 completed chemotherapy. Has now presented with cytopenia(s).

CBC: *Hb*: 7.3 g/dl, TLC: 1510/cu mm and Platelets: 27,000/cu mm.

DLC: Neutrophils: 4.0%, Lymphocytes: 83.4%, Eosinophils: 0.0%, Monocytes: 12.6% , Basophils: 0.0%

Peripheral Blood Smear and Bone Marrow
Smear showed Leukopenia with marked neutropenia (ANC~60 cells/cu mm). An occasional blast was seen. Bone marrow was hypocellular and showed 30% blasts among the total nucleated cells.

Flow Cytometric Immunophenotyping
Blasts 25% of acquired events in the blast region which were CD34+, CD13+, CD33+, HLA-DR+, CD117+, CD38+

Cytogenetics
Conventional Karyotype

42~44,X,-Y,-4,del(5)(q15q33)-7,del(17)(p13),-18,-19,-20,add (21)(q22),-22,+mar1,+mar2,+mar3, [cp11]/46,XY[9]

Cytogenetic analysis of 20 metaphases showed (hypodiploid complex) karyotype with structural and numerical abnormalities. Structural abnormalities showed deletion of the long arm of chromosome 5 (5q), deletion of short arm of chromosome 17, i.e., TP53 gene deletion confirmed by FISH, and addition of unknown material on the long arm of chromosome 21. Numerical abnormalities were monosomies of chromosomes 4, 7, 18, 19, 20, and 22. There was also loss of sex chromosome Y and gain of 1–3 marker chromosomes of unknown origin in 11 of the 20 metaphases analyzed. The above cytogenetic abnormalities indicate a poor prognosis.

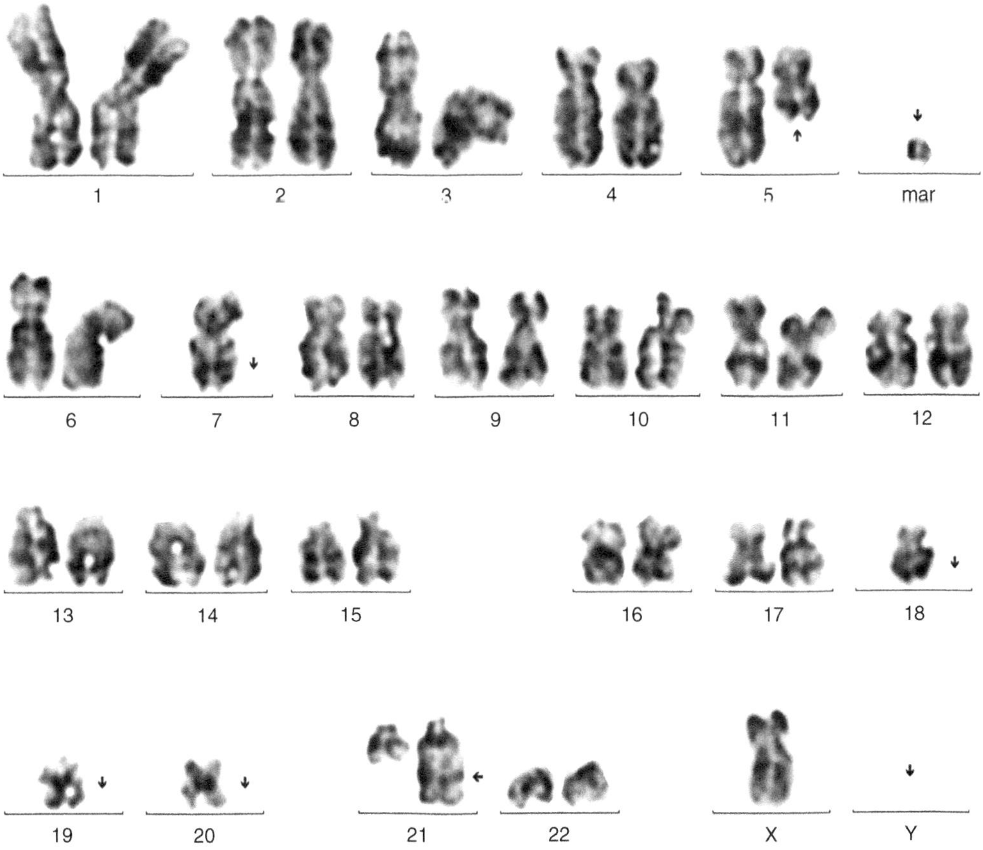

Case Study 3.22 (1) Complex karyotype showing deletion 5q, monosomy 7, monosomy 18, 19, and 20, along with loss of chromosome Y and derivative 21

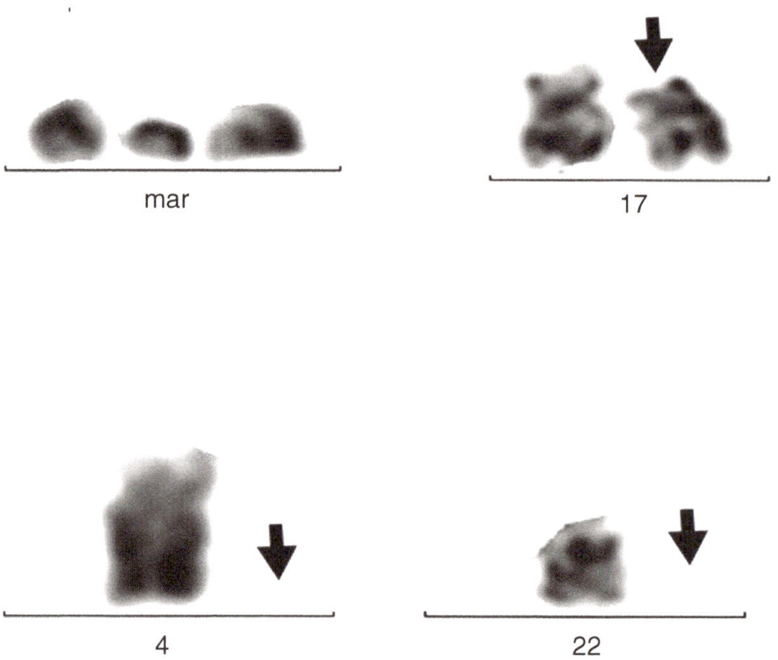

Case Study 3.22 (2) Partial karyotype showing monosomy 4, monosomy 22, derivative 17, and three different marker chromosomes of unknown origin

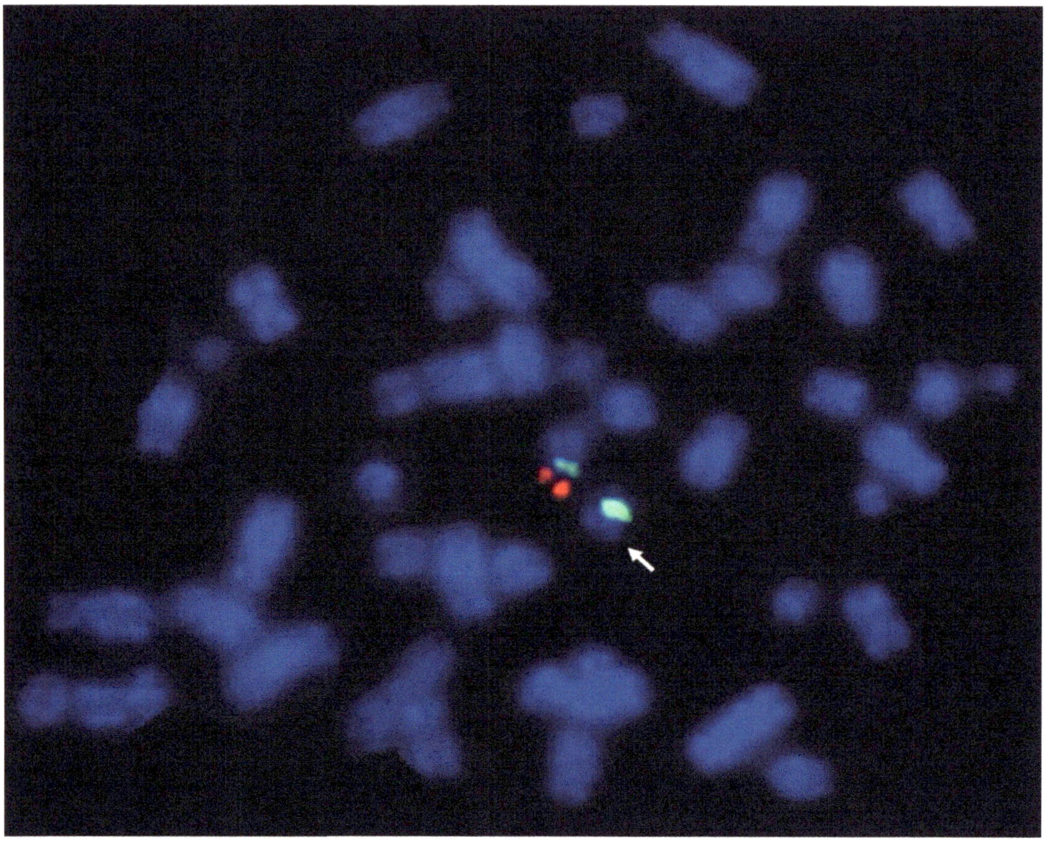

Case Study 3.22 (3) Metaphase FISH image showing 2G1R signal pattern using *TP53* DC deletion probe indicating *TP53* deletion

Discussion

Therapy-related MDS and AML (t-MDS/t-AML) are thought to be a direct consequence of mutations induced by chemotherapy, radiation therapy, or immunotherapy. The frequency of unfavorable cytogenetics such as complex karyotype is considerably higher in t-AML [34]. This patient of AML had a hypodiploid complex karyotype associated with TP53 gene deletion together with del (5q) and monosomies of chromosomes 4, 7, 18, 19, 20, and 22. He developed AML following completion of therapy for non-Hodgkin's lymphoma. Christiansen et al. [35], in a study of 52 t-MDS and 25 t-AML patients, reported TP53 mutations in 21 of 77 cases of t-MDS or t-AML and 19 of these 21 patients had received alkylating agents. Mutations with loss of function of TP53 are significantly associated with del (5q) in t-MDS and t-AML as seen in this patient, who had received alkylating agents for NHL.

3.23 Case Study 3.23: s-AML with Monocytic Differentiation with der(1) t(1;15)(q11;q13) Evolving from Polycythemia Vera

Brief Clinical History

The patient is a known case of Polycythemia vera, treated with hydroxyurea and phlebotomy in 2001. Now presented with weakness and loss of appetite since the past 4 months. On examination, multiple subcutaneous fleshy nodules with massive splenomegaly were noted. Peripheral blood reported as Acute Leukemia. Bone marrow was done for evaluation.

CBC: *Hb:* 6.6 g/dl, TLC: 131,080/cu mm and Platelets: 45,000/cu mm

DLC: Neutrophils: 62.0%, Myelocytes: 1.0%, Blasts: 21.0%, Lymphocytes: 3.0%, Eosinophils: 0.0%, Monocytes: 13.0% , Basophils: 0.0%

Peripheral Blood Smear and Bone Marrow

The Peripheral blood showed hyperleukocytosis with 21% blasts and promonocytes together with 13% abnormal monocytoid cells. Myeloid and monocytic cells showed significant dyspoiesis. Bone marrow aspirate showed similar blasts and promonocytes (23%). Erythroid cells and megakaryocytes showed dyspoiesis.

The peripheral blood, bone marrow aspirate, and trephine biopsy were consistent with Acute Myeloid Leukemia with monocytic differentiation in a known case of Polycythemia Vera.

Flow Cytometric Immunophenotyping

Blasts (19%) in CD45 blast region with 34+, CD117+, CD13+, CD33+, HLA-DR+.

Cytogenetics

Conventional Karyotype

47,XY,+der(1)t(1;15)(q11;q13)[20]

Cytogenetic analysis of 20 metaphases showed a neoplastic clone showing both structural and numerical abnormalities. Numerical abnormality showed gain of derivative chromosome 1 and structural abnormality of unbalanced translocation between the long arm of derivative chromosome 1 and the long arm of chromosome 15. These changes showed as partial trisomy of the long arm of chromosome 1 (1q) in all 20 metaphases analyzed.

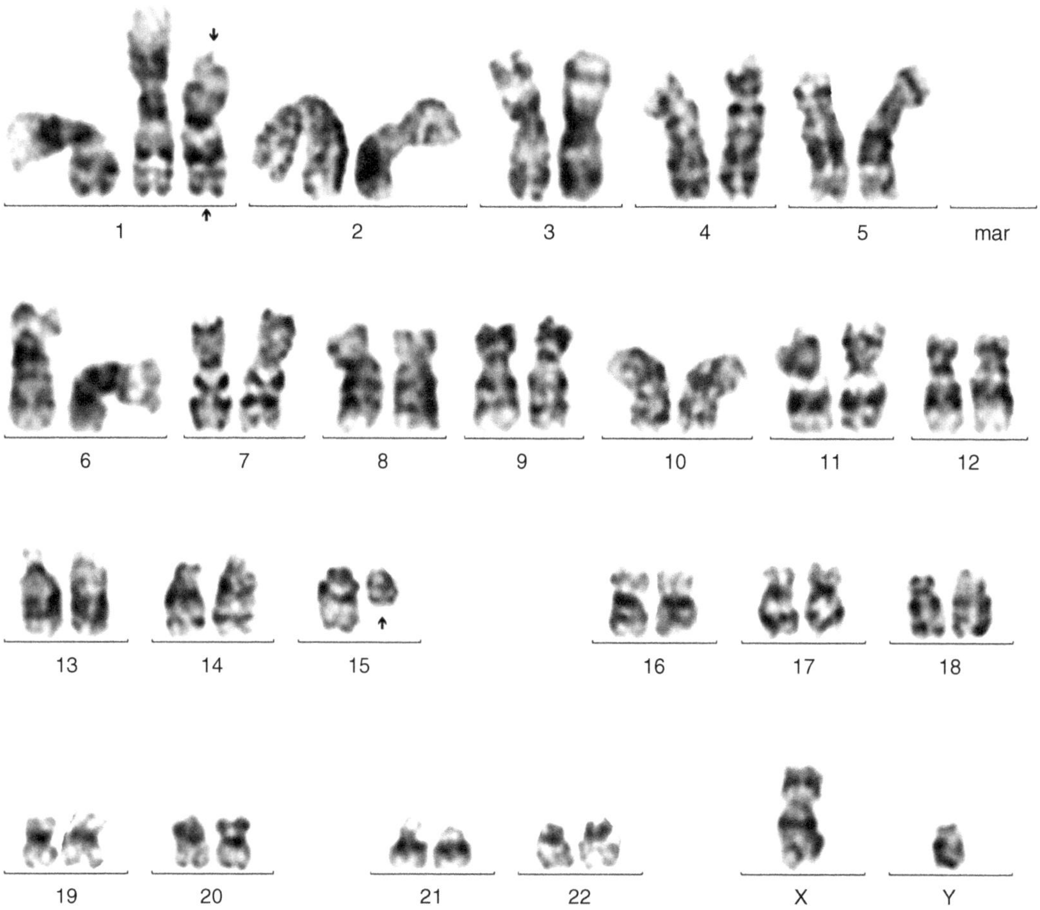

Case Study 3.23 Karyotype showing derivative chromosome 1 formed by an unbalanced translocation between the short arm of chromosome 1 and the long arm of chromosome 15

Molecular Findings

The Real-Time Quantitative PCR for *AML comprehensive panel was negative.*

Discussion

This patient evolved into acute myeloid leukemia from preexisting polycythemia vera. Myeloproliferative neoplasms are a heterogeneous group of disorders of hematopoietic stem cells with an intrinsic risk of evolution into AML. This risk is highest in PMF (10%–20% at 10 years) followed by PV (2.3% at 10 years and 7.9% at 20 years). In ET, this evolution is uncommon. *Leukemic evolution is generally associated with advanced age, leukocytosis, exposure to myelosuppressive therapy, cytogenetic abnormalities, and a number of gene mutations* [36]. In the Mayo cohort, the cumulative incidence of blast transformation (BT) was 3.8% for ET, 6.8% for PV, and 14.2% for PMF [37]. In this patient treated on hydroxyurea, a number of structural and numerical abnormalities were found. Karyotyping has an important role in prognosis. In this patient, the prognosis would be poor.

3.24 Case Study 3.24: Acute Myeloid Leukemia (s-AML) from Primary Myelofibrosis with Complex Karyotype

Brief Clinical History

A 63-year-old male was diagnosed with JAK 2 positive primary myelofibrosis in June 2015. Has now presented with breathlessness for the last 4–5 days. Bone marrow was done for evaluation of the disease. No history of fever, lymphadenopathy, bleeding from any sites, or any other physical signs of clinical importance were found.

CBC: *Hb:* 7.5 g/dl, TLC: 1240/cu mm and Platelets: 20,000/cu mm.

DLC*:* Neutrophils: 58.0%, Lymphocytes: 28.0%, Eosinophils: 0.0%, Monocytes: 7.0%, Basophils: 0.0%, Metamyelocytes: 1.0%, Myelocytes: 5.0%, Blast: 1.0%.

Peripheral Blood Smear and Bone Marrow

Smear showed pancytopenia with leukoerythroblastosis and 1% blasts. Platelets were markedly reduced. Bone marrow aspirate was hemodilute, paucicellular and imprint smears showed only a few hematopoietic cells. Bone marrow biopsy showed an overall cellularity of ~90% and was replaced by sheets of myeloblasts. There was pericellular fibrosis.

Cytogenetics

Conventional Karyotype and FISH

46,XY,t(1;11)(p32;q23),dup(3)(q25q29),del(20)(q11.2)[8]/51,XY,idem,+8,+13,+18,+19,+20[12]

Cytogenetic analysis of 20 metaphases showed a neoplastic clone showing complex karyotype with structural and numerical abnormalities. The structural abnormalities were balanced reciprocal translocation between the short arm of chromosome 1 (1p) and the long arm of chromosome 11(11q), duplication of the long arm of chromosome 3 (3q), and deletion of the long arm of chromosome 20 (20q) in all 20 metaphases analyzed. Twelve of the 20 metaphases also showed hyperdiploidy with gain of chromosomes 8, 13, 18, 19, and 20 with the existing structural abnormalities.

The above structural abnormalities were also confirmed by FISH which indicated that there was MLL gene rearrangement and duplication of (3q) EVI1(MECOM) gene. Collectively, all these abnormalities indicate a poor prognosis.

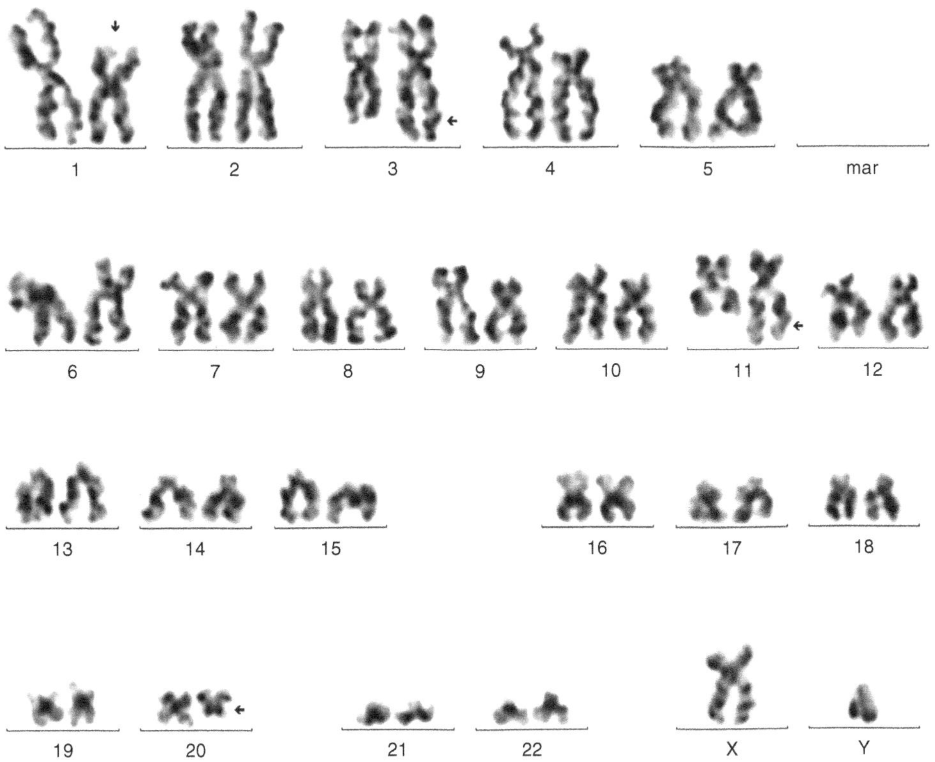

Case Study 3.24 (1) Karyotype showing t(1;11) along with derivative chromosome 3 which is formed by duplication of 3q along with deletion of 20q

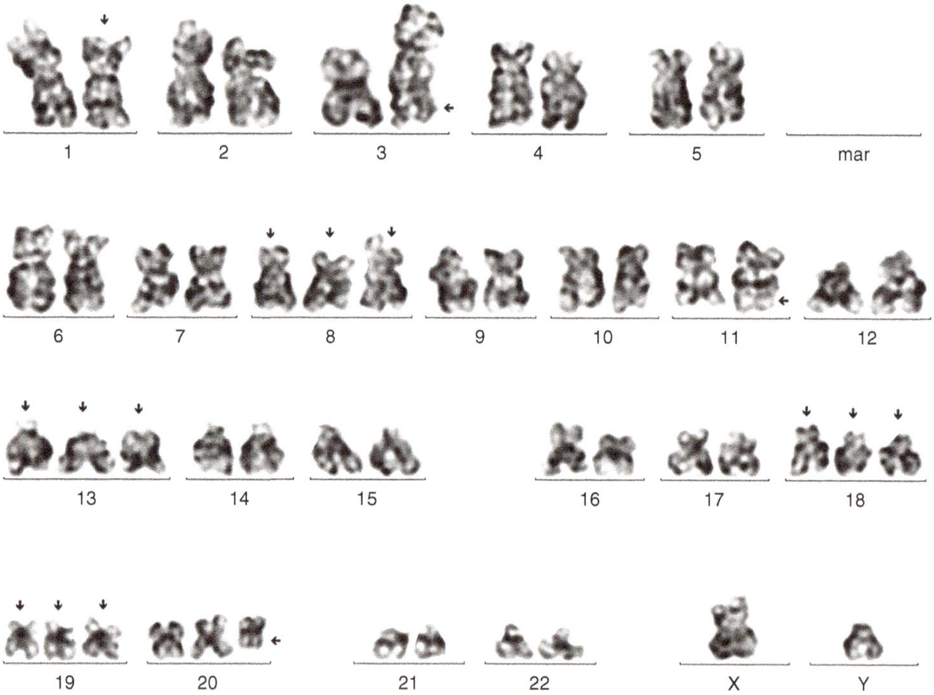

Case Study 3.24 (2) Complex karyotype showing t(1;11), duplication of 3q, deletion of 20q, and trisomy of chromosomes 8, 13, 18, and 19

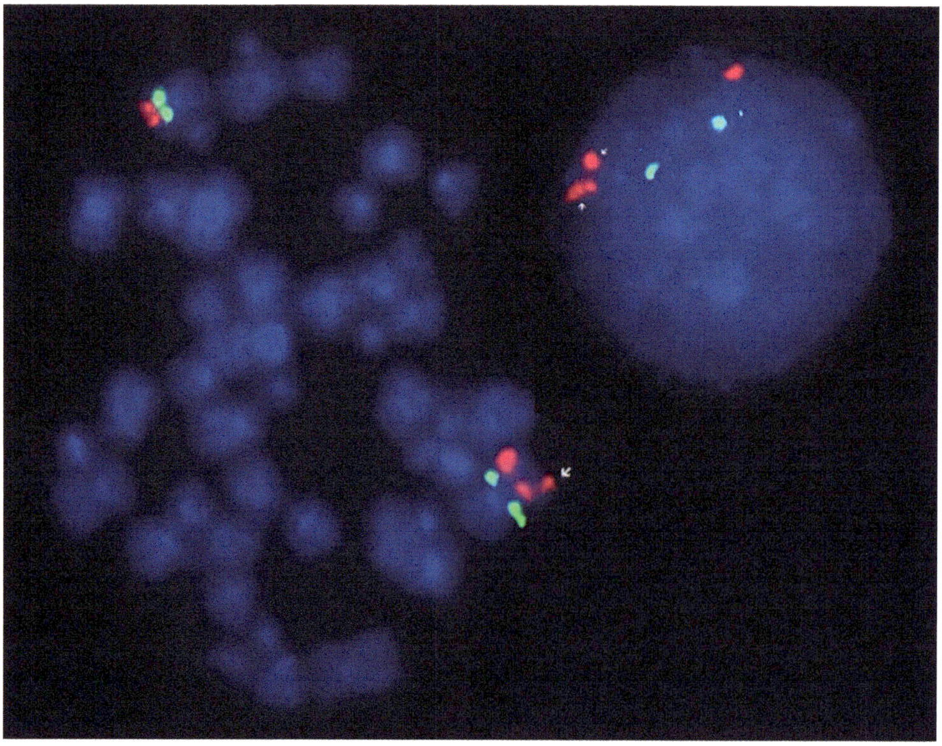

Case Study 3.24 (3) Metaphase FISH image showing *MECOM/GATA2* duplication using dual-color *MECOM/GATA2* probe

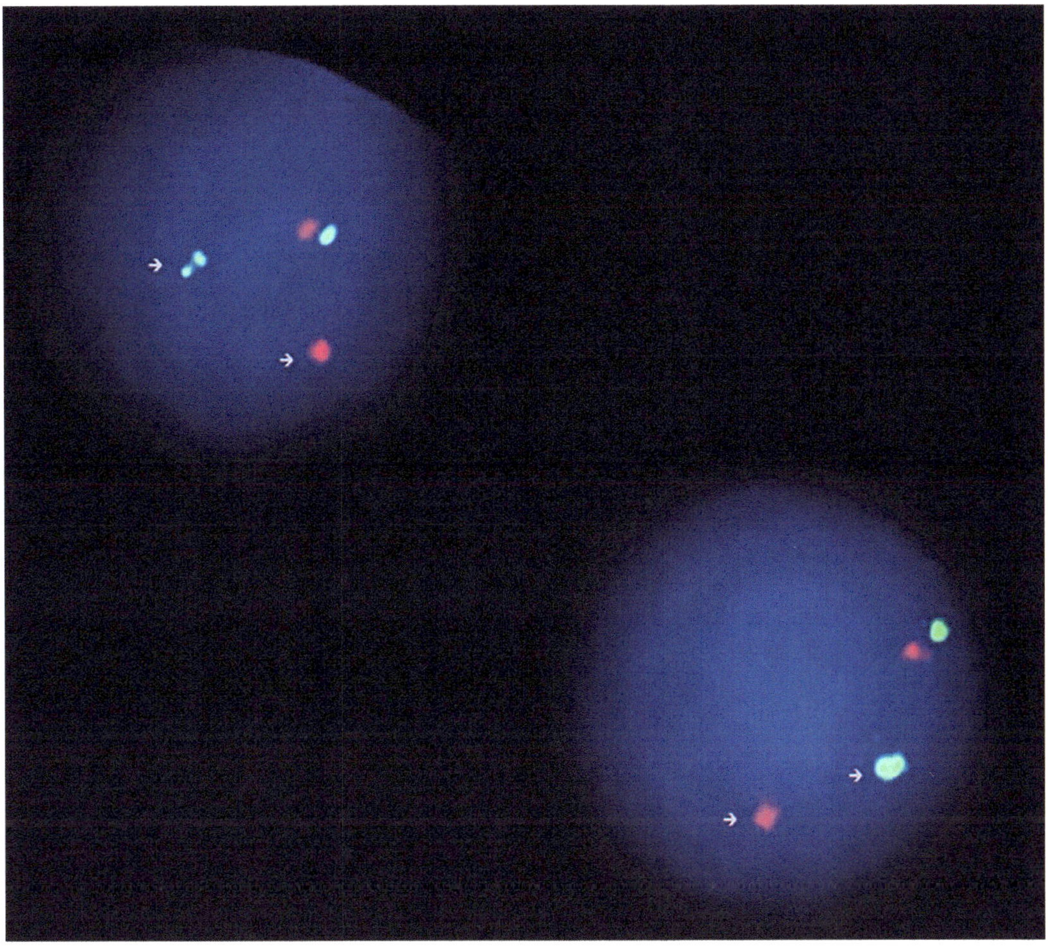

Case Study 3.24 (4) FISH image showing 1F1G1R signal pattern using *MLL(=KMT2A)* breakapart probe indicating MLL*(=KMT2A)* rearrangement

Molecular Findings

The Real-Time Quantitative PCR for *AML panel: negative.*

Discussion

Secondary AML (s-AML) is defined as AML occurring after an antecedent myeloid disease (e.g., MDS or MPN, excluding CML) regardless of prior cytotoxic therapy for these disorders. The WHO classification Swerdlow et al. [1] defines therapy-related AML (t-AML) as AML occurring as a late complication to prior cytotoxic chemotherapy and/or radiotherapy for a neoplastic or non-neoplastic disease. Myeloproliferative neoplasms such as ET, PV, and PMF might progress into blast phase disease (MPN-BP). The reported risk factors for leukemic transformation for PMF include increased blast percentage, thrombocytopenia, abnormal karyotype, triple-negative driver mutational status, and sequence mutations such as SRSF2, RUNX1, CEBPA, and SH2B3 [38]. In a Danish national population-based study of 3055 unselected patients with AML, the frequencies of s-AML and t-AML were 19.8% and 6.6%, respectively [39]. Similarly a Swedish Acute Leukemia Registry included 3363 adult patients of which 73.6% had de novo AML, 18.7% s-AML, and 7.7% t-AML [40].

3.25 Case Study 3.25: Acute Megakaryoblastic Leukemia with Isolated Deletion 9q

Brief Clinical History

A 58-year-old female presented with complaints of dyspnea on exertion and high-grade fever for 2 weeks. No comorbidities or history of regular medication was found. No organomegaly, lymphadenopathy, bleeding from any sites was seen. General physical examination was unremarkable.

CBC: *Hb*: 6 g/dl, TLC: 2100/cu mm and Platelet count: 1,50,000/cu mm.

DLC: Blasts: 62%, Neutrophils: 12%, Lymphocytes: 25%, Monocytes: 1%

Peripheral Blood Smear and Bone Marrow

Smear showed leukopenia with approximately 62% blasts. Blasts were medium to large in size with a high N:C ratio, 2–3 prominent nucleoli, and scant agranular cytoplasm with cytoplasmic blebs. Bone marrow aspirate was hemodilute but the biopsy was hypercellular showing fibrosis with marked megakaryocytic hyperplasia with atypia and interstitial clusters of blasts.

Flow Cytometric Immunophenotyping

A population of blasts (~80% of acquired events) was seen in the CD45-negative region with SSC low expressing CD41+, CD7+ and negative for HLA-DR, CD34, TDT, cMPO, CD117, CD13, CD33, CD56, CD38, CD11c, CD64. The immunophenotype was suggestive of megakaryoblasts with aberrant CD7.

Cytogenetics

Cytogenetic Karyotype and FISH

46XX,del 9(q21q22)[20]

Cytogenetic analysis of 20 metaphases showed a neoplastic clone characterized by an interstitial deletion of the long arm of chromosome 9 in all 20 metaphases.

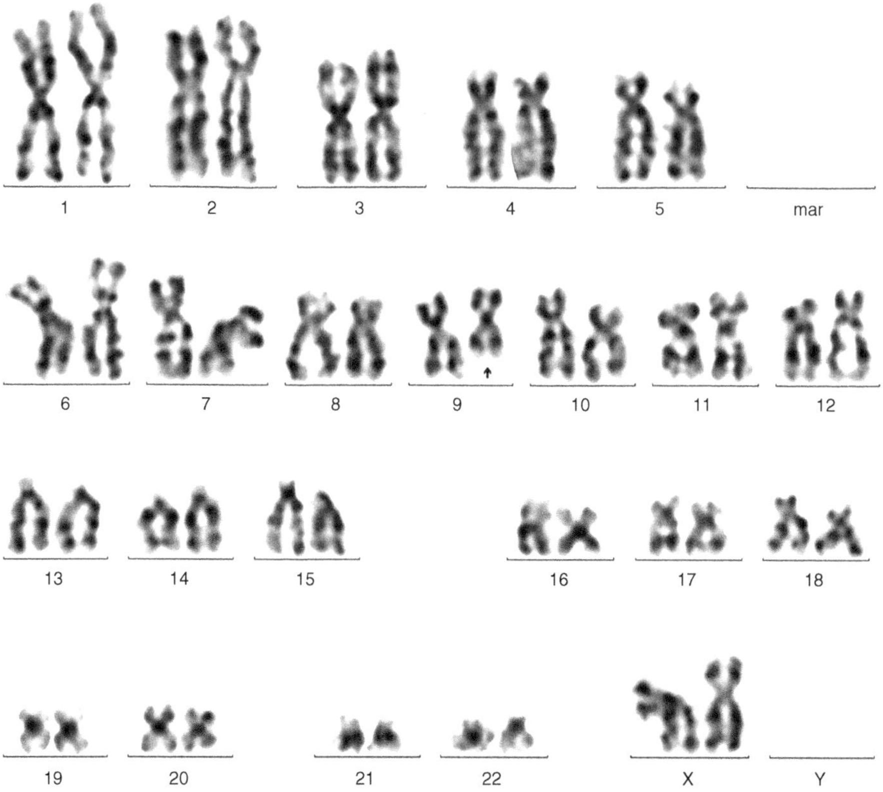

Case Study 3.25 Karyotype showing interstitial deletion of the long arm of chromosome 9, i.e., del(9q)

Molecular Findings

The Real-Time Quantitative PCR was *negative for the AML panel.*

Discussion

The case described above is morphologically and immunophenotypically acute megakaryoblastic leukemia with isolated deletion of 9q. Molecularly, RQ-PCR showed no markers of AML.

Deletions on the long arm of chromosome 9 (del (9q) is a rare cytogenetic abnormality observed in 2% of AML patients. In a quarter of these patients, it presents as an isolated abnormality. The rest of these patients have additional abnormalities such as t(8;21), inv(16), PML-RARA [41].

Various studies have shown that 9q21.32 is the most frequently deleted region. A minimal deleted region (MDR) within the commonly deleted region (CDR) to 9q21.32-9q21,33 including seven annotated genes (GKAP1, KIF27, C90RF64, HNRNPK, RMI1, SLC28A3, NTRK2) has been described. Expression of the MDR genes is significantly reduced in del(9q) patients [42].

AMLs with del (9q) in general have been described to have various FAB morphologies including M1, M2, M4, M5, M6. AMLs with isolated del (9q). They have characteristic myeloid lineage vacuolation, erythroid dysplasia, and single long and slender Auer rods. However, the case described above was Acute megakaryoblastic leukemia [FAB (AMLM7)] and had no evidence of dyshematopoiesis or presence of Auer rods.

3.26 Case Study 3.26: Acute Myeloid Leukemia, NOS, with Isolated Trisomy 11

Brief Clinical History

A 45-year-old female presented with a case of incisional hernia and complaints of high-grade fever for 10 days. No comorbidities or history of regular medication, organomegaly, lymphadenopathy, bleeding from any sites were found. General physical examination was unremarkable.

CBC: *Hb*: 8 g/dl, TLC: 31,6100/cu mm and Platelet count: 150,000/cu mm

DLC: Blasts: 90%, Neutrophils: 1%, Lymphocytes: 9%

Peripheral Smear and Bone Marrow

Smear showed leukocytosis with approximately 90% blasts. Blasts were large in size with a high N:C ratio with 1–2 prominent nucleoli and agranular cytoplasm. Occasional blasts showed Auer rods. Similar blasts (85%) were seen in the bone marrow.

Flow Cytometric Immunophenotyping

This confirmed a population of blasts ~88% of acquired events in CD45 moderate, SSC low expressing HLA-DR+, CD34+, CD117+, CD13+, CD33+, CD38+.

Cytogenetics

Conventional Karyotype and FISH

47XX, +11[16]/46XX[3]

Cytogenetic analysis of 19 metaphases showed a neoplastic clone characterized by a gain of chromosome 11 in 16 metaphases.

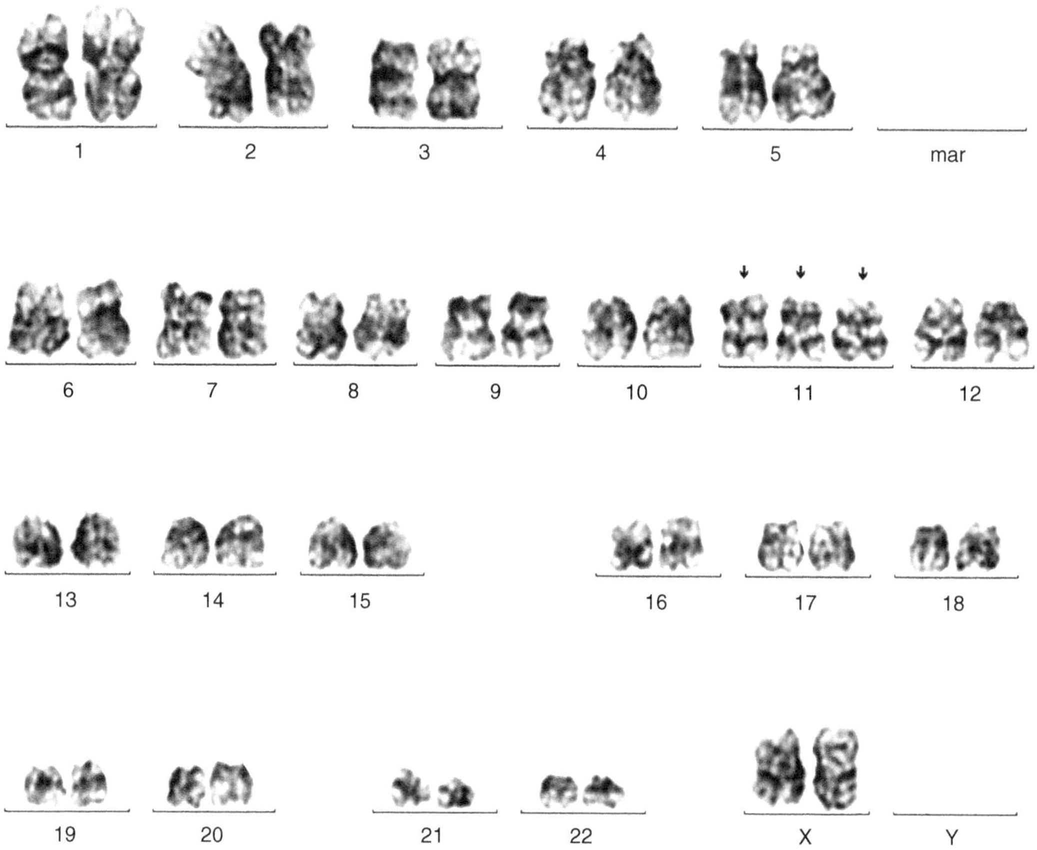

Case Study 3.26 Karyotype showing gain of chromosome 11, i.e., trisomy 11

Molecular Findings

The Real-Time Quantitative PCR for the AML panel was negative

Discussion

The case described above is morphologically and immunophenotypically AML without maturation with isolated trisomy 11. Molecularly, RQ-PCR showed no markers of AML.

Recurrent chromosomal abnormalities are identified in 55%–60% of newly diagnosed adults with AML. Trisomies occur as a sole abnormality in 7%–8% of adult patients with AML involving chromosomes 4, 8, 11, 13, and 21 which comprise the majority of sole trisomies in AML (approximately 89%–95% of all sole trisomies) [43]. An isolated trisomy 11 in AML is a very rare abnormality occurring in <1% cases. In a large study Eisfeld et al. [44] including 1496 AML patients, AMLs with isolated trisomy 11 are most often morphologically AML without maturation (FABM1) and have a stem cell phenotype with CD34, HLA-DR, CD117 expression similar to what was seen in this case.

Approximately 90% of these cases have partial tandem duplication of the MLL (KMT2A) gene and FLT3 gene mutations (20%). U2AF1, IDH2, DNMT3A mutations are also frequent. They have a very low frequency of mutations of the NRAS, KRAS, NPMI, and KIT genes.

AML with isolated trisomy 11 has been consistently shown to have poor prognosis in many studies and is classified as intermediate-risk AML by ELN classification [7, 45].

3.27 Case Study 3.27: Acute Myeloid Leukemia, NOS with Isolated Trisomy 14

Brief Clinical History

A 52-year-old male presented with complaints of easy fatigability and fever for 2 days. There were no comorbidities or any history of regular medications. No organomegaly, lymphadenopathy, bleeding from any sites were found. General physical examination was unremarkable.

CBC: *Hb*: 9.45 g/dl, TLC: 30,040/cu mm and Platelet count: 24,700/cu mm

DLC: Blasts: 70%, Neutrophils: 2%, Lymphocytes: 27%, Monocytes: 1%

Peripheral Blood Smear and Bone Marrow Findings

Smear showed leukopenia with approximately 70% blasts. Blasts were large in size with a high N:C ratio, homogenous nuclear chromatin with 1–2 prominent nucleoli, and moderate pale blue cytoplasm. Occasional blast showed the presence of an Auer rod. Similar blasts constituted 88% of marrow nucleated cells.

Flow Cytometric Immunophenotyping:

A large population of myeloblasts (~90% of acquired events), in CD45-moderate blast region with SSC moderate with CD34+, HLA-DR+, cMPO+, CD117+, CD13+, CD33+, *CD56+*, CD38+, CD11c+, CD64-dim were seen.

Cytogenetics

Conventional Karyotype

47,XY,+14[5]/46,XY[15]

Cytogenetic analysis of 20 metaphases showed a neoplastic clone characterized by gain of chromosome 14 as trisomy in 5 metaphases.

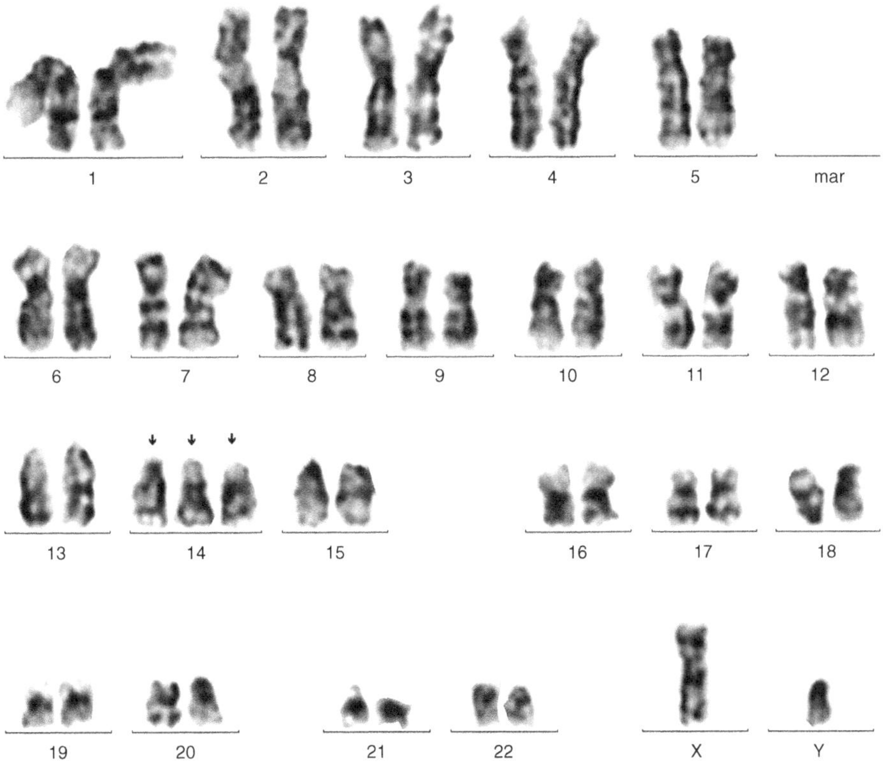

Case Study 3.27 Karyotype showing gain of chromosome 14, i.e., trisomy 14

Molecular Findings
The Real-Time Quantitative PCR was *negative for the AML panel.*

Discussion
The case described above is morphologically and immunophenotypically AML with monocytic differentiation with a sole cytogenetic abnormality of trisomy 14. Molecularly, RQ-PCR showed no markers of AML.

Trisomy 14 as an isolated chromosome abnormality is a rare non-random cytogenetic abnormality in myeloid neoplasms. It is more common in MDS followed by MDS/MPN and AML. These patients are usually elderly with a median age of 70.5 years and have a male predominance [46].

Molecularly RAS, c-KIT, NPM1 mutations have not been observed in these cases. FLT3 mutation has been described.

Trisomy 14-positive AMLs, frequently have trilineage dyspoietic changes. The case described above was AML with monocytic differentiation [FAB AML(M4)] and had no evidence of dyshematopoiesis.

AML with trisomy 14 is classified as intermediate-risk AML, by ELN classification [7]. In addition, the immunophenotyping finding of CD56 positivity in this case may add to the adverse prognosis.

3.28 Case Study 3.28: Acute Myeloid Leukemia, NOS with Sole Abnormality of Trisomy 8

Brief Clinical History
A 65-year-old male presented with complaints of fever on and off for 1 week. There were no comorbidities or history of regular medications. No organomegaly, lymphadenopathy bleeding from any sites were found. General physical examination was unremarkable.

CBC: *Hb*: 9 g/dl, TLC: 1000/cu mm, Platelet count: 16,000/cu mm

DLC: Blasts: 32%, Neutrophils: 8%, Lymphocytes: 58%, Monocytes: 2%

Peripheral Smear and Bone Marrow
Smear showed leukopenia with approximately 32% blasts. Blasts were large in size with a high N:C ratio, round nucleus with 1–2 prominent nucleoli, and scant to moderate pale blue cytoplasm. No Auer rods were seen. Similar blasts constituted 70% of the bone marrow.

Flow Cytometric Immunophenotyping
A large population of blasts (~90% of acquired events) were seen in CD45-moderate blast region with SSC moderate with CD34+, HLA-DR+, cMPO+, CD117+ , CD13+, CD33+, *CD56+*, CD38+, CD11c+, CD64+, CD123+.

Cytogenetics
Conventional Karyotype
47,XY,+8[11]/46,XY[9]

Cytogenetic analysis of 20 metaphases showed a neoplastic clone characterized by gain of chromosome 8 as trisomy in 11 metaphases.

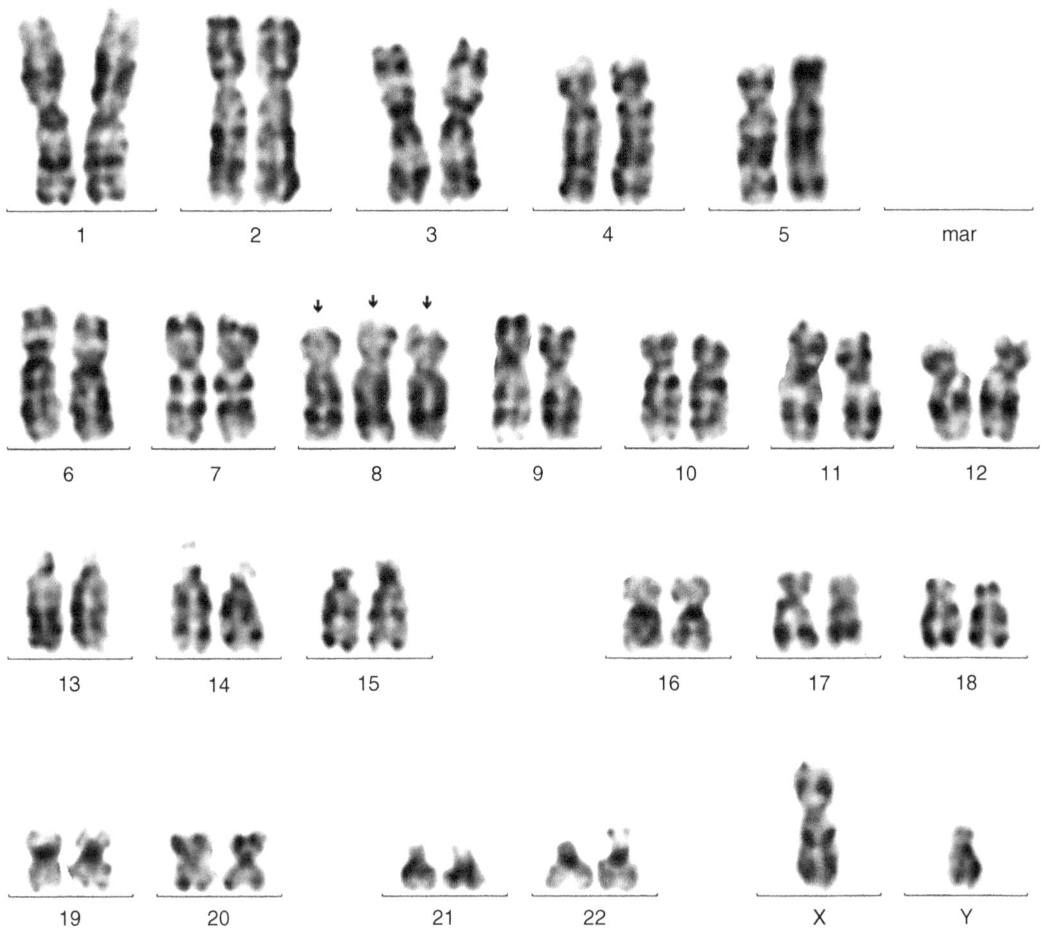

Case Study 3.28 Karyotype showing gain of chromosome 8, i.e., Trisomy 8

Molecular Findings

The Real-Time Quantitative PCR was *negative for the AML Panel.*

Discussion

The case described above is morphologically and immunophenotypically AML with monocytic differentiation with a sole cytogenetic abnormality of trisomy 8. Molecularly, RQ-PCR showed no markers for AML. Trisomy 8 is one of the most common cytogenetic alterations in AML, with a frequency between 10% and 15% [1]. The median age of trisomy 8 AML is approximately 70 years. The frequency of trisomy 8 is lower in t-AML compared to de novo AML. Trisomy 8-positive AMLs are a heterogeneous group with regards to morphology and can fall into any FAB AML category. However, in the pediatric age group, it is commonly associated with AML with monocytic differentiation [FAB AML (M5)]. Trisomy 8 is seen as a sole cytogenetic alteration in 30%–40% of cases and is seen in association with other cytogenetic alternations in 60%–70% of cases and 25%–30% of AML with complex karyotype [47]. Various somatic mutations have also been seen with trisomy 8 in varying frequencies such as IDH1 and IDH2 (45%), NPM1 (20%–25%), WTI (5%), FLT3-TKD (7%–10%), FLT3-ITD (25%), RUNX1 (25%–35%), ASXL1 (up to 50%). ASXL1 (up to 50%), ASXL1 and RUNX1 mutations are found to be co-occurring in AML with sole trisomy 8. AML with trisomy 8 is classified as intermediate-risk AML by ELN classification [7]. In addition, the immunophenotyping finding of CD56 positively in this case may add to the adverse prognosis.

3.29 Case Study 3.29: Acute Myeloid Leukemia: AML-M7 (NOS) with Complex Karyotype

Brief Clinical History

A 51-year-old male patient presented with fever, epistaxis, and oral mucosal bleeds. He was on Ayurvedic medicines and supportive therapy with blood transfusions.

CBC: *Hb:* 6.1 g/dl, TLC: 1810/cu mm and Platelets: 8000/cu mm.

DLC: Neutrophils: 80.0%, Lymphocytes: 15.0%, Eosinophils: 0.0%, Basophils: 0.0%, Monocytes: 5.0%

Peripheral Smear and Bone Marrow

Peripheral smear showed pancytopenia. BM was dry tap. Bone marrow aspirate was hemodilute. Bone biopsy showed large immature nucleated cells separated by fibrous strands. The large cells (blastoid) on IHC were CD45+, CD33+,CD34+, CD117+, and MPO+. Factor VIII was positive on occasional blasts and megakaryocytes. Reticulin stain showed marrow fibrosis (grade 2). The picture was suggestive of AML-M7.

Cytogenetics

Conventional Karyotype

47,XY,del(5)(q12q33),del(13)(q12q22),-16,-17, del(19) (p13), add(19) (q13.4), +21, +mar1, +mar2 [10]/46,XY[10]

Cytogenetic analysis of 20 metaphases done on the peripheral blood showed a neoplastic clone characterized by a complex karyotype showing structural and numerical abnormalities. Numerical abnormalities showed monosomy of chromosomes 16 and 17 along with trisomy of chromosome 21 with gain of two marker chromosomes of unknown origin. Structural abnormalities showed deletion of the long arm of chromosome 5 (5q) and deletion of the long arm of chromosome 13 (13q) along with deletion of the short arm of chromosome 19 (19p). This chromosome (19) also showed the addition of extrachromosomal material of unknown origin on the long arm (19q) in 10 metaphases analyzed.

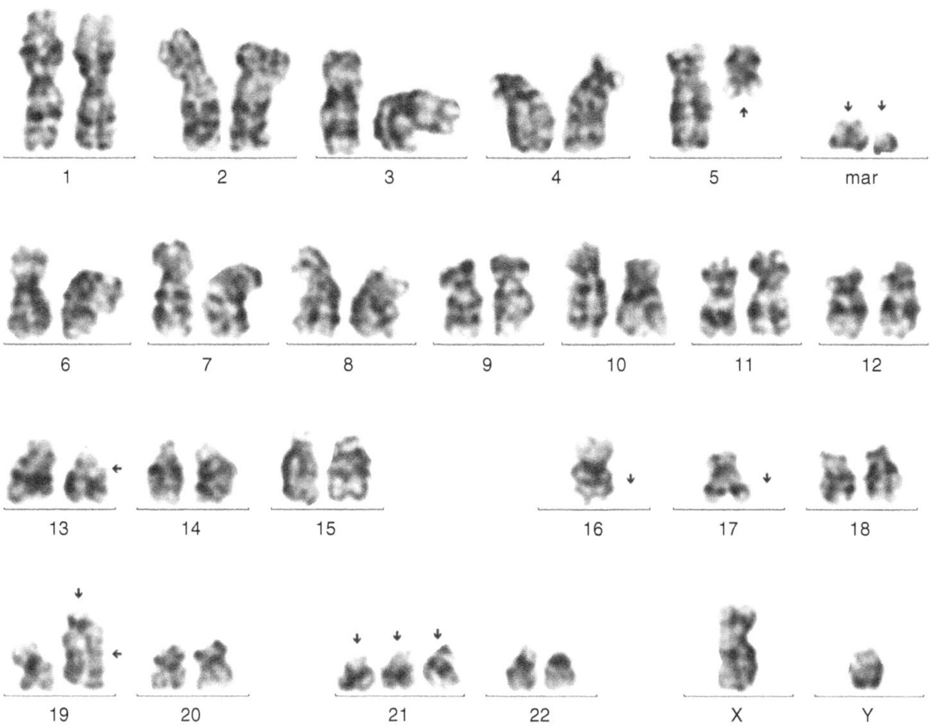

Case Study 3.29 Complex karyotype showing deletion of the long arm of chromosome 5, interstitial deletion 13q, monosomy 16, 17, derivative (19), gain of chromosome 21 along with two different marker chromosomes of unknown origin

Molecular Findings

The Real-Time Quantitative PCR for *AML comprehensive Panel: Negative*

Discussion

This patient of AML-M7 (NOS) showed a complex karyotype with both numerical and structural abnormalities. No cytogenetic abnormality is specific for AML-M7 [48]. In the vast majority, karyotype is complex. -5/Del(5q) and or -7/del(7q) are found in almost all cases with complex karyotype (as in this case). Dastugue et al. reported the cytogenetic profile of 53 patients (30 children, 23 adults) with AML-M7 evaluated by the Groupe Francais de Cytogenetique Hematologie and reported 7 patients with trisomy 19, 1 patient with hyperdiploid karyotype and an extra copy of chromosome 19, 1 patient with a t(4;19) (p12;?) translocation. Trisomy 21 and del (19P) are also found together with several numerical and structural abnormalities [49].

The prognosis is poor in these patients.

3.30 Case Study 3.30: Acute Myeloid Leukemia with Lineage Switch from ALL with TP53 alterations and Complex Karyotype

Brief Clinical History

A 63-year-old female diagnosed case of B-ALL on maintenance chemotherapy. Peripheral blood smear showed blasts. Bone marrow was done for suspected relapse.

CBC: *Hb*: 8.5 g/dl, TLC: 1190/cu mm and Platelets: 68,000/cu mm

DLC: Neutrophils: 27.0%, Lymphocytes: 51.0%, Eosinophils: 5.0%, Monocytes: 1.0%, Basophils: 1.0%, Blasts: 15.0%,

Peripheral Blood Smear and Bone Marrow

Smear showed leukopenia with neutropenia and blasts 15%. Blasts in BM constituted 48.6% of the total nucleated cells. The blasts were large with high N:C ratio, scant to moderate cytoplasm, open nuclear chromatin and indistinct nucleoli, few with cytoplasmic vacuolations.

Flow Cytometric Immunophenotyping
Blasts 27% of acquired events were seen in CD45 moderate blast region. They were CD34+, CD13+, CD117+, HLA-DR+, CD38+, CD19+.

Molecular Findings
The Real-Time Quantitative PCR showed *AML Comprehensive Panel: Negative*

Cytogenetics
Conventional Karyotype

46,XX,-4,+8,der(17)t(15;17)(q24;q23),del(17)(p13),der(20)t(10;20)(q11.2;q13.3),+der(11)t(5;11) (q31;q13)[20]

Cytogenetic analysis of 20 metaphases showed a neoplastic clone characterized by a complex karyotype showing monosomy for chromosome 4 and trisomy of chromosomes 8 and 11. There was unbalanced translocation between the long arm of chromosome 5 and the long arm of chromosome 11. Chromosome 17 showed deletion of the short arm at breakpoint p13 which was confirmed by FISH using TP53 deletion probe. The same chromosome showed balanced translocation between the long arm of chromosome 15q24 and the long arm of chromosome 17q23. This, however, is different from t(15;17) of APML which is seen at breakpoints 15q24 and 17q21. There was an additional abnormality showing balanced translocation between the long arm of chromosome 10 and the long arm of chromosome 20.

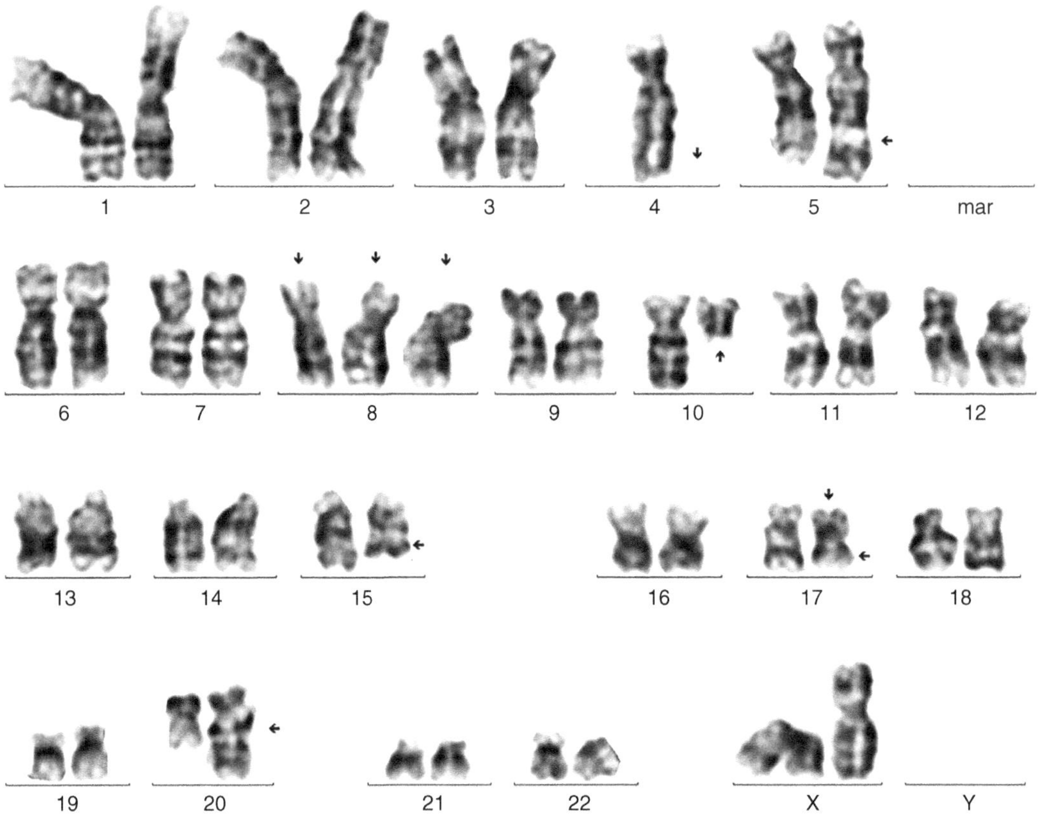

Case Study 3.30 (1) Complex karyotype showing monosomy 4, derivative 5, trisomy 8, deletion 10q, and add (20q)

Case Study 3.30 (2)
Metaphase FISH image showing 3F signal pattern using *MLL(=KMT2A)* DC BA probe indicating t(5;11)

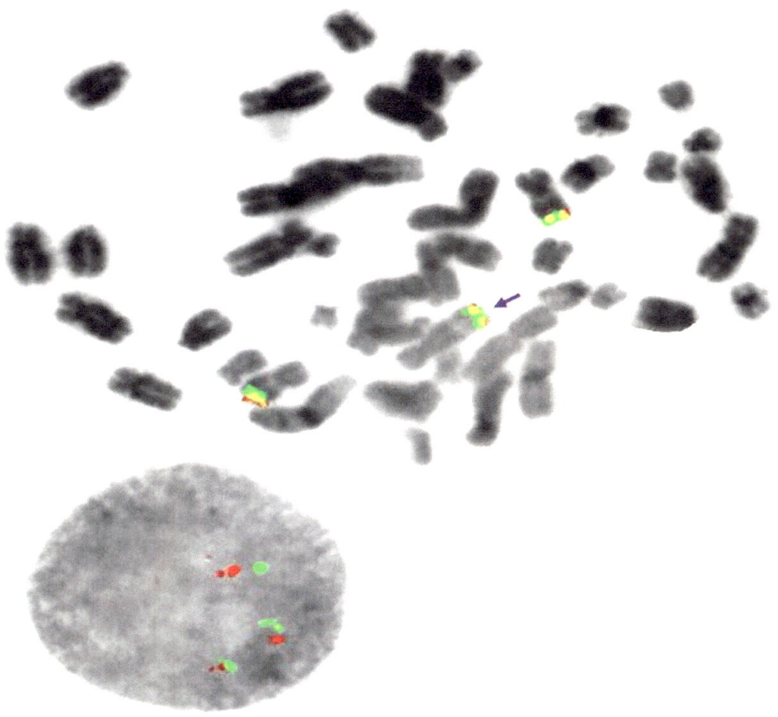

Discussion

This patient of ALL on maintenance therapy showed a lineage switch to AML in relapse with complex karyotype and TP 53 alterations. Rucker et al. [50] performed an integrative analysis using TP53 mutational screening in 234 complex karyotype AMLs (CK-AMLs). TP53 mutations were found in 60%. They are often associated with copy number alterations such as -5/del(5q), -7/del(7q), -16/del(16q), -18/del(18q), +1/+1p and +11/+11q/amp11q 13~25. In this patient, in addition to TP53 change, there was monosomy 4, trisomy of chromosome 8 and 11 t(5;11) t (10;20). Interestingly, chromosome 17 showed deletion of the short arm at breakpoint p13 and on FISH, a balanced translocation was seen between the long arm of 17q23 and the long arm of 15q24. This finding is different from that seen in APML t(15;17) where the breakpoints are 15q24 and 17q21.

These cytogenetic alterations signify poor prognosis. The lineage switch seen in this patient has rarely been reported [50]. It should be determined whether blast cells arise from some leukemic clone or be considered therapy-related malignancy. This patient was on maintenance therapy and, therefore, not therapy-related, also because the cytogenetic profile was of myeloid and not original lymphoid lineage. The mechanism of lineage switch is unclear, but in this case, it could be a selection of the preexistent chemotherapy insensitive minor population of cells of myeloid lineage. The prognosis in these cases is dismal [51].

3.31 Case Study 3.31: Acute Myeloid Leukemia with Normal Karyotype and NPM1 Mutation on (RT-PCR) with NPM1 and FLT3-ITD Mutation on NGS

Brief Clinical History

A 20-year-old female was evaluated for peripheral blood picture suggestive of Acute Leukemia.

CBC: *Hb:* 7.7 g/dl; TLC: 1,11,470/cu mm, Platelets: 18,000/cu mm.

DLC: Neutrophils: 1.0%, Lymphocytes: 2.0%, Eosinophils: 1.0%, Monocytes: 2.0%, Basophils: 0.0%, Blasts and promonocytes: 94.0%.

Peripheral Blood and Bone Marrow

Smear showed Hyperleukocytosis with blasts and promonocytes constituting ~94%. Blasts were large in size with a high N:C ratio, opened up chromatin with irregular nuclear contours, 1–2 prominent nucleoli, and moderate basophilic cytoplasm. Occasional blasts showed Auer rods and Phi bodies.

The peripheral blood and bone marrow findings were morphologically consistent with acute myeloid leukemia with monocytic differentiation.

Flow Cytometric Immunophenotyping

Total Blasts constituted ~77% of acquired events on the CD45 vs SSC plot. Two distinct blast populations were seen on CD45 vs SSC plot. One constituted ~67% of total blast population with CD117+, CD13+, CD33+, CMPO+, HLA-DR+, CD34−, CD64+, CD4+, CD11C+. The other population of blasts constituted ~10% of total blast population CD33+, CD38+, CD117+, CD71−, CD34−.

Peripheral blood immunophenotyping was consistent with Acute Myeloid Leukemia with monocytic differentiation.

Cytogenetics

Conventional Karyotype

46,XX[20]

Cytogenetic analysis of unstimulated bone marrow aspirate culture showed normal karyotype in all the analyzed 20 metaphases, and none of the myeloid neoplasia associated abnormalities were seen.

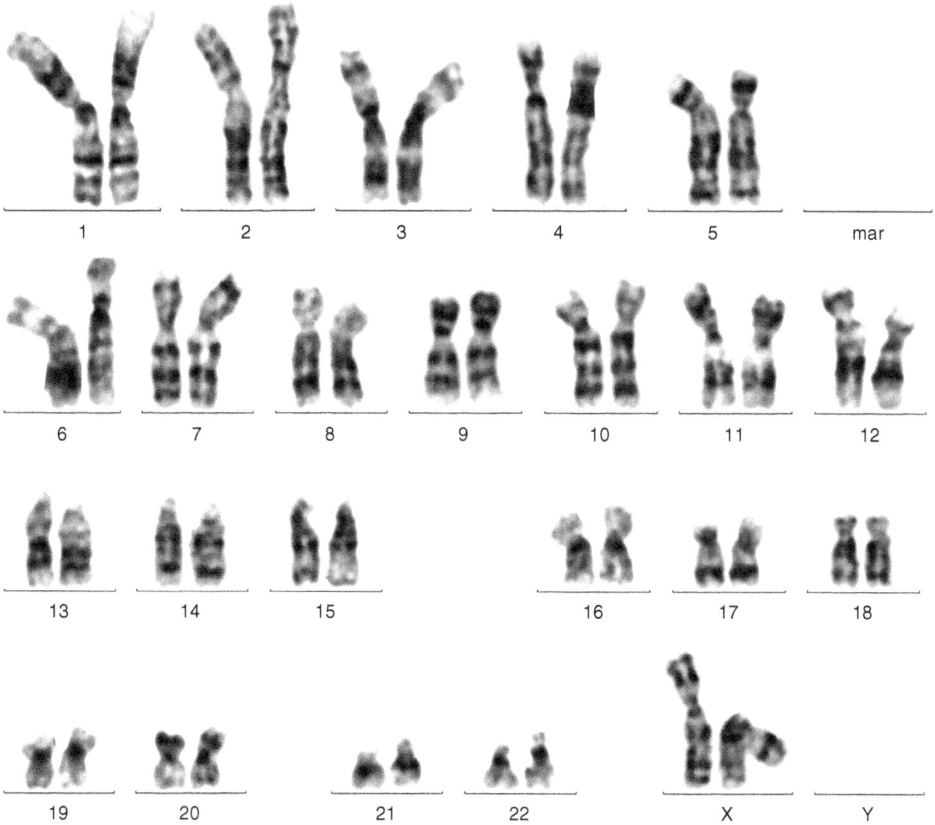

Case Study 3.31 Karyotype showing normal female karyotype

Molecular Studies
The Real-Time Quantitative PCR for Acute Myeloid Leukemia panel was *positive for NPM1 gene mutation.*

Discussion
This patient was morphologically AML without maturation, cytogenetically normal with absent CD34 and HLA-DR, an immunophenotype characteristically associated with NPM1 mutated AML [30]. Mutations of the nucleophosmin gene have been reported as the most frequent mutation in AML, especially in the presence of a normal karyotype. It accounts for nearly one-third of AML patients and shows mutual exclusion, of recurrent genetic abnormalities, high frequency of FLT3–ITD mutations, frequent M4 or M5 morphology, multilineage involvement, and distinctive gene expression signature.

In this patient, there were no secondary mutations such as FLT3, DNMT3A, IDH1, KRAS, or NRAS genes. NPM1 mutation is associated with good prognosis in both young and older adults as well as in children with normal karyotype and no associated FLT3 mutation or FLT3–ITD allele ratio low. As per ELN risk stratification, this patient would fall in the favorable prognosis group [7].

Next-Generation Sequencing (NGS) on PGM (Ion Torrent)
Oncomine Myeloid Panel
The panel identified the NPM1 driver variant confirming the positive NPM1 by Real-Time PCR. In addition, it identified the FLT3-ITD variant which was not captured on the gel images showing the importance of next-generation sequencing in this case, which had a normal karyotype and mutation of NPM1 clone. The additional finding of FLT3-ITD mutation put this case in the intermediate-risk category as per ELN classification and not favorable category which would otherwise have been the case with NPM1 mutation clone showing the significance of NGS analysis in this case.

3.32 Case Study 3.32: Acute Myeloid Leukemia with t(8;21) and RUNX1-RUNX1T1 Gene Rearrangement on RT-PCR and NGS

Brief Clinical History
A 14-year-old male investigated for relapsed AML.

CBC: *Hb:* 12.7 g/dl, TLC: 1,32,660/cu mm, Platelets: 7910/cu mm.

DLC: Neutrophils: 1.0%, Blasts: 96%; Lymphocytes: 3.0%, Eosinophils: 0.0%, Monocytes: 0.0%, Basophil: 0.0%,

Peripheral Smear and Bone Marrow
Peripheral blood smear showed Hyperleukocytosis with the presence of ~96% blasts which were large in size, having fine homogenous chromatin, 1–2 nucleoli, and scant to moderate amount of cytoplasm. Few blasts showed nuclear cleaving, cytoplasmic granules, and an Auer rod. The picture was consistent with acute myeloid leukemia. Bone marrow was not done.

Cytogenetics
Conventional Karyotype
45,X,-Y,t(8;21)(q22;q22)[18]/46,XY[2]

Cytogenetic analysis of 20 metaphases showed a neoplastic clone characterized by a balanced translocation between the long arms of chromosomes 8 and 21. The clone also showed the loss of sex chromosome (loss of Y) as a secondary change in 18 metaphases analyzed.

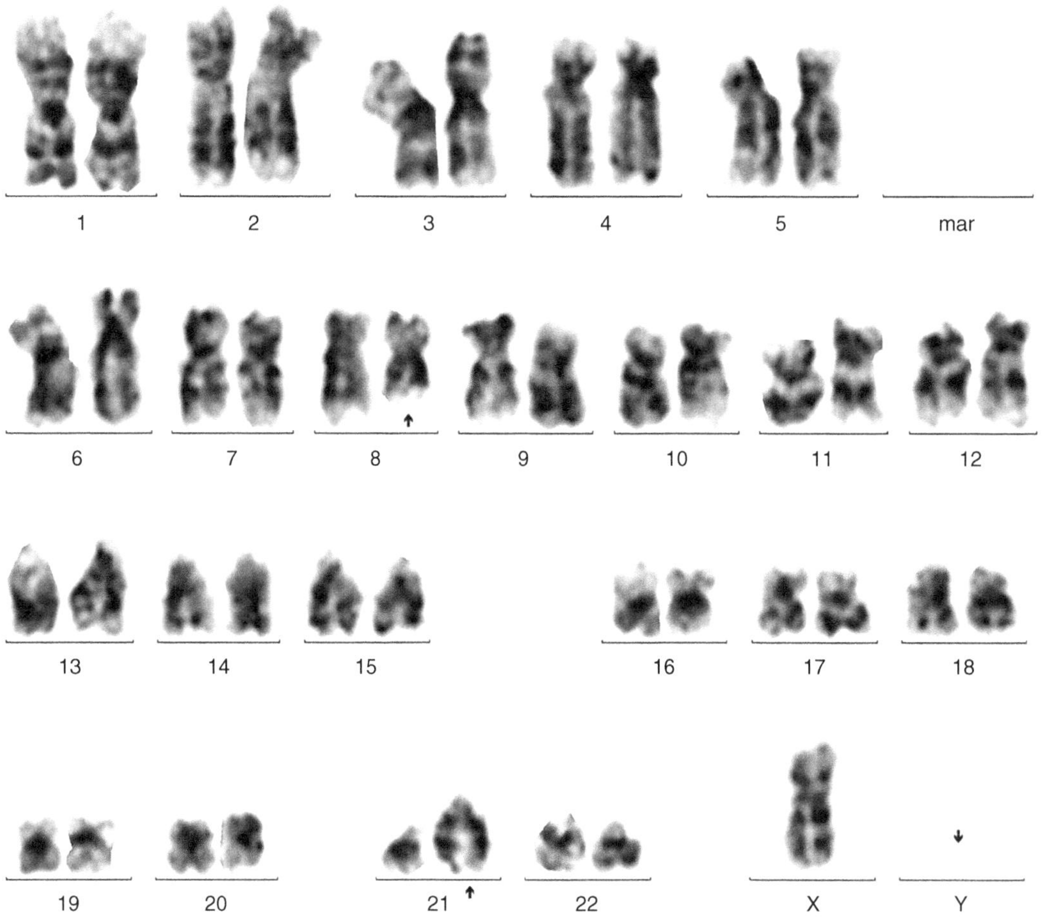

Case Study 3.32 Karyotype showing t(8;21)(q22;q22) along with loss of chromosome Y.

Flow Cytometric Immunophenotyping
Blasts (~92% of acquired events); present in the blast region. CD34+, MPO+, CD13+, HLA-DR+, CD117+, CD38+, CD11c+. The immunophenotype was that of Acute Myeloid Leukemia.

Molecular Findings
The Real-Time Quantitative PCR showed (*RUNX1-RUNXT1) GENE REARRANGEMENT.*

Discussion
As per the 2017 WHO classification, the presence of t(8;21) is diagnostic of acute myeloid leukemia irrespective of the percentage of blasts in the marrow [1]. It is the commonest recurrent cytogenetic abnormality occurring in acute myeloid leukemia and is associated with a good prognosis.

Additional cytogenetic abnormalities are seen in nearly 70% of cases such as loss of a sex chromosome or del(9q) [2]. The presence of additional cytogenetic abnormalities does not alter the good prognosis. The loss of the sex chromosome is the most frequent secondary cytogenetic abnormality in t(8;21) AML [3].

Next-Generation Sequencing (NGS) on PGM (Ion Torrent)
Oncomine Myeloid Panel: confirmed the presence of (RUNX1-RUNX1T1) gene rearrangement in concordance with its presence on Real-Time PCR and t (8;21) on karyotyping.

In this case, NGS showed the same molecular change, t(8;21)(RUNX1-RUNXT1) as the Real-time PCR.

3.33 Case Study 3.33: Acute Myeloid Leukemia with Normal Karyotype and NPM1 Mutation on RT-PCR and NGS

Brief Clinical History
A 68-year-old male, follow-up case of AML post four cycles of decitabine. Evaluated for disease status.

CBC: *Hb:* 12.0 g/dl, TLC: 1890/cu mm and Platelets: 2,15,000/cu mm.

DLC: Neutrophils: 26.1%, Lymphocytes: 67.2%, Eosinophils: 1.7%, Monocytes: 4.4%,
Basophils: 0.6%.

Peripheral Blood Smear and Bone Marrow
Smear showed leukopenia with neutropenia, ANC: 470/uL. There were no blasts. Bone marrow aspirate was hemodilute and imprint smears showed a few erythroid and myeloid cells with 2% blasts. Bone marrow biopsy was cellular, showed erythroid hyperplasia. No blast clusters were seen.

Flow Cytometric Immunophenotyping
Blasts 28% of acquired events, moderate CD45 with moderate SSC

MPO+, CD13+, CD33+, CD64+, CD4+, CD11C+, HLDR+, CD117+, CD34−

The picture showed the immunophenotype of Acute Myeloid Leukemia with monocytic differentiation.

Cytogenetics
Conventional Karyotype

46,XY[20]

Cytogenetic analysis of unstimulated peripheral blood culture showed normal karyotype in all the analyzed 20 metaphases, and none of the myeloid neoplasia associated abnormalities were seen.

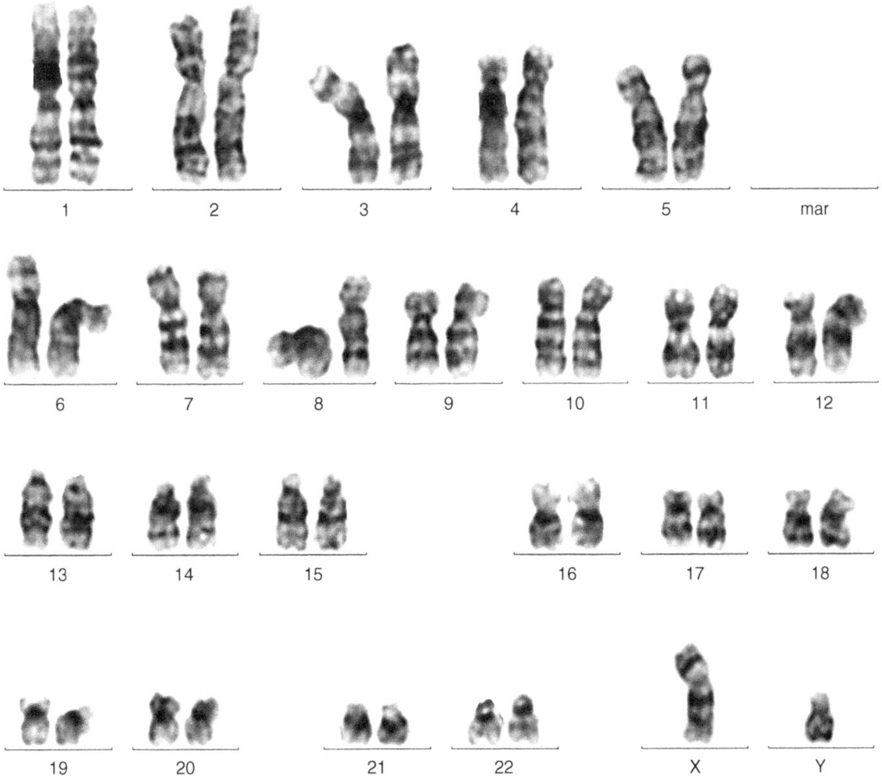

Case Study 3.33 Karyotype showing normal male karyotype.

Molecular Studies
The Real-Time Quantitative PCR for the AML panel was *positive for NPM1 gene mutation.*

Discussion
This patient was morphologically AML without maturation, cytogenetically normal with absent CD34 and HLA-DR, an immunophenotype characteristically associated with NPM1 mutated AML. Mutations of the nucleophosmin gene have been reported as the most frequent mutation in AML, especially in the presence of a normal karyotype [7]. It accounts for nearly one-third of AML patients and shows mutual exclusion, of recurrent genetic abnormalities, high frequency of FLT3-ITD mutations, frequent M4 or M5 morphology, multilineage involvement, and distinctive gene expression signature [14].

In this patient, there were no secondary mutations such as FLT3, DNMT3A, IDH1, KRAS, or NRAS genes. NPM1 mutation is associated with good prognosis in both young and older adults as well as in children with normal karyotype and no associated FLT3 mutation or FLT3–ITD allele ratio low. As per ELN risk stratification, this patient would fall in the favorable prognosis group.

Next-Generation Sequencing (NGS) on PGM (Ion Torrent)
Oncomine Myeloid Panel

Positive for NPM 1 gene mutation with no additional mutations identified.

In this case, there was a concurrence between Real-Time PCR and NGS findings.

Annexure

- Case Study 3.1: Acute Myeloid leukemia with t(8;21)(q22;q22.1);RUNX1-RUNX1T1 and KIT Gene Mutation
- Case Study 3.2: Acute Myeloid leukemia with a variant 3 way translocation t(8;17;21)(q22;q23;22.1);RUNX1-RUNX1T1
- Case Study 3.3: Acute Myeloid Leukemia, t(8;21)(q22;q22) with rare duplication of der (21)
- Case Study 3.4: AML with inv(16)(p13;q22)/t(16;16)p(13;q22);CBFB–MYH11
- Case Study 3.5: Acute Myeloid Leukemia with t(9;11)(p21;q23.3)(KMT2A-MLL3) with Trisomies of Chromosome 8 and 21
- Case Study 3.6: Acute Promyelocytic leukemia with PML-RARA, t(15;17) with Mutated FLT3-TKD (D835)
- Case Study 3.7: Acute Promyelocytic leukemia with variant (11;17)(q23;q21.1)(PLZF/RARA)
- Case Study 3.8: Acute Promyelocytic Leukemia with t(15;17) and t(9;22) Translocations, Trisomy 8 and multiple molecular abnormalities
- Case Study 3.9: Acute Promyelocytic Leukemia with PML-RARA t(15;17) with mutated FLT3-ITD
- Case Study 3.10: Acute Myeloid Leukemia with Normal Karyotype, mutated NPM1 and FLT3-ITD
- Case Study 3.11: Acute Myeloid Leukemia with normal Karyotype and NPM 1 Gene Mutation Mutation
- Case Study 3.12: Acute Myeloid Leukemia with Normal Karyotype and FL3-ITD Gene Mutation
- Case Study 3.13: Acute Myeloid Leukemia with t(11;19) and FLT3-ITD Mutation
- Case Study 3.14: Acute Myeloid Leukemia with Myelodysplasia related changes (AML-MRC) transformed from Aplastic Anaemia
- Case Study 3.15: Acute Myeloid Leukemia with Myelodysplasia related changes with Monosomy 7
- Case Study 3.16: Acute Myeloid Leukemia with Myelodysplasia Related Changes with Monosomal Karyotype
- Case Study 3.17: Acute Myeloid Leukemia (s-AML) from (MDS-EB2) with Myelodysplasia Related Changes with Complex Karyotype
- Case Study 3.18: Acute Myeloid Leukemia (t-AML) with Myelodysplasia Related Changes with Monosomy 7
- Case Study 3.19: Acute Myeloid Leukemia with Myelodysplasia Related changes with Complex Karyotype

- Case Study 3.20: Acute Myeloid Leukemia with Myelodysplasia Related changes (MDS-EB1) with complex Karyotype
- Case Study 3.21: Acute Myeloid Leukemia (t-AML) with t(11;19)(q23;p13.1) (KMT2A-ELL)
- Case Study 3.22: Therapy Related Acute Myeloid Leukemia (t-AML) with Hypodiploid Complex Karyotype
- Case Study 3.23: s-AML with Monocytic Differentiation with der(1) t(1;15)(q11;q13) Evolving from Polycythermia Vera
- Case Study 3.24: Acute Myeloid Leukemia (s-AML) from Primary Myelofibrosis with Complex Karyotype
- Case Study 3.25: Acute Megakaryoblastic Leukemia with Isolated deletion 9q
- Case Study 3.26: Acute Myeloid Leukemia, NOS, with Isolated Trisomy 11
- Case Study 3.27: Acute Myeloid Leukemia, NOS with Isolated Trisomy 14
- Case Study 3.28: Acute Myeloid Leukemia, NOS with sole abnormality of Trisomy 8
- Case Study 3.29: Acute Myeloid Leukemia: AML-M7 (NOS) with Complex Karyotype
- Case Study 3.30: Acute Myeloid Leukemia with Lineage Switch from ALL with TP53 alterations and Complex Karyotype
- Case Study 3.31: Acute Myeloid Leukemia with Normal Karyotype and NPM1 mutation on (RT-PCR) with NPM1 and FLT3-ITD mutation on NGS
- Case Study 3.32: Acute Myeloid Leukemia with t(8;21) and RUNX1-RUNX1T1 Gene Rearrangement on RT-PCR and NGS
- Case Study 3.33: Acute Myeloid Leukemia with Normal Karyotype and NPM1 mutation on RT-PCR and NGS

References

1. Swerdlow SH, Campo E, Harris NL, Jaffe ES, Pileri SA, Stein H, Thiele J, editors. WHO classification of tumours of haematopoietic and lymphoid tissues. 4th ed. Lyon: IARC; 2017.
2. Lagunas-Rangel FA, et al. Acute myeloid leukemia-genetic alterations and their clinical prognosis. Int J Hematol Oncol Stem Cell Res. 2017;11(4):328–39.
3. Yohe S. Molecular genetic markers in acute myeloid leukemia. J Clin Med. 2015;4:460–78.
4. Vundinti BR, Kerketta L, Madkaikar M, et al. Three way translocation in a new variant of t(8;21) acute myeloid leukemia involving Xp22. Indian J Cancer. 2008;4(1):30–2.
5. Mikulasovich M, LeBlanc A, Scalise A, Manwani D, Keyzner A, Najfeld V. Duplication and triplication of der(21)t(8;21)(q22;q22) in acute myeloid leukemia. Cancer Genet Cytogenet. 2009;188:83–7.
6. Kuykendall A, Duployez N, Boissel N, Lancet JE, Welch JS. Acute myeloid leukemia: the good, the bad. and the ugly. Am Soc Clin Oncol Educ Book. 2018 May;23:555–73.
7. Döhner H, Estey E, Grimwade D, et al. Diagnosis and management of AML in adults: 2017 ELN recommendations from an international expert panel. Blood. 2017;129:424–47.
8. Mao L, Wang H, Cheng Y, et al. Occurrence of t(15;17) (q22;q21) and t(9;22) (q34;q11) in a patient with acute promyelocytic leukemia. Leuk Lymphoma. 2009;50(3):466–70. https://doi.org/10.1080/10428190802676728.
9. Pettijohn EM, et al. Article commentary coexisting of PML-RARa and BCR-ABL in acute promyelocytic leukemia. Leuk Lymphoma. 2013;55(2):238–9.
10. Sun X, He Y, Mao C, et al. BCR-ABL fusion gene detected in acute promyelocytic leukemia: a case study of clinical and laboratory results. Leuk Lymphoma. 2014;55:435–8.
11. Takahashi H, Sakai R, Hattori Y, et al. Biclonal existence t(15;17) and t(9;22) chromosomal abnormalities in acute promyelocytic leukemia. Rinsho Ketsueki. 2011,52.37–40.
12. Sanz MA, Fenaux P, Tallman MS, Estey EH, et al. Management of acute promyelocytic leukemia: updated recommendations from an expert panel of the European leukemia net. Blood. 2019;133:1630–43.
13. Haferlach C, Mecucci C, Schnittger S, et al. AML with mutated NPM carrying a normal or aberrant karyotype show overlapping biologic, pathologic, immunophenotypic and prognostic features. Blood. 2009;114(14):3024–32.
14. Yoshimoto G, Nagafuji K, Miyamato T, et al. FLT3 mutations in normal karyotype acute myeloid leukemia in first complete remission treated with antologans peripheral blood stem cell transplantation. Bone Marrow Transplant. 2005;36:977–83.
15. Yung Shih L, Fuang Huang C, Hou Wu J, et al. Heterogeneous patterns of FLT3 Asp 835 mutations in relapsed de novo acute myeloid leukemia. Clin Cancer Res. Feb. 2004;10:1326–32.
16. Maciejewski J, consultqd cleveland clinic.org. A paradigm shift in conception of aplastic anaemia; 2016.
17. Takai K, Sananda M, Ketsueki R. Transformation of severe aplastic anemia to myelodysplastic syndrome with monosomy 7 in a patient who achieved transfusion independence after immunosuppressive therapy. Rinsho Ketsueki. 2000;41(4):347–53.

18. Desangles F. 7/del(7q) in adults. Atlans Genet Cytogenet Oncol Haematol. 1999;3(3):139–40.
19. Vardiman J, Reichard K. Acute myeloid leukemia with myelodysplasia related changes. Am J Clin Pathol. 2015;144(1):29–43.
20. Montalban Bravo G, Kanagal-Shamanna R, Sasaki K, et al. Outcome of patients with acute myeloid leukemia (AML) with myelodysplasia related changes (AML-MRC) are dependent on diagnostic criteria and therapy. Blood. 2019;134:1312.
21. Greenberg PL, Tuechler H, Schanz J, et al. Revised international prognostic scoring system (IPSS-R) for myelodysplastic syndrome. Blood. 2012;120:2454–65.
22. Anelli L, Pasciolla C, Zagaria A, et al. Monosomal karyotype in myeloid neoplasias: a literature review. Onco Targets Ther. 2017;10:2163–71.
23. Patnaik MM, Hanson CA, Hodnefield JM, et al. Monosomal karyotype in myelodysplastic syndromes, with or without monosomy 7 or 5, is prognostically worse than an otherwise complex karyotype. Leukemia. 2011;25(2):266–70.
24. Solary E, Vainchenker W. Encyclopedia of cancer. 3rd ed. New York: Elsevier; 2019.
25. Miesner M, Haferlach C, Bacher U. Multilineage dysplasia (MCD) in acute myeloid leukemia (AML) correlates with MDS-related cytogenetic abnormalities and a prior history of MDS or MDS/MPN but has no independent prognostic relevance: a comparison of 408 cases classified as "AML not otherwise specified" (AML-NOS) or "AML with myelodysplasia-related changes". Blood. 2010;116(15):2742–51.
26. Pezeshki A, Podder S, Kamel R, et al. Monosomy 7/del (7q) in inherited bone marrow failure syndromes: a systematic review. Pediatr Blood Cancer. 2017; https://doi.org/10.1002/pbc.26714.
27. Kayser S, Dohner K, et al. The impact of therapy-related acute myeloid leukemia (AML)an outcome in 2853 adult patients with newly diagnosed AML. Blood. 2017;117(7):2137–45.
28. Schoch C, et al. Karyotype is an independent prognostic parameter in therapy related acute myeloid leukemia (t-AML) an analysis of 93 patients with t-AML in comparison to 1091 patients with de novo AML. Leukemia. 2004;18(1):120–5.
29. Sarova I, Zemanova Z. Molecular cytogenetic analysis of dicentric chromosomes in acute myeloid leukemia. Leuk. Res. 2016;43:51–7. https://doi.org/10.1016/j.leukres.2016.01.001.
30. Arber DA, Orazi A, Hasserjian R, et al. The 2016 revision to the World Health Organization classification of myeloid neoplasms and acute leukemia. Blood. 2016;127(20):2391–405.
31. Nagy A, Neubauer A. Acute myeloid leukemia with myelodysplasia related changes. Atlas Genet Cytogenet Oncol Haematol. 2017;21(11):404–8.
32. Huret JL. t(11;19)(q23;p13.1)KMT2A/ELL. Atlas Genet Cytogenet Oncol Hematol. 1997;1(2):104–5.
33. Moorman AV, Hagemeijer A, Charrin C. The Translocations t(11;19) (q23;p13.1): A cytogenetic and clinical profile of 53 patients. European 11q workshop participants. Leukemia. 1998;12(5):805–10.
34. Godley LA, Larson RA. Therapy related myeloid leukemia. Semin Oncol. 2008 Aug;35(4):418–29.
35. Christiansen DH, Anderson MK, Pedersen-Bjergaard J. Mutations with loss of heterozygosity of p53 are common in therapy related myelodysplasia and acute myeloid leukemia after exposure to alkylating agents and significantly associated with deletion or loss of 5q, a complex karyotype and a poor prognosis. J Clin Oncol. 2001;19(5):1405–13.
36. Lurlo A, Cattaneo D, Gianelli U. Blast transformation in myeloproliferative neoplasm: risk factors, biological finding and targeted therapeutic options. Int J Mol Sci. 2019;20(8):1839.
37. Tefferi A, Guglielmelli P, Larson D, et al. Long term survival and blast transformation in molecularly annotated essential thrombocythemia, polycythemia vera and myelofibrosis. Blood. 2014;124(16):2507–13.
38. Yogarajah M, Tefferi A. Leukemic transformation in myeloproliferative neosplasms. a literature review on risk characteristics and outcome. Mayo Clin Proc. 2017 July;92(7):1118–28.
39. Hulegardh E, Nilsson C, Lazarevic V, et al. Characterization and prognostic features of secondary acute myeloid leukemia (AML) in a population based setting: a report from the Swedish Acute Leukemia Registry. Am J Hematol. 2015;90:208–14.
40. Ostgard LSG, Medeiros BC, Sengelov H, et al. Epidemiology and clinical significance of secondary and therapy related acute myeloid leukemia: a national population based cohort study. J Clin Oncol. 2015;33(31):3641–9.
41. Peniket A, Wainscoat J, Side L, et al. Del (9q) AML: clinical and cytological characteristics and prognostic implications. Br J Haematol. 2005 Apr;129(2):210–20.
42. Naarmann-de Vries IS, Sackmann Y, Klein F, Ostareck Lederer A, Ostareck DH, Jost E, Ehninger G, Brummendorf TH, Marx G, Rollig C, Crysandt M. Characterization of acute myeloid leukemia with del(9q) - impact of the genes in the minimally deleted region. Leuk Res. 2019 Jan;76:15–23.
43. Bhatnagar B, Eisfeld AK, Kohlschmidt J, et al. Clinical and molecular characterization of patients with acute myeloid leukemia and sole trisomies of chromosomes 4,8,11,13 or 21. Leukemia. 28 Aug 2019;34(2):358–68.
44. Eisfeld AK, Kohlscjmidt J, Mrozek K, Blachly JS, Nicolet D, Kroll K, et al. Adult acute myeloid leukemia with trisomy 11 as the sole abnormality is characterized by the presence of five distinct gene mutations: MLL-PTD, DNMT3A, U2AF1, FLT3-ITD and IDH2. Leukemia. 2016;30:2254–8.
45. Alseraye FM, Zuo Z, Bueso-Ramos C, Wang S, Medeiros LJ, Lu G. Trisomy 11 as an isolated abnormality in acute myeloid leukemia is associated with unfavorable prognosis but not with an NPM1 or KIT mutation. Int J Clin Exp Pathol. 2011;4:371–7.
46. Cui W, Bueso Ramos CE, YIN CC, Sun J, Chen S, Muddasani R, Lu G. Trisomy 14 as a sole chromo-

some abnormality is associated with older age, a heterogenous group of myeloid neoplasms with dysplasia, and a wide spectrum of disease progression. J Biomed Biotechnol. 2010; https://doi.org/10.1155/2010/365318.
47. Hemsing AL, Hovland R, Tsykunova G, Reikvam H. Trisomy 8 in acute myeloid leukemia. Expert Rev Hematol. 2019; https://doi.org/10.1080/17474086.2019.1657400.
48. Cuneo A, Cavazzini F, Castoldi GL. Acute megakaryoblastic leukemia AML-M7. Altas Genet Cytogen Oncol Haematol. 2004;8(1):29–30.
49. Nimer SD, MacGrogan D, Jhanwar S, et al. Chromosome 19 abnormalities are commonly seen in AML-M7. Blood. 2002;100(10):3838–9.
50. Rucker FG, Schlenk RF, Bullinger L, et al. TP53 alterations in acute myeloid leukemia with complex karyotype correlate with specific copy number alterations, monosomal karyotype, and dismal outcome. Blood. 2012;119(9):2114–21.
51. Ruiz Delgado NGJ, Corteza AK, Gazca O, et al. Lineage switch from acute lymphoblastic leukemia to myeloid leukemia. Med Universitaria. 2017;19(74):27–31.

4 Myelodysplastic Syndrome

4.1 Case Study 4.1: Myelodysplastic Syndrome (MDS-EB2) with Rare Interstitial Deletion of the Long Arm of Chromosome 9

Brief Clinical History
A 66-year-old male presented with pancytopenia for evaluation for suspected MDS.

CBC: *Hb:* 8.6 g/dl, TLC: 1620/cu mm and Platelet count: 98,000/cu mm.

DLC: Blasts: <1%, Neutrophils: 39%, Lymphocytes: 50%, Eosinophils: 3%, Monocytes: 8%, Basophils 0.0%.

Peripheral Blood Smear and Bone Marrow
Smear showed leukopenia with neutropenia (ANC: 700/ml) and an occasional blast. Bone marrow showed erythroid prominence with trilineage dysplasia with blasts constituting 16% of all nucleated cells. No ring sideroblasts were seen. The features were suggestive of MDS-EB2.

Immunophenotype
Blasts 4% in moderate CD45 region in a hemodilute BM Positive for CD34+, CD117+, HLA-DR+, CD13+, CD33+, CD38+.

Cytogenetics
Conventional Karyotype

46,XY,del(9)(q12q22)[5]/46,XY[15]

Cytogenetic analysis of 20 metaphases showed the presence of an abnormal clone characterized by an interstitial deletion of the long arm of chromosome 9 in 5 metaphases.

M. Bhargava, *Hematologic Malignancies*, https://doi.org/10.1007/978-981-33-4799-1_4

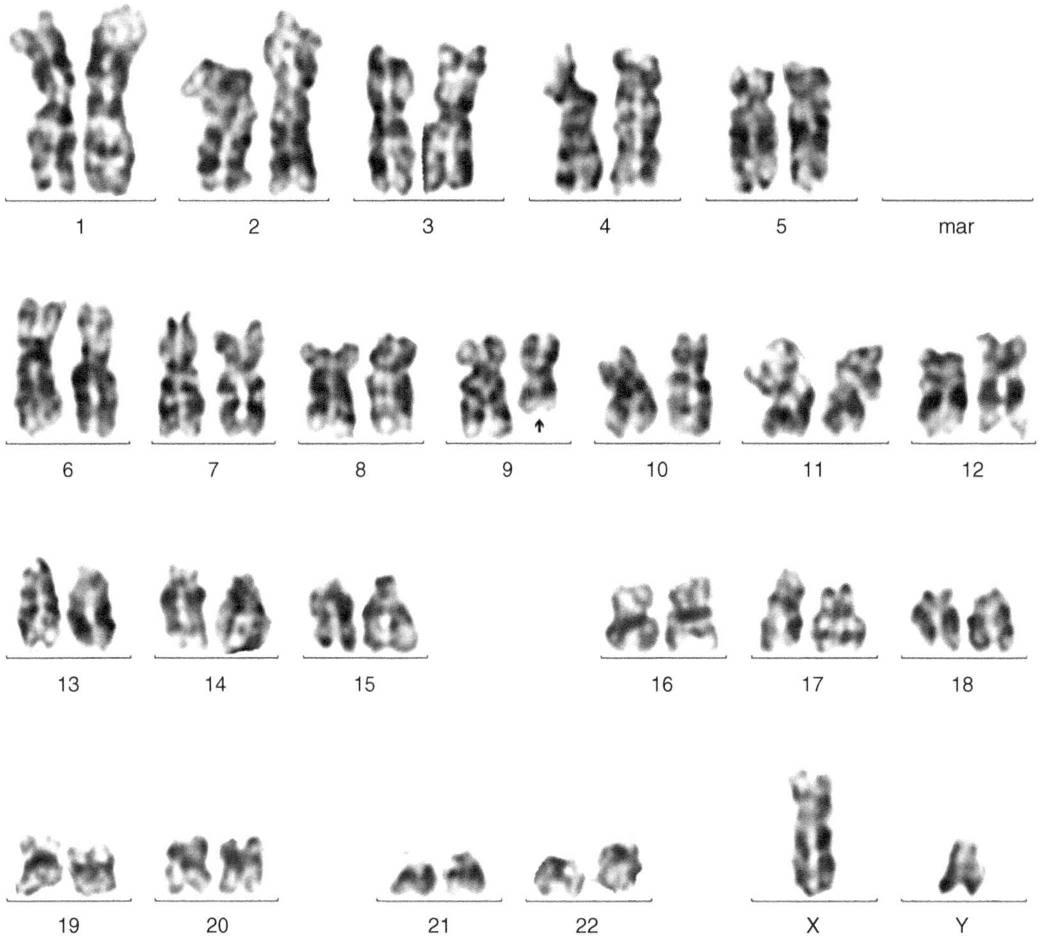

Case Study 4.1 Male karyotype showing interstitial deletion of the long arm of Chromosome 9, i.e., del(9q)

Discussion

Mecucci et al studied 16 patients of hematological malignancies with interstitial deletions of the long arm of chromosome 9 (9q-). From their observations and 10 other studies, it was deduced that this anomaly is almost exclusively found in myeloproliferative disorders and rarely occurs as sole abnormality. In more than one-third of cases, it was associated with t(8;21) and occurred as a secondary event.

Del (9q) is seen mainly in AML and rarely observed in MDS (1–2%) [1]. In our case, it was seen in MDS-EB2. When del (9q) is the unique chromosomal abnormality, the prognosis is variable. In a study of 34 patients, no significant correlation was found between del(9P) and progression of the disease to AML [2].

4.2 Case Study 4.2: Myelodysplastic Syndrome with Excess Blasts (MDS-EB2) and Trisomy 8

Brief Clinical History

A 72-year-old male presented with a weakness for 15 days. He is a known hypertensive. Reported as MDS-EB at an outside facility and put on Azacytidine. Has received two blood transfusions. Bone marrow is done to rule out Acute Leukemia.

CBC: *Hb:* 7.3 g/dl, TLC: 1100/cu mm and Platelets: 40,000/cu mm.

DLC: Neutrophils: 2.0%, Lymphocytes: 57.0%, Eosinophils: 14.0%, Monocytes: 26.0, Basophils: 1.0%.

Peripheral Blood Smear and Bone Marrow:
Smear showed leukopenia with neutropenia (ANC: 22/μL). Bone marrow showed trilineage dysplasia with a blast count of 12.0% of total nucleated cells. Peripheral blood and bone marrow features were consistent with myelodysplastic syndrome with excess blasts (MDS-EB-2).

Flow Cytometric Immunophenotyping
Blasts constituting ~4% of acquired events were present in the moderate CD45 blast region. They were CD34+, CD117+, HLA-DR+, CD13+, CD33+, CD38+. Granulocytic cells showed features of dysgranulopoiesis in form of underexpression of CD10 and CD16 and abnormal maturation pattern of CD16/CD11b.

Cytogenetic
Conventional Karyotype

47,XY,+8[18]/46,XY[2]

Cytogenetic analysis of 20 metaphases showed a neoplastic clone characterized by gain of chromosome 8 as trisomy in 18 metaphases.

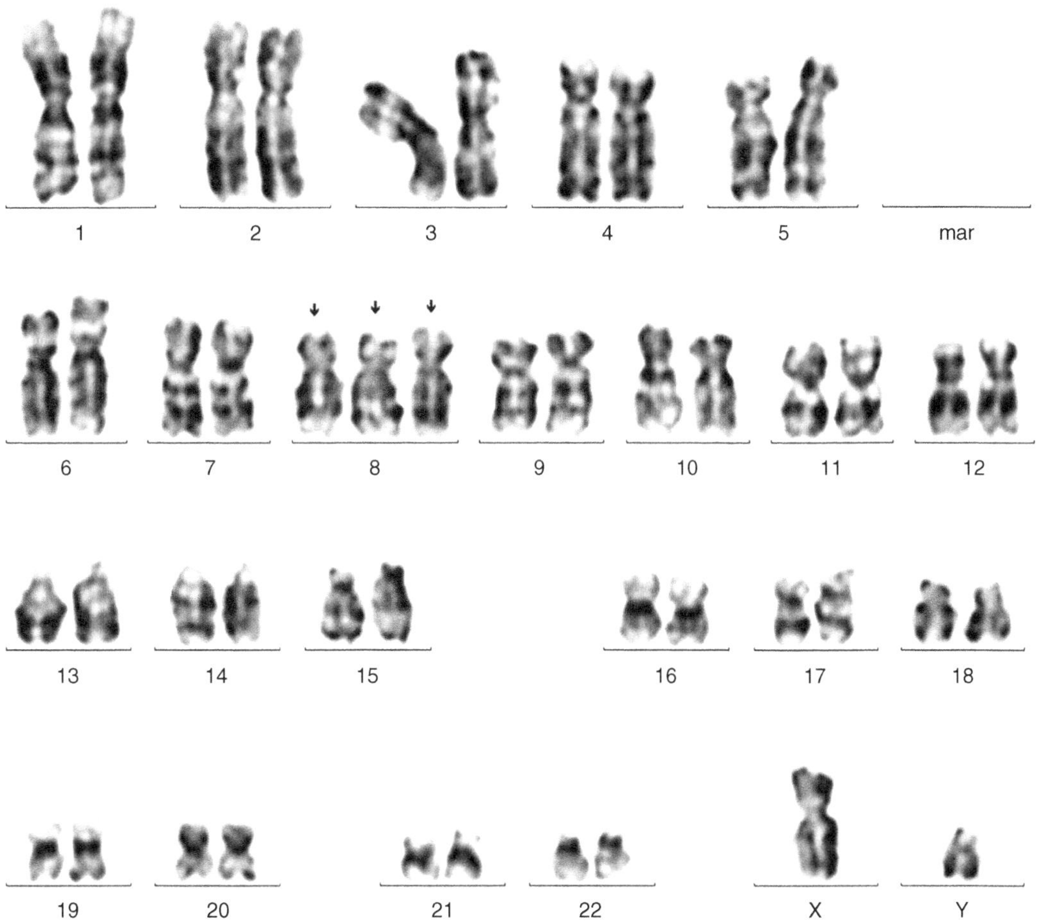

Case Study 4.2 Karyotype showing three copies of chromosome 8, i.e., Trisomy 8

Molecular Studies
The Real Time Quantitative PCR showed *AML Comprehensive panel: Negative.*

Discussion
Isolated trisomy 8 is a frequent cytogenetic abnormality in MDS representing 9-11% of de novo MDS with cytogenetic abnormality and 5% of all MDS. It is, however not specific to MDS. It is also frequent in CMML, AML, MPN, PMF, PV. In a study by Drevon et al. [3], 45% of patients had MDS-EB-1or-2, 40% MDS-SLD or MDS-MLD, 11% MDS-RS, and 5% MDS-U. This patient had MDS-EB-2 with isolated trisomy 8. Progression towards AML occurs in about half of the cases when trisomy 8 is the sole abnormality. In the same retrospective study of 138 MDS patients, 54 (39.1%), had myeloproliferative features (MP), 28 at diagnosis, Median event free survival (EFS), and overall survival (OS) were 25 and 27 months, respectively. In conclusion, MP features are frequent in MDS with isolated+8 and they have more E2H2, ASXL1, and STAG2 mutations.

As per IPSS-R, these patients fall in the intermediate category [4].

4.3 Case Study 4.3: Myelodysplastic Syndrome with Multilineage Dysplasia (MDS-MLD) with Monosomal Karyotype

Brief Clinical History
A 60-year-old male was diagnosed as ITP at an outside facility and started on oral steroids for one year. Has presented with Hemoptysis.

CBC: *Hb:* 10.9 g/dl, TLC: 4460/cu mm and Platelets: 75,000/cu mm.

DLC: Neutrophils: 74.6%, Lymphocytes: 16.4%, Eosinophils: 0.7%, Monocytes: 8.3%, Basophils: 0.0%.

Peripheral Blood Smear and Bone Marrow
Smear showed Neutrophilic predominance with left shift with prominent granulations. No abnormal cells were seen.

Bone marrow showed trilineage dysplasia with a variable blast count of 3%–4%.

Cytogenetics
Conventional Karyotype
45,XY,t(3;21)(q26;q22),-7[14]/46,XY[6]

Cytogenetic analysis of 20 metaphases showed a monosomal karyotype with the presence of one autosomal monosomy and one structural abnormality. There was Monosomy of chromosome 7 and structural abnormality was a balanced reciprocal translocation between the long arm of chromosome 3 (3q) and long arm of chromosome 21 (21q) i.et(3;21)(q26;q22) RUNX1/MECOM in 14 metaphases analyzed. Monosomal karyotype is associated with poor prognosis.

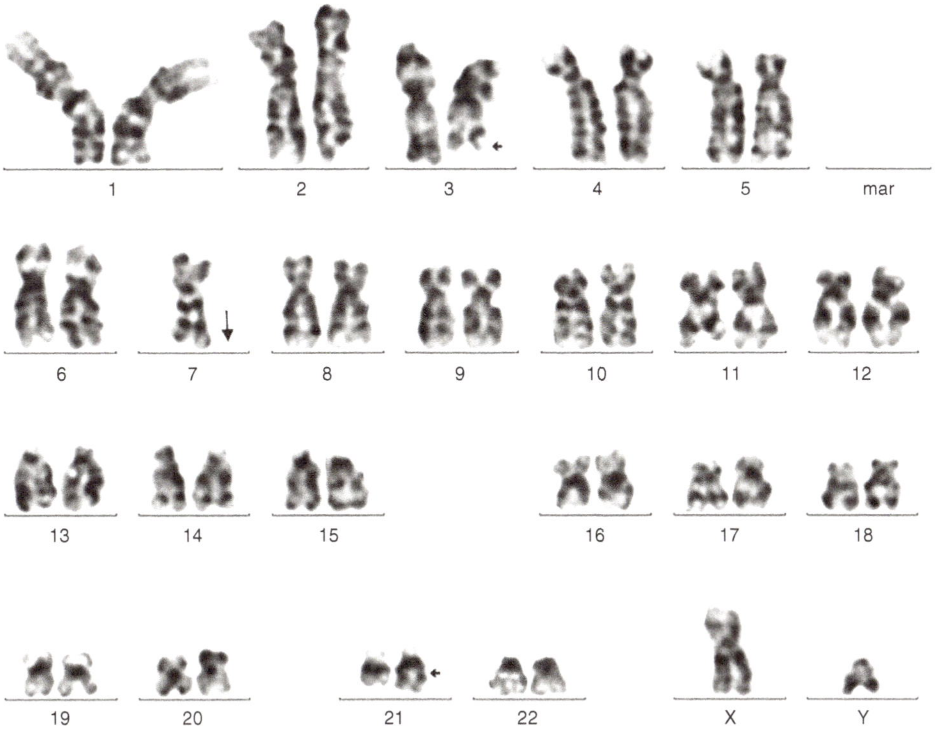

Case Study 4.3 (1) Karyotype showing 3q abnormality along with monosomy 7 which is secondary abnormality

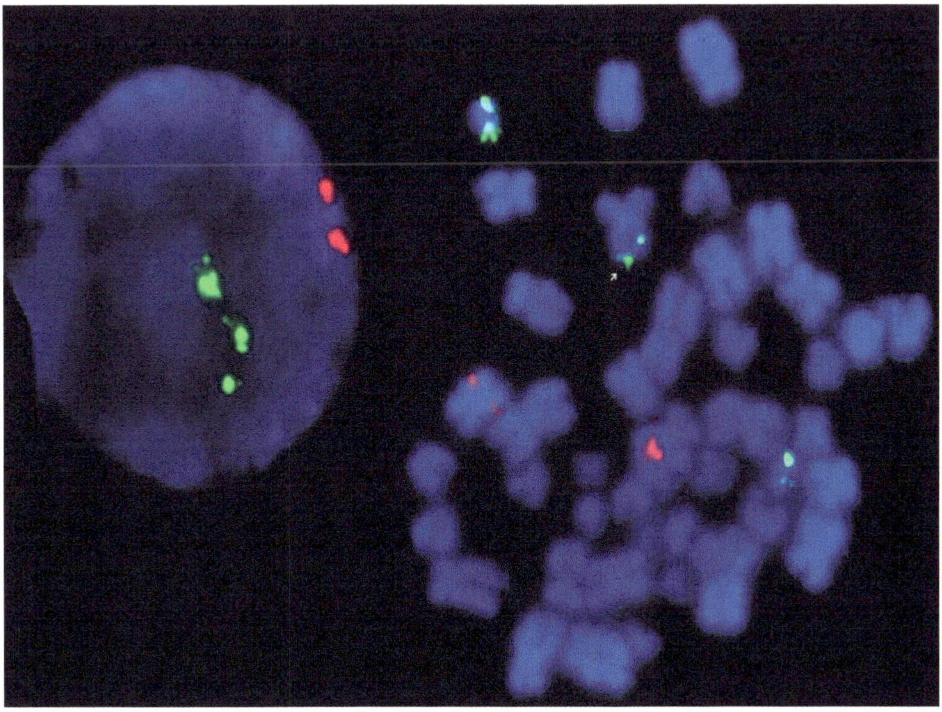

Case Study 4.3 (2) FISH image showing three copies of *RUNX1* allele by using *RUNX1/RUNX1T1* DC DF probe indicating *RUNX1* allele on chromosome 3 on metaphase FISH

Discussion

This case of MDS-MLD had a rare t(3;21) translocation together with a monosomal karyotype (monosomy 7) Monosomal karyotype is defined as the presence of at least 2 autosomal monosomies or a single autosomal monosomy associated with at least one structural abnormality (as in this patient) and can be seen in AML, MDS, PMF [5]. Monosomal karyotype does not include AML with t(15;17),t(8:21) or inv(16). The incidence of 3q abnormalities in AML is 4–5%. The inversion 3 t(3;3) is frequently associated with monosomy 7 and N-RAS mutations. AML with diverse 3q abnormalities such as unbalanced 3q26 or unbalanced 3q21 present with monosomal or complex karyotypes and show dismal survival [6].

The most frequent monosomies include Chr 7,5,17 and 18. Cases of t(3;21)(q26;q22) are mostly treatment-related blast crisis of CML and treatment-related MDS/AMLs. A very few cases of chronic myeloproliferative disease without t(9;22) have been documented (3.4%). Of the 146 cases of t(3;21) in the database, the gene involved was RUNX1-MECOM (EVI1) which means MDS1and EVI1 complex locus [7].

The prognosis of these patients is poor. In MDS and PMF, the risk of leukemic transformation is higher.

4.4 Case Study 4.4: Myelodysplastic Syndrome with Multilineage Dysplasia (MDS-MLD) with Deletion 13q

Brief Clinical History

A 59-year-old male presented with bilateral lid swelling and weakness. No history of bleeding. Had received three units of blood at an outside facility where he was diagnosed as MDS-MLD and started on erythropoietin.

CBC: *Hb:* 7.7 g/dl, TLC: 4040/cu mm and Platelets: 8000/cu mm.

DLC: Neutrophils: 46.3%, Lymphocytes: 22.5%, Eosinophils: 6.2%, Monocytes: 25.0%, Basophils: 0.0%.

Peripheral Smear and Bone Marrow

Smear showed absolute monocytosis. No blasts were seen. The bone marrow was hypercellular showing trilineage dysplasia in >10% of cells with <5% blasts. In a known case of MDS, overall features were suggestive of Myelodysplastic Syndrome with Multilineage dysplasia (MDS-MLD).

Cytogenetics

Conventional Karyotype

46,XY,del(13)(q12q14)[9]/46,XY[11]

Cytogenetic analysis of unstimulated bone marrow aspirate culture showed neoplastic clone characterized by the deletion of the long arm of chromosome 13 (13q) in 9 of the 20 metaphases analyzed.

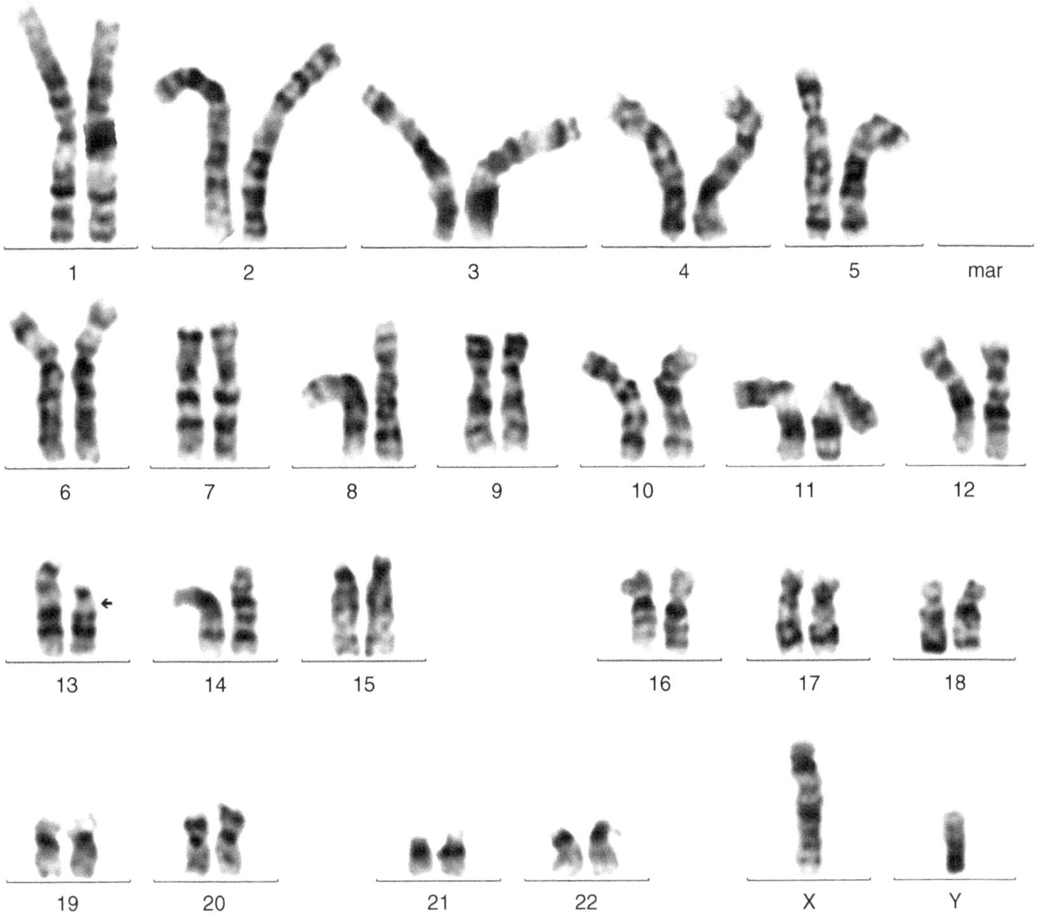

Case Study 4.4 Karyotype showing deletion of the long arm of chromosome 13, i.e., del(13q)

Discussion

The International Prognostic Scoring System (IPSS) by Greenberg et al. [4] is an important standard for assessing the prognosis of untreated MDS. Bone marrow cytogenetics, marrow blast percentage, and cytopenias continue to be the basis. In cytogenetics, five rather than three prognostic subgroups are included. This system included numerous clinical features towards analyzing prognosis in MDS into very good, good, intermediate, poor, and very poor categories based on cytogenetic features.

In MDS, the frequency of 13q deletion is seen in around 3% cases. In the IPSS-R, Greenberg et al. [4] included five rather than three cytogenetic prognostic groups with specific and new classifications of a number of less common cytogenetic subsets. As per this classification, our patient with del (13q) fell into the intermediate category with a median survival of 2.7 years (any other single or double independent clones).

Hosokawa et al. [8] in a review of bone marrow failure with del(13q) diagnosed as MDS (U) due to absence of apparent dysplasia showed good response to immunosuppressive therapy and suggested that since MDS-U with del(13q) is a benign bone marrow failure subset responsive to immunotherapy, it should not be considered as intermediate risk category.

4.5 Case Study 4.5: Therapy Related Myelodysplastic Syndrome with Complex Karyotype

Brief Clinical History

A 67-year-old male diagnosed as a case of Squamous Cell Carcinoma Left Lung in 2017. Received chemotherapy and radiotherapy. Presented in August 2019 with pancytopenia and again in January 2020 following therapy. BM was done for evaluation of the disease both times.

CBC: *Hb:* 13.9g/dl, TLC: 2200/cu mm and Platelets: 51,000/cu mm.

DLC: Neutrophils: 27.7%, Lymphocytes: 65.9%, Eosinophils: 3.2%, Monocytes: 2.7%, Basophils: 0.5%.

Peripheral Blood and Bone Marrow

Smear showed leukopenia with neutropenia, ANC: 610/μL. No abnormal cells were seen. Bone marrow done in September 2019 showed it to be hemodilute. Imprint smears showed a few erythroid cells and dyspoietic myeloid cells with occasional blasts. A few megakaryocytes showed dysmegakaryopoiesis. Bone marrow biopsy was hypocellular (cellularity 30%) but showed megaloblastic erythroid cells, few blasts (CD34+) and megakaryocytes showing dysmegakaryopoiesis. A follow-up bone marrow done in March 2020 was cellular and showed trilineage dyshematopoiesis with blast count of 15%.

Flow Cytometric Immunophenotyping

Cells of interest: Blasts constituted ~8% of acquired events, expressing moderate CD45.

CD34+, CD117+, HLA-DR+, CD13+, CD33+, cMPO+, CD38+, CD56+.

Cytogenetics

Conventional Karyotype

Karyotype: (August 2019)

47~49,XY,-Y,del(5)(q31q35), del(6)(q12q25),+8,+8,-20,-21,+mar1,+mar2,+mar3[cp12]/46,XY[8]

Karyotype: (January 2020)

45~47,-4,del(5)(q12q33),-6, -6,-7,+8+8,+10,+add(12)(q24.3),-16,+17, -20,-21,-21. +mar1, +mar2[cp7]/46,XY[13]

In addition to the changes in August 2019, of hyperdiploid complex karyotype with del(5q), del(6q), tetrasomy 8, monosomies of chromosome 20 and 21, loss of chromosome Y and 2-3 marker chromosomes, in 2020 January, some more changes occurred which were monosomy 7 and 16 trisomy of 10, 12 and 17, complete pair loss of chromosomes 6 and 21 and addition of unknown material on the long arm of chromosome 12.

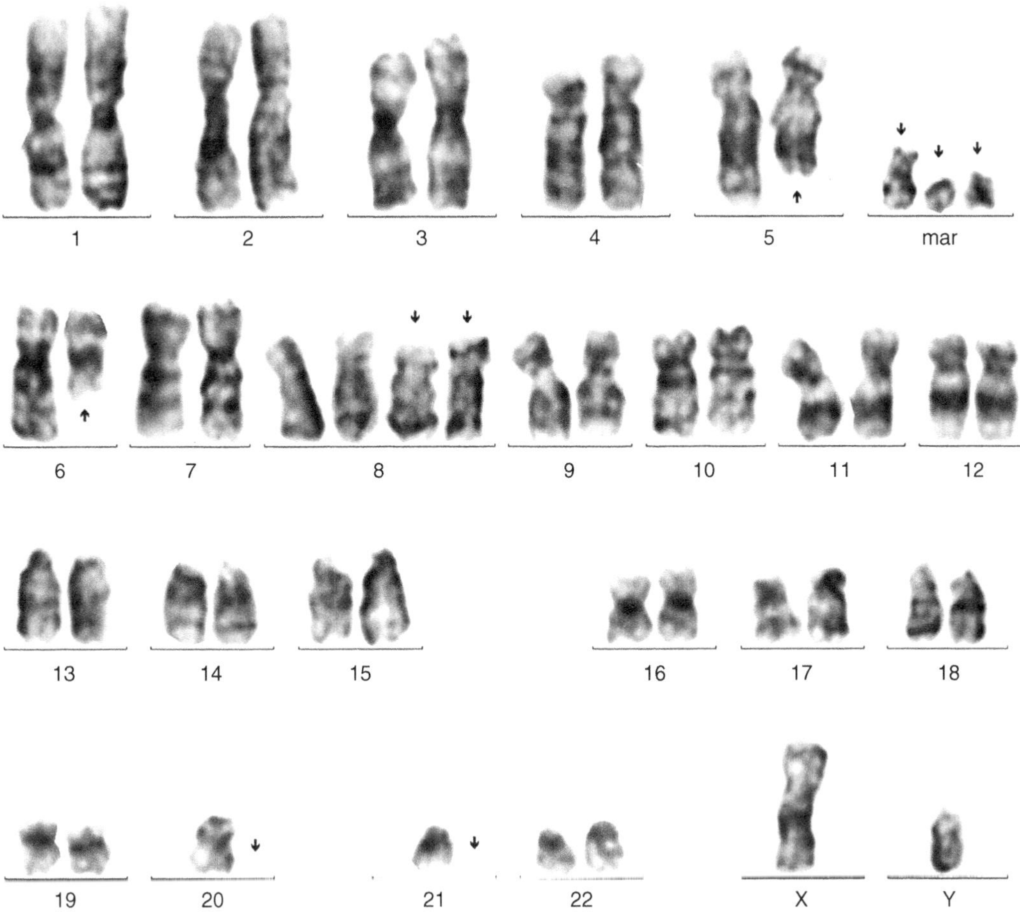

Case Study 4.5 Complex Karyotype showing deletion of the long arm of chromosome 5, 6, with tetrasomy 8 along with other abnormalities of chromosome 20, 21 and three different marker chromosomes of unknown origin

Discussion

This patient first presented in August 2019 following therapy for Squamous Cell Carcinoma Lung. Karyotyping showed a complex karyotype and on follow-up in March 2020 showed additional cytogenetic mutations indicating a progression of the disease.

As per the IPSS-R [4], this case was categorized as poor risk. Complex karyotypes are heterogeneous with prognosis worsening with each additional aberration rather than by the chromosome involved (most frequently 5, 7, and 17. They are highly chemo resistant. Therapy-related myeloid neoplasms (t-MNs) have a significantly high frequency of high-risk karyotypes (complex karyotypes abnormalities of chromosome 5 and/or 7 are seen in ~70% vs 20% in AML de novo). TP53 mutations are associated with del(5q) in 80% cases, complex karyotypes, resistance to therapy, and poor overall survival. -7/del(7q) defines another major subgroup of t-MN occurring in 49% patients. The remaining t-(MN) myeloid neoplasms cases are akin to de novo counterparts (balanced translocations, normal karyotype with NPM1 and FLT3 mutations, etc. [9].

4.6 Case Study 4.6: Myelodysplastic Syndrome with Excess Blasts (MDS-EB-2) with Dicentric Chromosome

Brief Clinical History

A 55-year-old male symptomatic since July 2019 with anemia needing recurrent transfusions. No organomegaly.

CBC: Hb: 6.7 g/dl, TLC: 940/cu mm Platelets: 40,000/cu mm.

DLC: Neutrophils: 21.0%, Lymphocytes: 69.0%, Eosinophils: 1.0%, Monocytes: 4.0%,

Basophils: 2.0%, Myelocytes: 2.0%, Blasts: 1.0%.

Peripheral Smear and Bone Marrow

Peripheral blood showed pancytopenia with ~1% blasts and mild dyspoiesis in neutrophils. The bone marrow examination showed significant trilineage dyspoiesis with an overall increase in myeloid blasts, constituting ~11.2% on the aspirate with no large aggregates on biopsy. Overall, features were consistent with Myelodysplastic syndrome with Excess blasts-2 (MDS-EB2).

Cytogenetics

Conventional Karyotype

45,XY,dic(7;11)(q10;p10)[3]/46,XY,idem,+mar[11]/46,XY[6]

Cytogenetic analysis showed a neoplastic clone characterized by a formation of dicentric chromosome between the long arm of chromosome 7 and the short arm of chromosome 11 in 14 of the 20 metaphases. The clone also showed the gain of a marker chromosome of unknown origin in 11 of the 14 metaphases analyzed.

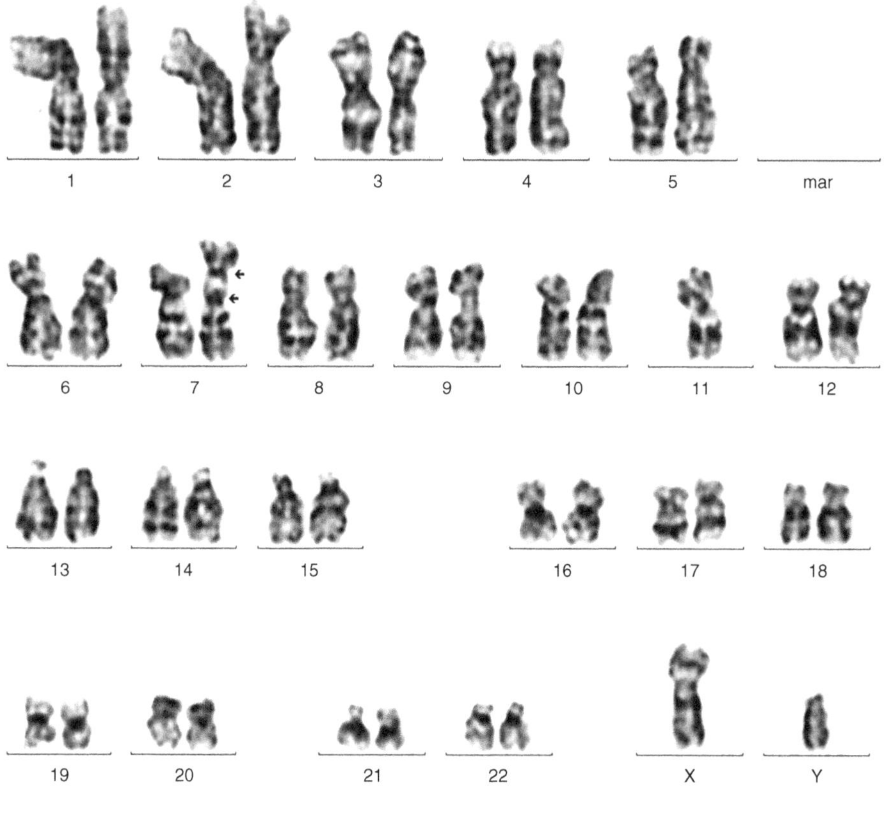

Case Study 4.6 Karyotype showing dicentric chromosome formed by translocation (7;11)

Discussion

This patient had a rare cytogenetic abnormality of a dic (7,11) formed between the long arm of Chromosome 7 and the short arm of Chromosome 11. Dicentric chromosomes are the result of translocations with a fusion of two chromosome arms and centromeric or pericentromeric sequences from both chromosomes involved [10]. Dicentric chromosomes are seen in MDS and AML but mostly in t-MDS and t-AML. In this patient, it was in de novo MDS. In a study of 180 cases of t-MDS and t-AML, 27 showed dic chromosome compared to only 7 out of 231 patients in de novo disease, indicating its rarity [10]. As per IPSS-R score cytogenetic abnormalities are placed into distinct prognostic subgroups including inv (3)/t(3q),del(11q) del(12p),i(17q),+19, double anomalies including de (5q),del(7q) or monosomy 7 or any other double changes. According to this classification, this patient falls in the intermediate category [4].

4.7 Case Study 4.7: Myelodysplastic Syndrome with Excess Blasts-EB1 (MDS-EB-1) with Composite Karyotype

Brief Clinical History

A 63-year-old male presented with weakness and bicytopenia in 2019. Bone marrow aspirate and biopsy done outside was suggestive of MDS-EB1. Was given Inj. Azacytidine 100 mg for 5 days. Bone marrow was done to assess disease status.

CBC: *Hb:* 6.5g/dl, TLC: 5440/cumm and Platelets: 8000/cumm.

DLC: Blasts: 3.0%, Myelocytes: 10.0%, Metamyelocytes: 2.0%, Neutrophils: 41.0%, Lymphocytes: 24.0%, Eosinophils: 12.0%, Monocytes: 7.0%, Basophils: 1.0%;

Peripheral Smear and Bone Marrow

Smear showed a shift to the left with 3% blasts. Bone marrow showed significant trilineage dyspoiesis with ~8.2% blasts and more than 15% ring sideroblasts.

Cytogenetics

Conventional Karyotype

44~48,XY,-Y, t(3;12) (p13;p11.2), del(5) (q12q33), add(6) (p25), del(6) (p21),+der(6) del(6) (p21), t(6;13)(q13;p13),+8,-10,i(11) (q11),t(11;12)(q13;p13),-16,-17,-18, +mar1, +mar2[cp20]

Cytogenetic analysis of 20 metaphases showed a neoplastic clone characterized by a composite karyotype showing numerical (44–48 modal number) and several structural abnormalities. Numerical abnormalities showed derivative chromosome as trisomy 6, trisomy 8, monosomy 10, monosomy 16, monosomy 17, monosomy 18, and loss of chromosome Y. Structural abnormalities showed balanced translocation between the short arm of chromosome 3 and the short arm of chromosome 12, interstitial deletion of the long arm of chromosome 5(5q), the addition of unknown material on the short arm of chromosome 6, deletion of the short arm of chromosome 6, unbalanced translocation between the long arm of chromosome 6 and the short arm of chromosome 13, there was also gain of (11q23) region as isochromosome 11, unbalanced translocation between the long arm of chromosome 11 and the short arm of chromosome 12, and also gain of 2 marker chromosomes of unknown origin.

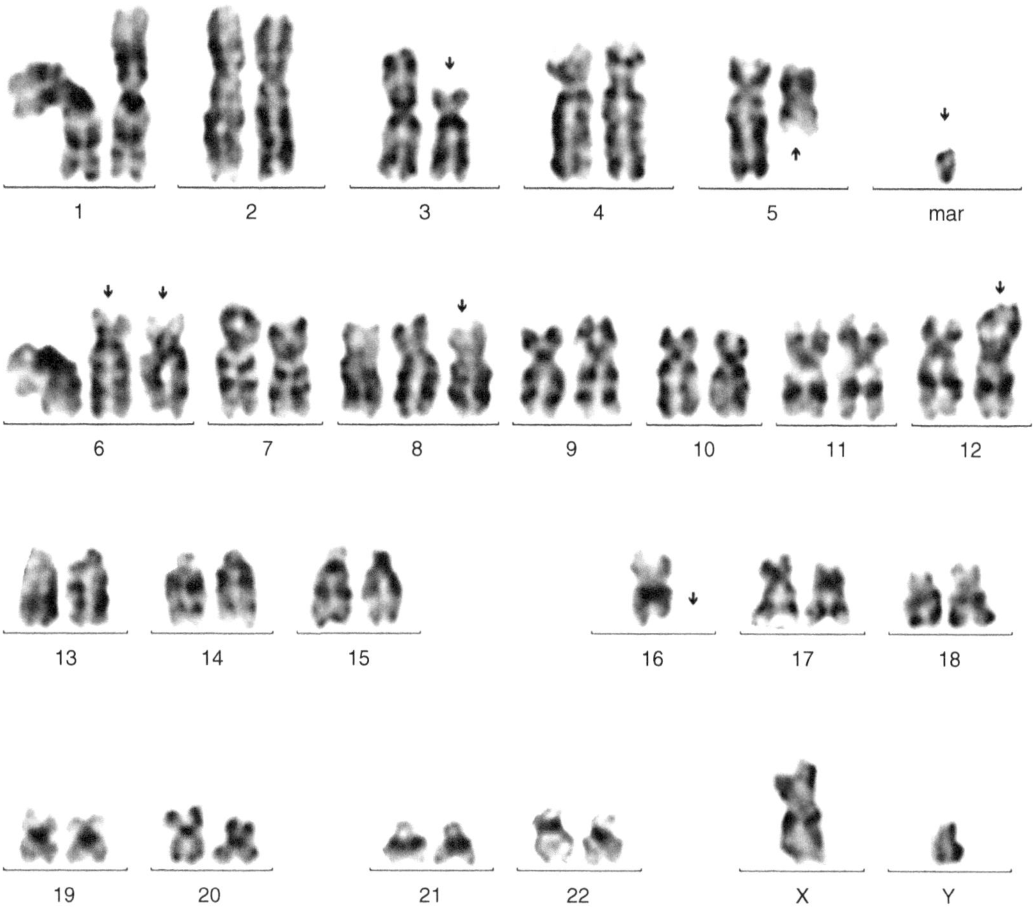

Case Study 4.7 (1) Composite Karyotype showing abnormalities of the short arm of chromosome 3, deletion of the long arm of chromosome 5, deletion of the short arm of chromosome 6 along with its gain, and trisomy 8, add(12p), monosomy 16 and one marker chromosome of unknown origin

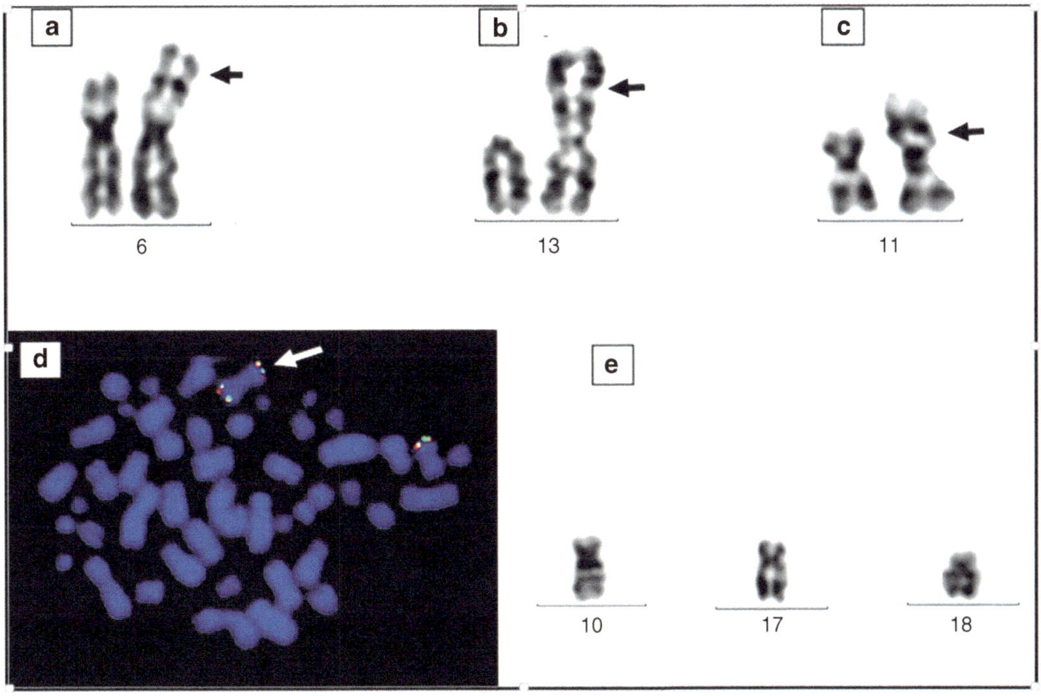

Case Study 4.7 (2) Partial karyotype showing abnormalities of chromosome fig a)add(6)(p25), b) t(6;13)(q13;p13)13, c) t(11;12)(q13;p13), d) FISH image showing isochromsome 11 by using *MLL(=KMT2A)* break-apart probe e) Monosomy of 10, 17 and 18

Discussion

One of the points to be noted in this patient was the presence of 15% ring sideroblasts. It could be argued that this patient initially may have had MDS-RS-MLD before progressing to MDS-EB-1 Metaphase chromosome analysis of BM is the cornerstone of documenting clonal hematopoiesis for diagnosis of MDS. MDS-EB represents the most clinically aggressive end of MDS characterized by dysplasia, variable increase in the number of blasts in the blood and/or bone marrow, and an increased risk of progression to AML [11]. As per the number of blasts in the PB and BM, this patient fell in MDS-EB1 category. About 40% of patients of MDS-EB may have normal metaphase chromosomes. Targeted sequencing techniques may reveal mutations in up to 90% of patients. Commonly mutated genes include SF3B1, TET2, SRSF2, ASXL1, DNMT3A, RUNX1, U2AF1, TP53, and EZH2. IDH1 and IHD2 can be targeted by specific drug therapies [10]. As per the IPSS-R, this patient falls into a poor prognostic category based on composite karyotype [4].

4.8 Case Study 4.8: Myelodysplastic Syndrome (MDS-EB-2) with Monosomal Karyotype

Brief Clinical History

An 81-year-old male presented with severe anemia in critical condition to the emergency.

CBC: *Hb:* 6.5 g/dl, TLC: 2220/cu mm and Platelets: 19,000/cu mm.

DLC: Blasts: 4.0%; Myelocytes: 9.0%, Neutrophils: 32.0%, Lymphocytes: 32.0%, Eosinophils: 3.0%, Monocytes: 20.0%, Basophils: 0.0%.

Peripheral Smear and Bone Marrow

Smear showed Leukopenia with neutropenia with the presence of ~4% blasts. Granulocytes showed dyspoiesis. The bone marrow was cellular showing trilineage dysplasia with ~11.4% blasts. Picture was consistent with Myelodysplastic Syndrome with excess blasts-2 (MDS-EB-2).

Cytogenetics

Conventional Karyotype

45,XY,t(3;3)(q21.3;q26.2),-7[17]/46,XY[3]

Cytogenetic analysis of 20 metaphases showed a neoplastic clone characterized by a balanced reciprocal translocation between the long arms of chromosome 3 i.et(3;3) and Monosomy of chromosome 7 in 17 of the 20 metaphases analyzed. t(3;3) as well as Monosomy 7 indicate a poor prognosis.

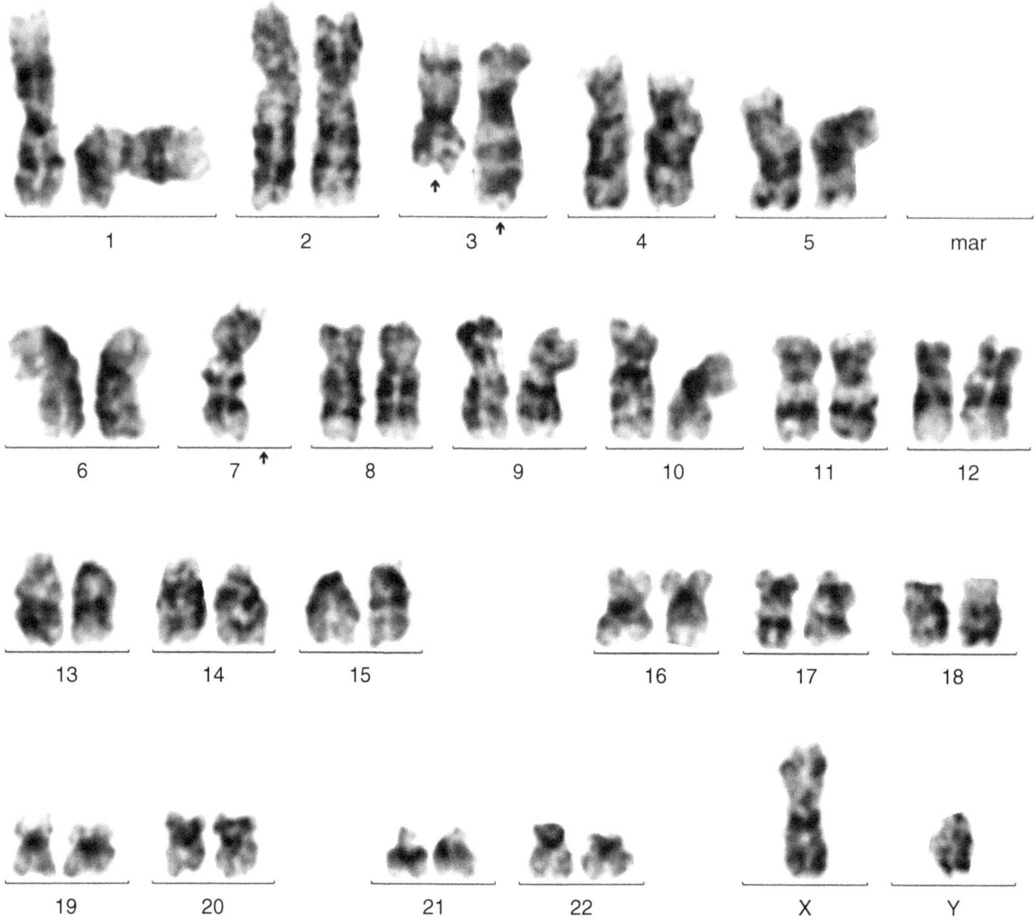

Case Study 4.8 (1) Karyotype showing translocation between long arms of chromosome 3, i.e, t(3;3)

Case Study 4.8 (2) FISH image showing 2F1R1G, i.e., t(3;3) using *MECOM/GATA2* DC BA probe

Molecular Findings

The Real Time Quantitative PCR for *MPN Mutation Panel: Negative.*

Discussion

This patient of MDS-EB-2 had t(3;3) as well as monosomy 7 qualifying for a monosomal karyotype.

Monosomal karyotype does not include recurrent genetic abnormalities in AML such as (t8;21),

t(15;17) and inv (16). inv(3) and t(3;3) have been documented in de novo AML, t-AML,S-AML and MDS. Patients with inv(3) or t(3:3) have an aggressive course with the short OS. OS is shorter if additional monosomy is present as in our case. The incidence of 3q abnormalities in AML is 4%–5%. The inversion 3 t(3;3) is frequently associated with monosomy 7 and N-RAS mutations. AML with diverse 3q abnormalities such as unbalanced 3q26 or unbalanced 3q21 present with monosomal or complex karyotypes and show dismal survival [12].

Monosomal karyotype (MK) is defined as the presence of at least 2 autosomal monosomies or a single autosomal monosomy with at least one structural abnormality as in this case. In MDS, MK among patients with complex karyotype in 83% [13]. The prognosis is very poor. In adult MDS patients with MK 2 yr survival is 6% OS is 23% for complex karyotype without monosomies [14].

4.9 Case Study 4.9: Myelodyaplastic Syndrome with Isolated 5q Deletion (5q-Syndrome)

Brief Clinical History
A 71-year-old female presented with persistent anemia with ascites, predominantly lymphocytic. She was on ATT and had received two units of blood transfusion.

CBC: *Hb:* 7.5 g/dl, TLC: 3.690/cu mm and Platelet count: 3,20,000/cu mm

DLC: Neutrophils: 76.1%, Lymphocytes: 14.1%, Eosinophils: 1.0%, Monocytes: 8.7%

Peripheral Smear and Bone Marrow
Smear showed bicytopenia with neutrophilia and thrombocytosis. Bone marrow showed dyserythropoiesis and dysmegakaryopoiesis with monolobate megakaryocytes Blasts were 3%.

Cytogenetics: Karyotyping
Conventional Karyotype

46,XX,del(5)(q22q34)[20]

Cytogenetic analysis of unstimulated bone marrow aspirate culture showed interstitial deletion of the long arm of chromosome 5 as an isolated abnormality in all the analyzed 20 metaphases.

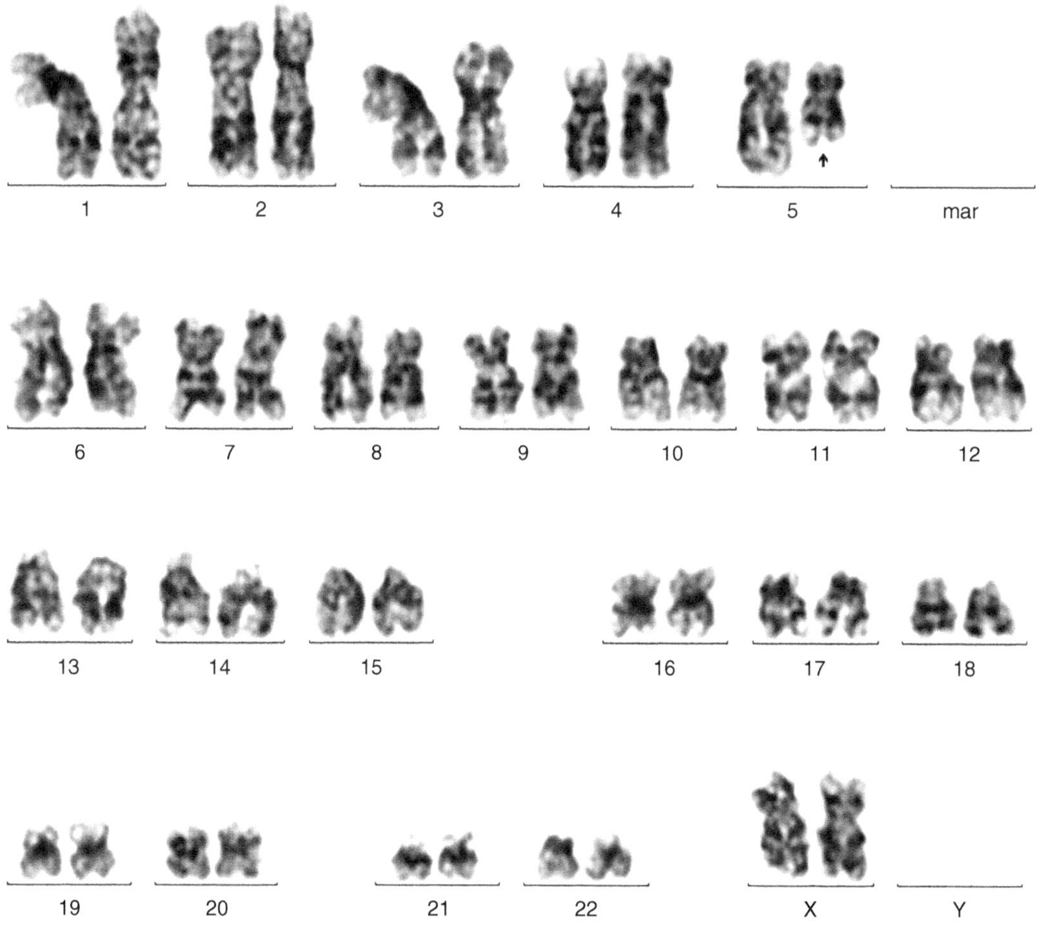

Case Study 4.9 Isolated 5q deletion (5q-syndrome)—Karyotype showing deletion of the long arm of chromosome 5

Discussion

MDS with isolated del(5q) is present in less than 10% of patients of MDS, more common in females with a median age at diagnosis of 65-70 years. These patients present with anemia and thromobocytosis (as in our case) but without neutropenia [15]. The bone marrow shows an increase in megakaryocytes with monobilobated or bilobulated nuclei. There are less than 1% and <5% blasts in the peripheral blood and bone marrow respectively, they respond well to lenalidomide. They generally have favourable prognosis unless associated with mutation in TP53 [4]. MDS with isolated del (5q) can still be diagnosed if there is one additional cytogenetic abnormality besides del (5q) [16].

Annexure

- Case Study 4.1: Myelodysplastic Syndrome (MDS – EB2) with rare interstitial deletion of the long arm of Chromosome 9
- Case Study 4.2: Myelodysplastic Syndrome with excess blasts (MDS-EB2) and Trisomy 8
- Case Study 4.3: Myelodysplastic Syndrome with Multilineage dysplasia (MDS-MLD) with monosomal karyotype
- Case Study 4.4: Myelodysplastic Syndrome with Multilineage dysplasia (MDS-MLD) with deletion 13q
- Case Study 4.5: Therapy Related Myelodysplastic Syndrome with Complex Karyotype
- Case Study 4.6: Myelodysplastic Syndrome with Excess Blasts (MDS-EB-2) with Dicentric Chromosome
- Case Study 4.7: Myelodysplastic Syndrome with excess blasts-EB1 (MDS-EB-1) with composite karyotype
- Case Study 4.8: Myelodysplastic Syndrome (MDS-EB-2) with Monosomal Karyotype
- Case Study 4.9: Myelodyaplastic Syndrome with Isolated 5q deletion(5q-syndrome)

References

1. Mecucci C, Vermaelen G, Kulling G, et al. Interstitial 9q—deletions in hematologic malignancies. Cancer Genetics Cytogenet. 1984;12(4):309–19.
2. Kerndrup G, Pedersen B, Hansen KB. Specific minor chromosome deletions in myelodysplastic syndromes: clinical and morphologic correlations. Cancer Genet Cytogenet. 1987; June;26(2):227–34.
3. Drevon L, Marceau A, Maarek O. Myelodysplastic syndrome with isolated trisomy 8: a type of MDS frequent associated with myelo-proliferative features? A report by the Groupe Francophone des Myelodysplasies. Br J Haematol. 2018 Sep;182(6):843–50.
4. Greenberg PL, Tuechler H, Schanz J, et al. Revised international prognostic scoring system for myelodysplastic syndromes. Blood. 2012;120:2454–65.
5. Perdigao J, Gomes da Silva M. Monosomal karyotype (MK) in myeloid malignancies. Atlas Genet Cytogenet Oncol Haematol. 2011;15(10):890–1.
6. Lughart S, Groschel S, Beverloo HB, et al. Clinical, molecular and prognostic significance of WHO type inv 3 (q21;q26.2)/t(3;3)(q21;q26.2) and various other 3q abnormalities in acute myeloid leukemia. J Clin Oncol. 2010;28:3890.
7. Huret JL. t(3;21)(q26;q22). Atlas Genet Cytogenet Oncol Haematol. 2014;18(1):53–6.
8. Hosokawa K, Katagiri T, Sugimore N, et al. Favorable outcome of patients who have 13q deletion: a suggestion for revision of the WHO MDS-U designation. Haematologica. 2012;97(12):1845–9.
9. McNemey ME, Godley LA, Le Beau MM. Therapy related myeloid neoplasms: when genetics and environment collide. Nat Rev Cancer. 2017;17(9):513–27.
10. Bejar R. Implications of molecular genetic diversity in myelodysplastic syndromes. Curr Opin Hematol. 2017 Mar;24(2):73–8.
11. Montalban BG, et al. Myelodysplastic syndromes: 2018 update on diagnosis, risk-stratification and management-NCBI. Am J Hematol. 2018 Jan;93(1):129–47.
12. Lughart S, Groschel S, Beverloo HB, et al. Clinical, molecular and prognostic significance of WHO type inv 3 (q21;q26.2)/t(3;3)(q21;q26.2) and various other 3q abnormalities in acute myeloid leukemia. J Clin Oncol. 2010 Aug;28(24):3890–8.
13. Patnaik MM, Hanson CA, Hodnefield JM, et al. Monosomal karyotype in MDS with or without monosomy 7 is prognostically worse than an otherwise complex karyotype. Leukemia. 2011 Feb;25(2):266–70.
14. Grimwade D, et al. Diagnosis and management of AML in adults: 2017 ELN recommenda-

tions from an international expert panel. Blood. 2017;129:424–47.

15. Arber D, Orazi A, Hassejcan R, et al. The 2016 revision to the WHO classification of myeloid meoplasms and acute leukemia. Blood. 2016;127(20):2391–405.

16. Mallo M, Cervera J, Schanz J, et al. Image impact of adjunct cytogenetic abnormalities for prognostic stratification in patients with myelodysplastic syndrome and deletion 5q. Leukemia. 2011 Jan;25(1):110–20.

5 Myelodysplastic/ Myeloproliferative Neoplasms (MDS/MPN)

5.1 Case Study 5.1: Myelodysplastic Syndrome: Myeloproliferative Neoplasm (CMML-1-RS)

Brief Clinical History

An 80-year-old male presented with giddiness and weakness. Received four units of blood transfusion. There was no response to Darbepoetin.

CBC: *Hb:* 7.1g/dl, TLC: 7970/cu mm and Platelets: 1,50,000/cu mm.

DLC: Neutrophils: 32.0%, Lymphocytes: 25.0%, Eosinophils: 0.0%, Monocytes: 33.0%, Basophils: 1.0%, Myelocytes: 6.0%, Metamyelocytes: 3.0%.

Peripheral Smear and Bone Marrow

Smear showed total count within normal limit with the shift to the left and absolute monocytosis (2630 cells/μL). Neutrophils showed dyspoiesis in the form of abnormal nucleation and granulations. Bone marrow smears were cellular showing trilineage dyspoiesis with erythroid hyperplasia, <5% blasts, and >15% ring sideroblasts. Peripheral blood and bone marrow features were suggestive of myelodysplastic/myeloproliferative neoplasms (MDS/MPN), Chronic Myelomonocytic Leukemia -1 with increased ring sideroblasts (CMML-1-RS).

Cytogenetics

Conventional Karyotype

45,XY,-7[3]/46,XY,del(7)(q21q34)[7]/46,XY,-7,+mar[6]/46,XY[4].

Cytogenetic analysis of unstimulated bone marrow aspirate culture showed neoplastic clone characterized by monosomy 7 in 9 metaphases along with one marker chromosome of unknown origin in 6 of these metaphases. Another clone in 7 metaphases showed interstitial deletion in the long arm of chromosome 7.

M. Bhargava, *Hematologic Malignancies*, https://doi.org/10.1007/978-981-33-4799-1_5

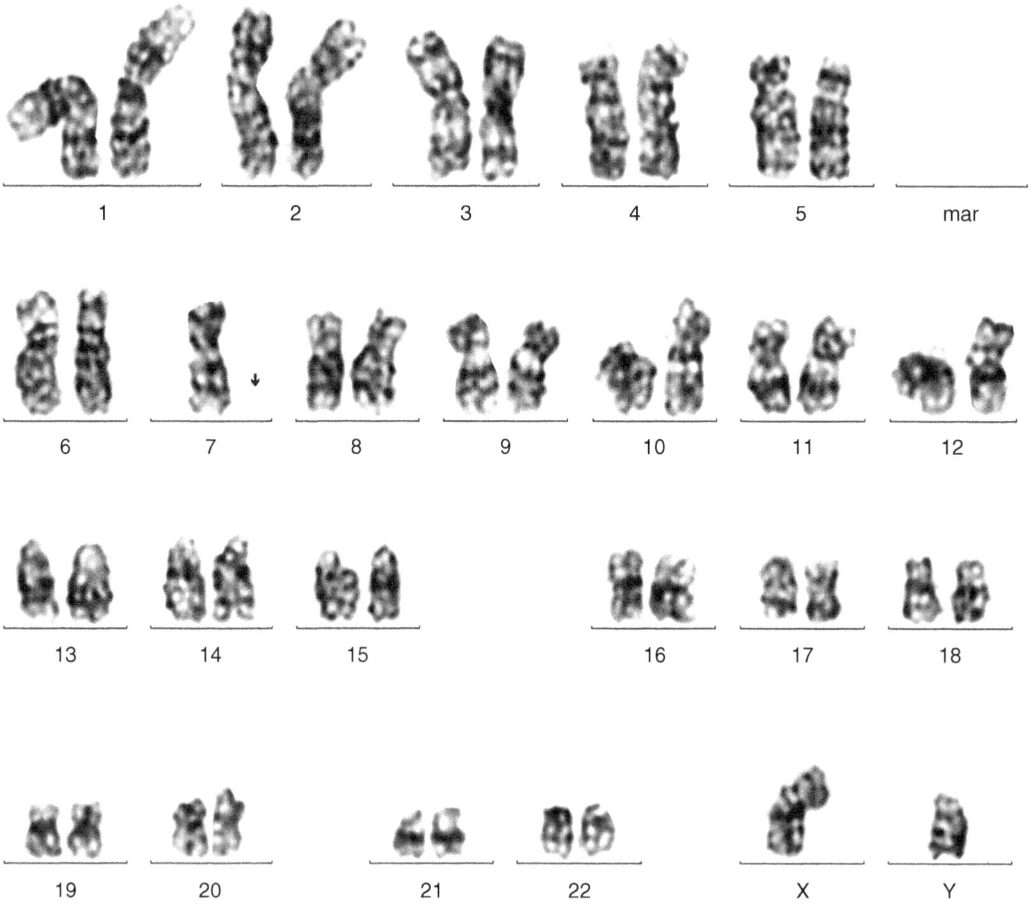

Case Study 5.1 (1) Karyotype showing loss of long arm of chromosome 7, i.e., monosomy 7

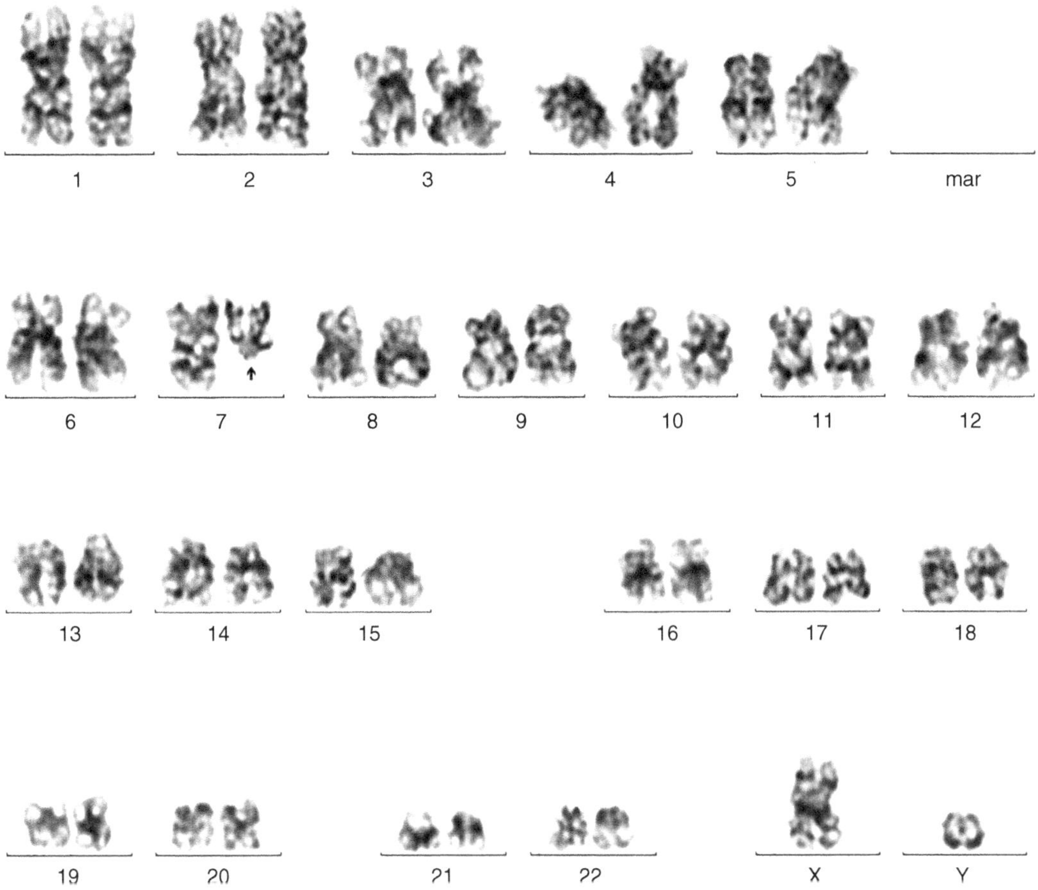

Case Study 5.1 (2) Karyotype showing deletion of long arm of chromosome 7, i.e., del(7q)

Discussion

This patient with severe anemia was transfusion dependent and had MDS-MPN-CMML-1-RS. CMML has overlapping features of MDS and MPN. Clonal cytogenetic changes are seen in ~30% and gene mutations are seen in more than 90% of patients [1]. Common abnormalities include trisomy 8, -Y,-7/del(7q) trisomy 21 and del(20q). Unlike MDS, del (5q) and monosomal karyotypes are infrequent [2]. Gene mutations are TET2 (~60%), (ASXL-1-40%), SRSF2(50%), RUNX-1(15%) RAS (30%) and CBL (15%) [3]. This patient also showed ring sideroblasts, thus making it CMML-1-RS. CMML has an inherent tendency to transform into AML and has poor prognosis [4].

In a study of 417 CMML patients, Tang et al. [5] using the Spanish risk stratification system, assigned 73% patients to low risk, 10.8% to intermediate risk, and 16.1% to high risk group. The median OS in low risk group was 33 months for the intermediate risk group was 33 months, for intermediate risk group 24 months, and for high risk group 14 months.

5.2 Case Study 5.2: Myelodysplastic Syndrome/ Myeloproliferative Neoplasm (CMML-2) with del (20q)

Brief Clinical History

A 65-year-old male presented with anemia and breathlessness. There was hepatosplenomegaly. He had received three units of blood transfusion at an outside facility where a BM examination had shown MDS/CMML.

CBC: *Hb:* 8.0 g/dl, TLC: 8070/cu mm and Platelets: 40,000/cu mm.

DLC: Neutrophils: 10.0%, Myelocytes: 1.0%, Blasts: 4.0%, Lymphocytes: 27.0%, Eosinophils: 0.0%, Monocytes: 58.0%, Basophils: 0.0%.

Peripheral Blood Smear and Bone Marrow

Smear showed absolute monocytosis (AMC ~ 4680/ml) with neutropenia (AMC ~807/ ml). Significant dyspoiesis was seen. Platelets were reduced. Bone marrow aspirate smears were hypercellular with myeloid preponderance. There was trilineage dysplasia with a blast count of ~12.0%. There was a marked monocytosis with few cells being promonocytes. The picture was consistent with Myelodysplastic/ Myeloproliferative neoplasm—Chronic myelomonocytic Leukemia-2 (CMML-2).

Immunotype

Blasts constituted ~3.8% of acquired events (gated on CD34-SSC), expressing dim CD45 with low side scatter. They were CD34+, CD117+, HLA-DR+, CD13+, CD33+, CD38+, CD11c+ (moderate), CD15+. Another population of cells in the myelomonocytic region (46.0%) were CD33+, CD64+ and CD13+ and CD14+, CD56+,CD16-.

The picture was suggestive of CMML.

Cytogenetics

Conventional Karyotype

46,XY,del(20)(q11.2)[20]

Cytogenetic analysis of 20 metaphases showed the presence of a neoplastic clone characterized by a deletion of the long arm of chromosome 20, i.e., del(20q). Deletion of the long arm of chromosome 20 is associated with a good prognosis in myelodysplastic syndrome (MDS International prognostic scoring system).

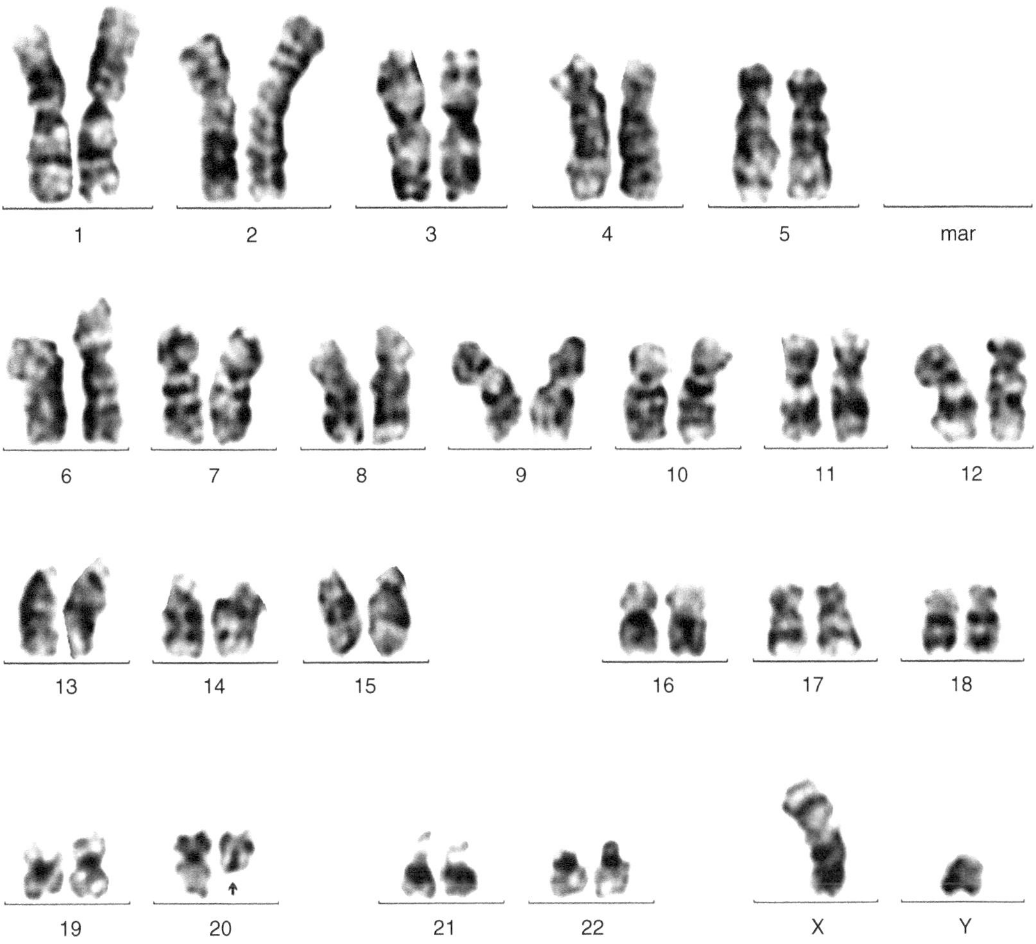

Case Study 5.2 Karyotype showing deletion of the long arm of chromosome 20, i.e., del(20q)

Molecular Findings

The Real Time Quantitative PCR *MPN mutation panel was negative.*

Discussion

This patient of MDS-MPN had CMML-2. In an international study from Mayo clinic and French consortium 409 patients with CMML were studied. Thirty percent showed an abnormal karyotype, including 72% sole, 16% two, and 11% complex abnormalities. The most common were +8(23%), -Y(20%), -7/7q-(14%), 20q-(8%), +21(8%), and der 3q (8%). Patients with abnormal karyotype were most likely to be elderly, be anemic, have leukocytosis with neutrophilia, show peripheral blasts are BM blasts. Complex and monosomal karyotypes are associated with a poor prognosis.

Clonal cytogenetic abnormalities are seen in nearby 30% patients with CMML, common being trisomy 8, monosomy 7/del(7q). Trisomy 21, -Y and complex karyotypes del(5q) or monosomal karyotypes are infrequent in CMML. CMML has overlapping features of MDS and MPN. In an analysis of 409 patients (Mayo clinic and French consortium [6], common abnormalities were +8 (23%), -Y(20%), -7del(7q)-(14%), 20q-(8%),+21(8%) and der (3q) (8%).

NGS has identified multiple gene mutations such as TET2, ASXL1, SRSFE,RAS, CBL, and RUNX1 [7]. So far only ASXL1 has been shown to impact OS. As per IPSS-R, del(20q) in CMML has a good prognosis [8].

5.3 Case Study 5.3: Myelodysplastic: Myeloproliferative Neoplasm MDS-MPN-CMML-O with Rare t(2;11)

Brief Clinical History

A 70-year-old male presented with fever and cough. Found to have bilateral Pneumonia. Responded to I/V antibiotics. Has leukocytosis. Bone marrow done for suspected MDS-MPN.

CBC: *Hb:* 12.4 g/dl; Platelets: 60,000/cu mm and TLC: 39,590/cu mm.

DLC: Neutrophils: 75.0%, Lymphocytes: 1.0%, Eosinophils: 1.0%, Monocytes: 16.0%, Basophils: 0.5%, Myelocytes: 7.0%.

Peripheral Smear and Bone Marrow

Smear showed Neutrophilic leukocytosis with left shift and ~7% myelocytes. There was absolute monocytosis (AMC~6334 cells/cu mm). Neutrophils showed dysgranulopoiesis. Bone marrow was hypercellular, showed myelomonocytic hyperplasia with dyspoiesis and blasts 3%. The picture was consistent with MDS-MPN-CMML-O.

Cytogenetics

Conventional Karyotype

46,XY,t(2;11)(p21;q23)[17]

Cytogenetic analysis of unstimulated bone marrow aspirate culture showed a neoplastic clone characterized by balanced reciprocal translocation between the short arm of chromosome 2 and the long arm of chromosome 11 in all the 17 analyzed metaphases.

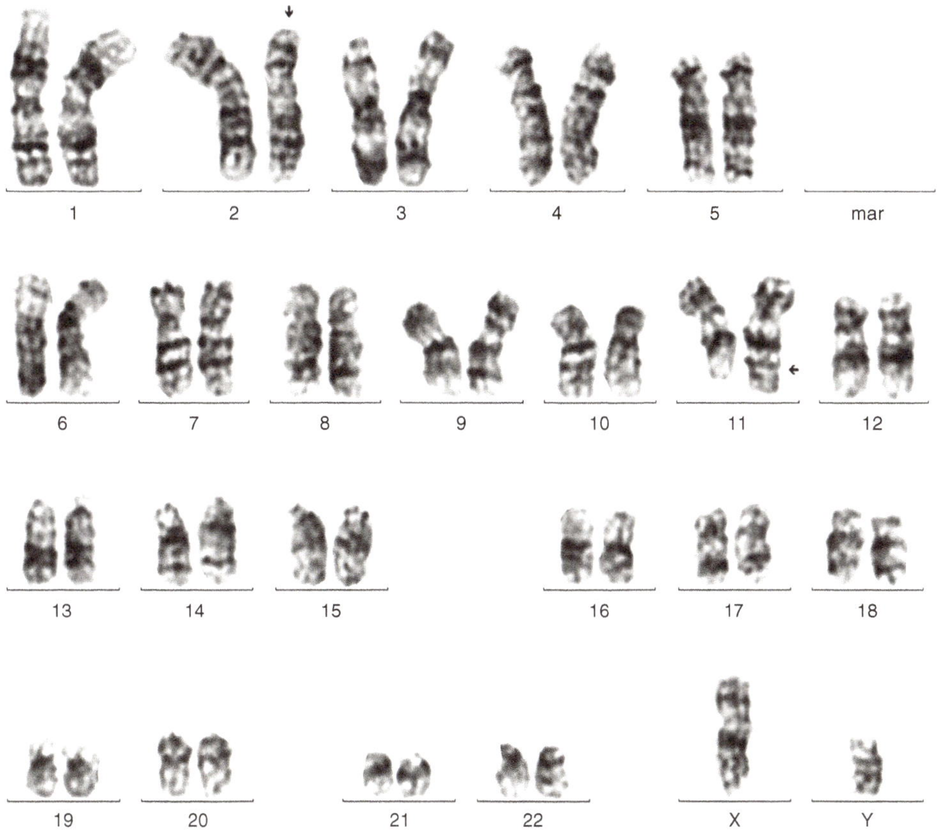

Case Study 5.3 (1) Karyotype showing reciprocal balanced translocation between short arm of chromosome 2 and the long arm of chromosome 11, i.e., t(2;11)

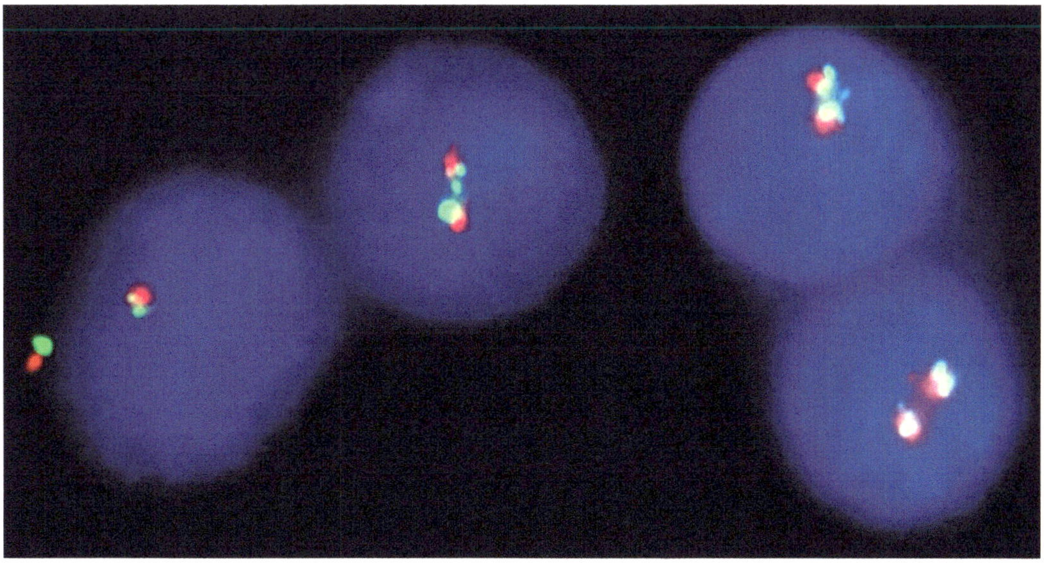

Case Study 5.3 (2) FISH image showing 2F, i.e., negative for *MLL(=KMT2A)* gene rearrangement using *MLL(=KMT2A)* dual-color break apart probe

Discussion

This patient had morphologic evidence of MDS-MPN-CMML-O with t(2;11)(p21;q23). This translocation has an overall frequency of ~1% in de novo MDS-MPN. It may or may not be associated with MLL/KMT2A gene rearrangement. A FISH test showed MLL/KMT2A gene rearrangement to be negative in this patient [9]. Dvorak et al. [10] reported 7 patients of MDS with t(2;11)(p21;p23), all of them without MLL gene rearrangement, and suggested that t(2;11)(P21;23) may be associated with good prognosis in MDS-MPN (median survival 72 months).

5.4 Case Study 5.4: Myelodysplastic Syndrome: Myeloproliferative Neoplasm CMML-1 with Isolated Trisomy 8

Brief Clinical History

A 73-year-old presented with history of thrombotic stroke and brain hemorrhage in 1999 and 2004 respectively. Previous marrow was reported as hypercellular with negative JAK2 and BCR-ABL1. Suspected MDS/MPN. Bone marrow for evaluation.

CBC: *Hb:* 8.3 g/dl; Platelets: 30,000/cu mm and TLC: 34,110/cu mm.

DLC: Neutrophils: 54.0%, Myelocytes: 5.0%, Blasts and promonocytes: 4.0%, Lymphocytes: 13.0%, Eosinophils: 0.0%, Monocytes: 21.0%, Basophils: 3.0%.

Peripheral Smear and Bone Marrow

Smear showed leukocytosis with the presence of ~4% blasts and promonocytes. There was absolute monocytosis (AMC ~ 7163 cells/cu mm) seen. There was significant dyspoiesis in myeloid and monocytic cells, including abnormal lobation of nuclei, hyposegmentation. Cellular bone marrow with myelomonocytic proliferation with myeloid and megakaryotic dysplasia with 4% blasts, features consistent with MDS-MPN-CMML-1

Cytogenetics

Conventional Karyotype

47,XY,+8[20]

Cytogenetic analysis of 20 metaphases showed a neoplastic clone characterized by trisomy 8.

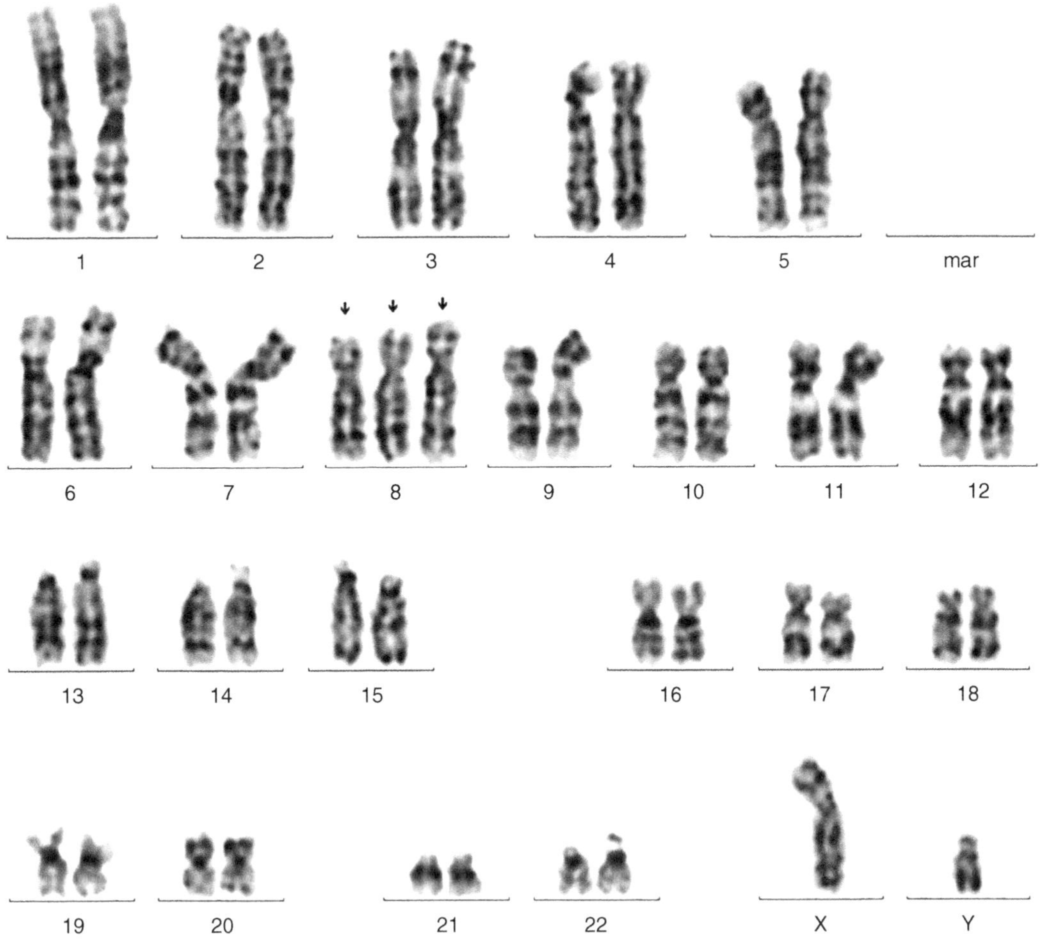

Case Study 5.4 Karyotype showing gain of chromosome 8, i.e., Trisomy 8

Discussion

This patient with morphologic evidence of MDS-MPN-CMML-1 showed isolated trisomy 8 on karyotyping. This is a frequent abnormality in MDS. In 138 patients of MDS, Drevon et al. [11] reported myeloproliferative features (MP) in 54. In these the median EFS and OS were 25 and 27 months. They conclude that MP features are frequent in MDS with isolated trisomy 8. MP forms had more EZH2, ASXL1, and STAG2 mutations. Trisomy 8 is also frequent in CMML as seen in this patient with MDS-MPN. In a study of 417 patients, Tang et al. [5], reported that patients with trisomy 8 showed a heterogeneous clinical course and outcome, similar to MDS. The median OS for patients with sole trisomy 8 was 22 months, similar to that in MDS intermediate category group [8]. Other single abnormalities included-7/del(7q),-Y,del(20q),-x, or del(xq)

5.5 Case Study 5.5: Myelodysplastic Syndrome: Myeloproliferative Neoplasm Atypical CML (a CML) with i(17q)

Brief Clinical History

A 69-year-old male presented with history of cough for 2–3 months. No comorbidities. No hepatosplenomegaly. Routine investigation showed high TLC. Provisional diagnosis was CML or MDS-MPN.

CBC: *Hb:* 7.6 g/dl; Platelets: 90,000/μL and TLC: 91,160/μL.

DLC: Blasts: 9.0%, Neutrophils: 15.0%, Metamyelocytes: 10.0%, Myelocyte: 49.0%, Lymphocytes: 10.5%, Eosinophils: 1.5%, Monocytes: 4.0%, Basophils: 1.0%. 24nRBC/100.

Peripheral Smear and Bone Marrow:

Peripheral smear showed leukoerythroblastic picture with ~9% blasts and ~49% myelocytes.

Hypercellular marrow showed significant trilineage dyspoiesis and myeloid hyperplasia with the predominance of myelocytes and ~6% blasts. There was grade 2 marrow fibrosis. There was no significant basophilia and absolute monocytosis. BCR-ABL done at this time was negative. In view

of leukocytosis with myelocyte predominance with increased blasts (9% on PS and 6% on BMA), significant trilineage dysplasia, absence of basophilia and absolute monocytosis, along with the absence of BCR-ABL by PCR, a possibility of MDS/MPN atypical chronic myeloid leukemia (*MDS-MPN-aCML, BCR-ABL negative*) was considered.

Cytogenetics

Conventional Karyotype

45,X,-Y,i(17)(q10)[4]/46,XY,i(17)(q10)[16]

Cytogenetic analysis of 20 metaphases showed a neoplastic clone characterized by the presence of isochromosome 17 resulting in a loss of the short arm (17p) and duplication of the long arm (17q) leading to a single copy of 17p and three copies of 17q in all the 20 metaphases analysed.4 of the 20 metaphases showed loss of chromosome "Y" a normal age-associated event.

The isochromosome (17q) is a rare event and is seen in atypical- chronic myeloid leukemia (CML) and is associated with poor prognosis.

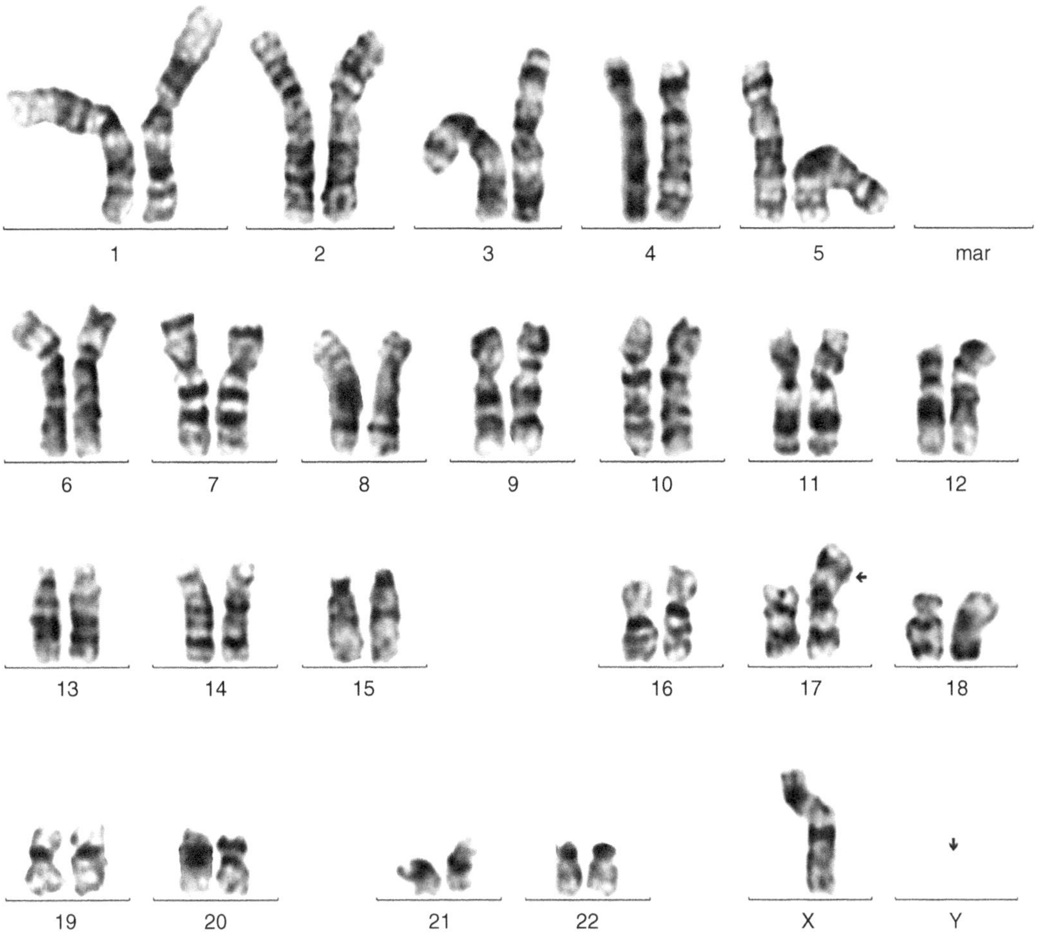

Case Study 5.5 Karyotype showing isochromosme of chromosome 17 formed by the long arm of chromosome 17, i.e.,(17q) and loss of chromosome Y.

Molecular Studies
The Real Time Quantitative PCR showed *BCR-ABL-IS (p210) as negative.*

Discussion
This patient with a high TLC and shift to the left in PBS was suggestive of CML. However, bone marrow was suggestive of a MDS-MPN picture. Morphology alone cannot differentiate between aCML or BCR-ABL positive CML. However, dysgranulopoiesis is prominent in aCML.

FISH studies as well as molecular studies did not show the presence of a BCR-ABL fusion transcript but the presence of i(17q) indicating that this patient had MDS-MPN-aCML. Orazi et al. [12] mentioned karyotypic abnormalities in 80% of aCML. The most common were gain of chromosome 8 and del(20q) but other common abnormalities involve chromosomes 13,14,17,19, and 12. Rarely, cases in which neoplastic cells have an isolated isochromosome 17q have features of atypical CML as was seen in this patient. In another study by McClure et al. [13], in a clinicopathologic study in 15 patients, i(17q) was the sole abnormality and was associated with chronic MDS-MPN disorders with an aggressive clinical course. In most studies, the median survival time is 14–29 months. In 30–40% patients, aCML evolves into AML.

5.6 Case Study 5.6: Myelodysplastic Syndrome Myeloproliferative Neoplasm (CMML-1) with t(1;15) with Derivative Chromosome 15

Brief Clinical History
A 64 year female presented with complaints of breathlessness for 25 days. There was evidence of hepatosplenomegaly, increased TLC, mild pericardial effusion, had been on antitubercular treatment. Suspected myeloproliferative neoplasm (MPN).

CBC: *CBC:* Hb: 8.5 g/dl, TLC: 9700/cu mm and Platelets: 1,50,000/cu mm.

DLC: Neutrophils: 52.0%, Myelocytes: 1.0%, Lymphocytes: 10.0%, Eosinophils: 0.0%, Monocytes: 36.0%, Basophils: 1.0%.

Peripheral Smear and Bone Marrow
Smear showed absolute monocytosis (AMC ~ 3492/cu mm). Neutrophils showed evidence of dyspoiesis with ~1.0% myelocytes. Bone marrow aspirate shows trilineage dysplasia with 2% blasts. Peripheral blood and bone marrow picture suggestive of MDS-MPN (CMML-1)

Cytogenetics
Conventional Karyotype

46,XX,der(15)t(1;15)(q12;p11)[20]

Cytogenetic analysis of 20 metaphases showed the presence of an abnormal clone characterized by an unbalanced translocation between the long arm of chromosome 1 and the short arm of chromosome 15 leading to the formation of derivative chromosome 15 in 20 metaphases analyzed.

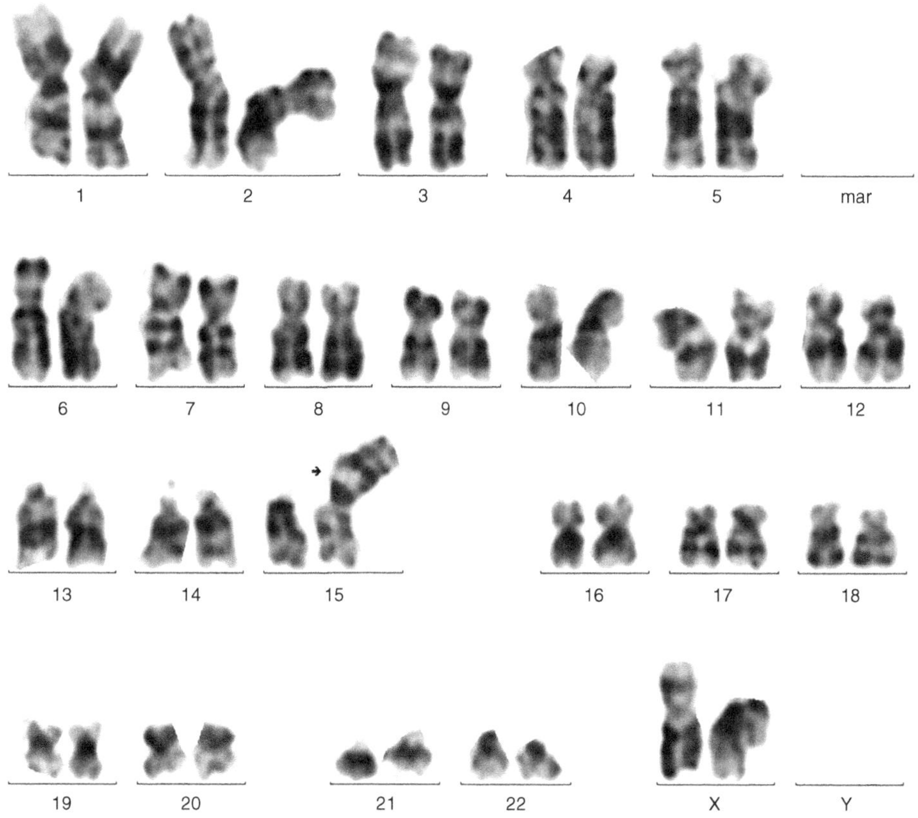

Case Study 5.6 Karyotype showing unbalanced translocation(1;15), i.e., gain of 1q or three copies of the long arm of chromosome 1, i.e., 1q.

Molecular Findings

The Real Time Quantitative PCR *MPN Mutation panel: Negative*

Discussion

Patients with der(15)t (1;15)(q11;p11-13) are diagnosed mainly with myeloid malignancies as in PV, MDS, AML [14]. This patient had morphologic evidence for MDS-MPN-CMML-1 with an unbalanced translocation between the long arm of chromosome 1 and the short arm of chromosome 15 with the formation of derivative chromosome 15 seen on karyotyping. This alteration of t (1;15) is seen in primary MDS involving 1q with partial trisomy for long arm of chromosome 1 [15]. Mecucci et al. reported two patients with an identical translocation between chromosome 1 and 15 in MDS. In the first patient, the translocation was similar to that seen in preleukemic states, i.e., a 5q-and a 20q-chromosome. In both patients, the long arm of chromosome 1 was trisomic. Follow-up of the second patient demonstrated a proliferative advantage of cells bearing t(1;15) over cells with trisomy 8 or normal cells but no evolution into acute leukemia. These alternations, if seen with complex karyotype have poor prognosis.

Annexure

- Case Study 5.1: Myelodysplastic Syndrome: Myeloproliferative Neoplasm (CMML-1-RS)
- Case Study 5.2: Myelodysplastic Syndrome/ Myeloproliferative Neoplasm (CMML-2) with del (20q)
- Case Study 5.3: Myelodysplastic: Myeloproliferative Neoplasm MDS-MPN-CMML-O with rare t(2;11)
- Case Study 5.4: Myelodysplastic Syndrome: Myeloproliferative Neoplasm CMML-1 with isolated trisomy 8
- Case Study 5.5: Myelodysplastic Syndrome: Myeloproliferative Neoplasm Atypical CML (a CML) with i(17q)
- Case Study 5.6: Myelodysplastic Syndrome Myeloproliferative Neoplasm (CMML-1) with t(1;15)with derivative Chromosome 15

References

1. Arber D, Orazi A, Hassejcan R, et al. The 2016 revision to the World Health Organization classification of myeloid Neoplasms and acute leukemia. Blood. 2016;127(20):2391–405.
2. Swerdlow SH, Campo E, Harris NL, Jaffe ES, Pileri SA, Stein H, Thiele J, editors. WHO classification of tumours of haematopoietic and lymphoid tissues. 4th ed. IARC: Lyon; 2017.
3. Patnaik MM, Tefferi A. Cytogenetic and molecular abnormalities in chronic myelomonocytic leukemia. Blood Cancer J. 2016;6(2):e393. https://doi.org/10.1038/bcj.2016.5.
4. Such E, Cervera J, Costa D, et al. Cytogenetic risk stratification in chronic myelomonocytic leukemia. Haematologica. 2011;96:375–83.
5. Tang G, Zhang L, Fu B, et al. Cytogenetic risk stratification of 417 patients with chronic myelomonocytic leukemia from a single institution. Am J Hematol. 2014 Aug;89(8):813–8.
6. Wassie EA, Itzykson R, Lasho TL. Molecular and prognostic correlates of cytogenetic abnormalities in chronic myelomonocytic leukemia: a Mayo clinic-French Consortium Study. Am J Hematol. 2014;89(12):1111–5.
7. Patnaik MM, Teffari A. Cytogenetic and molecular abnormalities in chronic myelomonocytic leukemia. Blood Cancer J. 2016;6(2):e393. https://doi.org/10.1038/bcj.2016.5.
8. Greenberg PL, Tuechler H, Schanz J, et al. Revised international prognostic scoring system for myelodysplastic syndromes. Blood. 2012;120:2454–65.
9. Ruano A, Shetty S. t(2;11)(p21;q23) without KMT2A(MLL) gene rearrangement. Atlas Genet Cytogenet Oncol Haematol. 2017;21(6):210–4.
10. Dvorak P, Lysak D, Vokurka S, et al. The translocation t(2;11)(p21;q23) without MLL gene rearrangement: a possible marker of good prognosis in myelodysplastic syndrome patients. Hematol Oncol. 2014;32:82–6.
11. Drevon L, Marceau A, Maarek O, et al. Myelodyplastic syndrome with isolated trisomy 8: a type of MDS frequently associated with myeloproliferative features? A report by the Groupe Francophone des Myelodysplasies. Br J Haematol. 2018; https://doi.org/10.1111/bjh.15490.
12. Orazi A, Bennett JM, Bain BJ, et al. WHO classification of tumours of haematopoietic and lymphoid tissues. 4th ed. IARC: Lyon; 2017. p. 87–9.
13. McClure RF, Dewald W, Hoyer JD, et al. Isolated isochromosome 17q: a distinct type of mixed myeloproliferative disorder/myelodysplastic syndrome with an aggressive clinical course. Br J Haematol. 1999;106:445–54.
14. Mecucci C, Tricot G, Boogaerts M, et al. An identical translocation between chromosome 1 and 15 in two patients with myelodysplastic syndromes. Br J Haematol. 1986 Mar;62(3):439–45.
15. Zamecnikova A. der (15)t(1;15)(q11-12;p11-13) and der(15)t(1;15)(q21-25;p10-13). Atlas Genet Cytogenet Oncol Haematol. 2017;23(1):8–13.

6 Myeloproliferative Neoplasms (MPNs)

6.1 Case Study 6.1: Myeloproliferative Neoplasm: Primary Myelofibrosis with Trisomy 9

Brief Clinical History

A 45-year-old male from Rwanda presented with splenomegaly for 3 years, was found to have massive splenomegaly with dilated portal veins. Bone marrow was done for evaluation.

CBC: *Hb:* 5.6 g/dl, TLC: 25,630/cu mm and Platelets: 70,000/cu mm.

DLC: Neutrophils: 62.0%, Lymphocytes: 7.0%, Myelocytes: 13.0%, Metamyelocytes: 12.0%, Monocytes: 3.0%, Basophil: 3.0%; ~3 nRBC/100 WBCs seen.

Peripheral Smear and Bone Marrow

Smear showed leukoerythroblastic blood picture with tear drop red cells. There was left shift in granulocytic series with basophilia. No blasts were seen in the peripheral smear. Bone marrow aspirate was hemodilute, imprint smears showed myeloid preponderance. Bone marrow biopsy showed myeloid predominance with increase in megakaryocytes with clustering and nuclear atypia and grade 2 fibrosis.

Overall, the features were consistent with *Myeloproliferative Neoplasm—Primary Myelofibrosis, Overt fibrotic stage.*

Cytogenetics

Conventional Karyotype

47,XY,+9,add(6)(p25)[19]/46,XY[1]

Cytogenetic analysis of 20 metaphases shows a neoplastic clone characterized by gain of chromosome 9 as trisomy and addition of unknown material at the short arm of chromosome 6 in 19 metaphases analyzed.

M. Bhargava, *Hematologic Malignancies*, https://doi.org/10.1007/978-981-33-4799-1_6

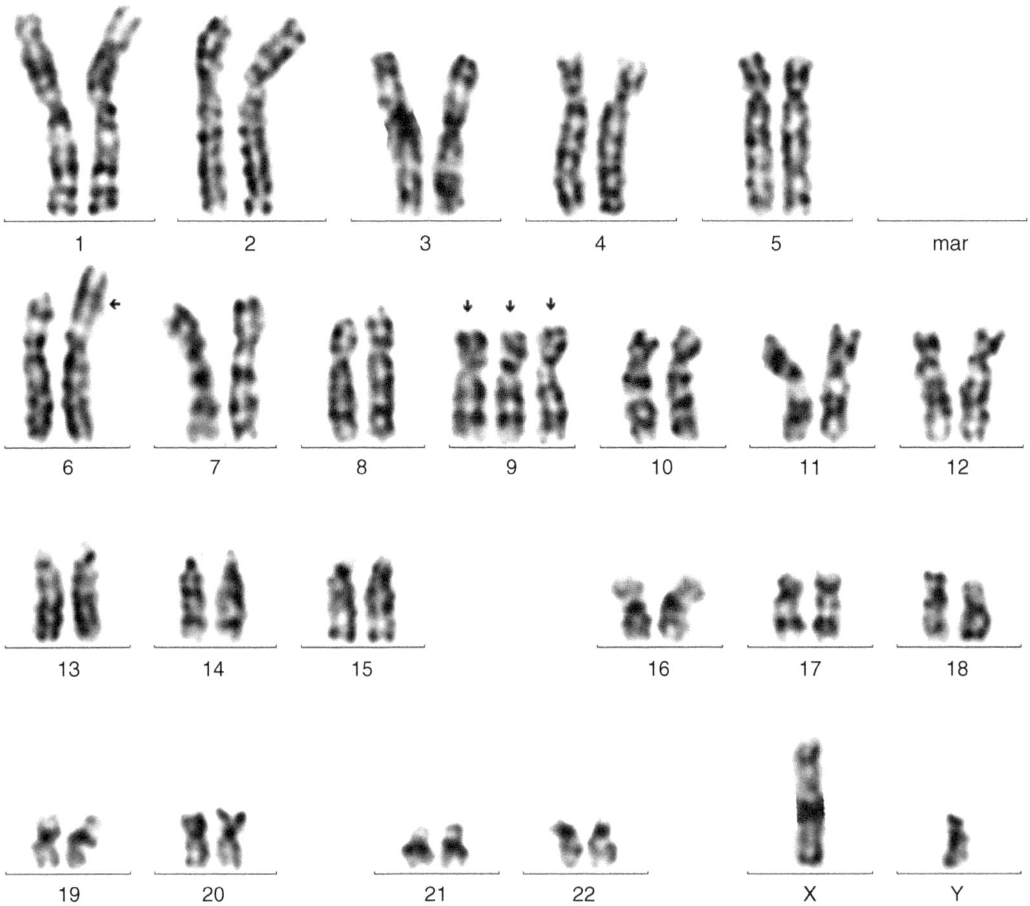

Case Study 6.1 Karyotype showing gain of chromosome 9, i.e., Trisomy 9 and add(6p)

Molecular Findings

The Real Time Quantitative PCR showed *JAK2 V617F mutation positive.*

Discussion

Unlike CML, there is no specific cytogenetic abnormality in PMF. There are however, recurrent abnormalities specific to the disease that carry significant prognostic relevance. Unfavorable include +8,-7/7q-, i(17q), inv(3), -5/5q-, 12p-, 11q23 rearrangement, complex, and monosomal karyotypes [1]. A number of Somatic mutations have also been described in PMF like JAK2 (58%), CALR(25%), MPL(8.3%), ASXL1(31%), EZHZ (6%), IDH1, or IDHZ (4%), SRSF2(12%), U2AF1(16%) and SF3B1 (7%) [2]. Some like CALR and ASXL1 were prognostically relevant.

This patient of primary myelofibrosis was molecularly JAK2 positive and had trisomy 9 on karyotyping. JAK2 positivity is seen in more than 95% of patients with PV and 50–60% of PMF and ET [3]. The most common recurrent abnormalities are gain of chromosome 8 or 9, del(20q), del(13q), and del(9p); gains of chromosomes 8 and 9 are sometimes found. In a study of 826 patients. Wassie et al. [4], reported 42.6% of PMF patients had abnormal karyotype with 68.2% showing sole abnormalities and 13.6% a complex karyotype. The most frequent abnormalities were 20q-(23.3%), 13q-(18.2%),+8 (11.1%), +9(9.9%), Chromosome 1q+(9.7%) and -7/7q(7.1%). In their study, -9 was seen in 17.9% together with JAK2 mutation. Higher frequency of trisomy 9 in JAK2 mutated PMF may reflect selective advantage through JAK2V617F dosage enhancement.

6.2 Case Study 6.2: Myeloproliferative Neoplasm: Polycythemia Vera with JAK 2 Mutation

Brief Clinical History
A 49-year-male old diagnosed case of intraductal papillary mucinous neoplasm of the pancreas with previous history of persistent high hemoglobin counts. Bone marrow was done for evaluation of a suspected myeloproliferative neoplasm? Polycythemia Vera.

CBC: *Hb:* 16.6 g/dl, TLC: 20,910/cu mm and Platelets: 17,84,000/cu mm.

DLC: Neutrophils: 59.0%, Lymphocytes: 19.0%, Eosinophils: 9.0%, Monocytes: 6.0%, Basophils: 3.0%, Myelocytes: 4.0%.

Peripheral Smear and Bone Marrow
Smear showed Neutrophilic leukocytosis with eosinophilia and basophilia. There were ~2% myelocytes. Platelets were markedly increased. Bone marrow aspirate was hemodilute. Bone biopsy showed panmyelosis with increase in myeloid cells and erythroid clusters. Megakaryocytes were increased with mild dysmyelopoiesis. There was no increase in Blasts or fibrosis.

Molecular Studies
The Real Time Quantitative PCR panel showed *JAK-2 mutation. BCR-ABL, CALR, and MPL mutations were negative.*

Discussion
This patient diagnosed as PV based on WHO criteria for the diagnosis of PV was found to have JAK2 mutation on molecular analysis [3]. In 2005, several groups identified a somatic mutation involving a protein kinase in patients with PV, ET, and PMF. This was a JAK2 point mutation at codon 617. It is observed in 95% patients of PV and 50% of patients with ET and PMF. Philadelphia negative MPNs (PV, ET, PMF) have overlapping clinical features and a common molecular basis. Three-quarters of these patients carry the unique JAK2 (V617F) mutations [5]. JAK2 positive mutated PV and ET form a biologic continuum in which the degree of erythrocytosis is determined by physiologic and genetic modifiers. The JAK 2 mutant allelic burden is significantly higher in PV. JAK 2 (V617F) positive patients have a 2 fold risk of developing thrombosis [6].

6.3 Case Study 6.3: Myeloproliferative Neoplasm: Primary Myelofibrosis with JAK2 Mutation

Brief Clinical History
A 57-year-old female presented with anaemia, generalized weakness and massive splenomegaly.

CBC:*Hb*: 9.2 g/dl, TLC: 14,560/cu mm and Platelets: 1,62,000/cu mm.

DLC: Myelocytes: 15.0%, Metamyelocytes: 4.0%, Neutrophils: 48.0%, Lymphocytes: 20.0%, Eosinophils: 3.0%, Monocytes: 7.0%, Basophils: 3.0%; 6nRBCs/100WBCs.

Peripheral Smear and Bone Marrow
Smear showed Neutrophilic leukocytosis with significant left shift. 3% basophils were seen. Platelets were adequate with megakaryocytic fragments. Bone marrow aspirate was hemodilute and aparticulate . Trephine biopsy showed large areas of fibrosis interspersed with few fairly cellular areas showing granulocytic hyperplasia. Megakaryocytes were increased and showed the formation of loose clusters with nuclear atypia, hyperchromasia, and cloud-like nuclei. Erythroid

cells were reduced. There was increased vascularity.

Stain for reticulin showed grade 2 fibrosis (WHO classification). Masson's trichrome showed focal collagen deposition.

Molecular Studies
The Real Time Quantitative PCR for MPN panel showed *JAK-2 mutation positive. BCR-ABL, CALR, and MPL mutations were negative.*

Discussion
This patient of primary myelofibrosis was positive for JAK2 mutation and negative for CALR and MPL mutations. The discovery of JAK2 V617F mutation opened a new era in the understanding of MPN biology, clinical evaluation and therapy and modified the robust classification of myeloproliferative disorders proposed by Dameshek in 1951 [7]. JAK 2 V617F mutation is found in ~50%–60% of patients. However, the presence of JAK2 mutation does not unequivocally diagnose PMF as distinct from ET in particular and PV since thrombocytosis can be seen in RARS-T as well (MDS-MPN)and a number of RARS-T patients also have JAK 2 mutation (50-80%). Also splenomegaly can be seen, apart from PMF, in 30% of PV and ET patients. JAK2 mutations, however, do provide a clear evidence for malignant clonal hematopoiesis. Increasing V617F mutant allele burden is shown to have the risk of developing MF but it is not clear about the risk for developing MDS/AML [8]. It is also not clear whether JAK 2 mutation can affect overall survival of patients with MPN. Some studies showed its association with poor survival in PMF but in a large series of PMF patients (n = 345). Cervvantes F et al. [9] which served as a basis for classification of International Prognostic Scoring System (IPSS) of PMF, JAK2 mutation was not associated with either prognostic score or survival.

6.4 Case Study 6.4: Myeloproliferative Neoplasm: Essential Thrombocythemia with CALR Mutation

Brief Clinical History
A 82-year-old female presented with vomiting and loose stools for a weak with Dehydration with dyselectrolytemia. A Status exploratory laparotomy with bowel resection and double barrel ileostomy for intestinal obstruction (omental band) with bowel gangrene was done in 2019.

CBC: *Hb:* 9.7 g/dl, TLC: 18,660/cu mm and Platelets: 10,30,000/cu mm.

Peripheral Smear and Bone Marrow
Peripheral blood smear showed marked thrombocytosis and neutrophilic leukocytosis.

Bone marrow aspirate and biopsy were not done.

Flow Cytometric Immunophenotyping
Was not done.

Cytogenetics
Was not done.

Molecular Studies
Real time PCR for MPN mutation panel showed *CALR mutation. BCR-ABL, JAK-2 mutation, and MPL mutation were negative.*

Discussion
This patient of splanchnic vein thrombosis was found to have ET and carried a CALR mutation but was negative for JAK2 as well as MPL mutation. Splanchnic vein thrombosis (SVT) is often associated with an underlying MPN (8%) and most often with JAK2V617F mutation [3]. The number of CALR positive patients is only 0–2.5% [6]. In a study of whole exome sequenc-

ing in 1107 patients of MPN, CALR and JAK2 and MPL mutations were mutually exclusive [10]. CALR mutations were detected in 67% of ET and 88% of PMF. Only 10% of patients with ET do not carry any of the three biomarkers. CALR mutated ET appears to be substantially different from JAK2(V617F) ET in terms of biologic, hematologic, and clinical features. The mechanism how the CALR mutant gene determines abnormal proliferation of megakaryocytes and overproduction of platelets is unclear. In CALR mutated ET, only few patients have allelic burden >75% (indicating homozygosity for the mutation). Phenotypically, CALR mutated ET is associated with platelet counts $>1000 \times 10^9$/L as was the case in this patient. The risk of thrombosis is however, lower than in JAK2 mutated ET. The absence of JAK2(V617F) and markedly raised platelet counts are the main features in CALR mutated ET. It is a distinct disease entity both at the molecular and clinical level [11].

6.5 Case Study 6.5: Myeloproliferative Neoplasm: Essential Thrombocythemia with JAK2 Mutation

Brief Clinical History

A 79 -year-old male with Parkinson's disease, with progressive generalized body weakness detected with leukocytosis and thrombocytosis was investigated further for Myeloproliferative neoplasm.

CBC:*Hb*: 13.6 g/dl, TLC: 20,620/cu mm and Platelets: 10,24,000/cu mm.

DLC: Neutrophils: 88.0%, Lymphocytes: 5.0%, Eosinophils: 2.0%, Monocytes: 4.0%, Basophils: 1.0%.

Peripheral Blood and Bone Marrow

Smear showed neutrophilic leukocytosis with no significant left shift. No blasts were seen.

Platelets were markedly increased with platelet anisocytosis and the presence of giant forms. Bone marrow showed megakarycytosis increased in number with megakaryocyte clusters. There was megakaryocytic atypia, variably sized with hyper lobulated forms. Platelet lakes were noted. Overall features of peripheral blood and bone marrows were suggestive of a Myeloproliferative Neoplasm, likely Essential thrombocythemia.

Molecular Studies

The Real Time Quantitative PCR studies of the MPN panel showed *JAK2 positive. BCR-ABL, CALR, and MPL mutations were negative.*

Discussion

This patient of ET on molecular analysis was positive for JAK2 mutation but negative for CALR and MPL mutations. JAK2 mutated ET and PV are different phenotypes in the evolution of a single disease. Rumi et al. [6] reported that the mutant allele burden was lower in JAK2 mutated than in CALR mutated ET. Patients with JAK2(VG17F) were older, had a higher Hb level and white blood cell count, and lower platelet count and serum erythropoietin than those with CALR mutation as in this case where hemoglobin and white cell count were also seen to be higher. While no polycythemia transformation was seen in CALR mutated patients, this risk was 29% at 15 years in those with JAK2 mutated ET No significant difference was seen in myelofibrotic transformation in these 2 categories. Patients with JAK2 mutated. ET and those with PV had similar risk for thrombosis which was twice as much as in CALR mutated ET. These findings are consistent with the notion that JAK2 mutated ET and PV are different phenotypes of a single myeloproliferative neoplasm whereas CALR mutated ET. Nangalia et al. [11] is a distinct entity.

Annexure

- Case Study 6.1: Myeloproliferative Neoplasm: Primary Myelofibrosis with Trisomy 9
- Case Study 6.2: Myeloproliferative Neoplasm: Polycythemia Vera with JAK 2 mutation
- Case Study 6.3: Myeloproliferative Neoplasm: Primary Myelofibrosis with JAK2 Mutation
- Case Study 6.4: Myeloproliferative Neoplasm: Essential Thrombocythemia with CAL-R Mutation
- Case Study 6.5: Myeloproliferative Neoplasm: Essential Thrombocythemia with JAK2 Mutation

References

1. Caramazza D, et al. Refined cytogenetic risk categorisation for overall and leukemia free survival in primary myelofibrosis: a single centre study of 433 patients. Leukemia. 2011;25:82–8.
2. Vannucchi AM, et al. Mutations and prognosis in primary myelofibrosis. Leukemia. 2013;27:1861–9.
3. Swerdlow SH, Campo E, Harris NL, Jaffe ES, Pileri SA, Stein H, Thiele J, editors. WHO classification of tumours of haematopoietic and lymphoid tissues. 4th ed. Lyon: IARC; 2017.
4. Wassie E, Finke C, Gangat N, et al. A compendium of cytogenetic abnormalities in myelofibrosis: molecular and phenotypic correlates in 826 patients. Br J Haematol. 2015 April;169(1):71–6.
5. Campbell PJ, Scott LM, Buck G, et al. Definition of sub types of essential thrombocythaemia and relation to polycythaemia vera based on JAK2 V617F mutation status: a prospective study. Lancet. 2005 Dec;366(9501):1945–53.
6. Rumi E, Pietra D, Ferretti V, et al. JAK 2 or CALR mutation status defines subtypes of essential thrombocythemia, with substantially different clinical course and outcomes. Blood. 2014;123(10):1544–51.
7. Tefferi A, Thiele J, Orazi A, et al. Proposals and rationale for revision of the World Health Organization diagnostic criteria for Polycythemia vera, essential thrombocythemia and primary myelofibrosis: recommendations from an ad hoc international expert panel. Blood. 2007;110:1092–7.
8. Kiladjian JJ. The spectrum of JAK2-positive myeloproliferative neoplasms. Hematology Am Soc Hematol. 2012;(1):561–6.
9. Cervvantes F, Dupriez B, Periera A, et al. New prognostic scoring system for primary myelofibrosis based on a study of the International Working Group for Myelofibrosis Research and Treatment. Blood. 2009;113(13):2895–901.
10. Klampfl T, Gisslinger H, Harutyunyan AS, et al. Somatic mutations of calreticulin, myeloproliferative neoplasms. N Engl J Med. 2013;369:2379–90.
11. Nangalia J, Massie CE, Baxter EJ, et al. Somatic CALR mutations in myeloproliferative neoplasms. N Engl J Med. 2013;369(25):2391–045.

7 Chronic Myeloid Leukemia

7.1 Case Study 7.1: Chronic Myeloid Leukemia with Lymphoid Blast Crisis with Rare dic (7;9)

Brief Clinical History
An 18-year-old male, a diagnosed case of CML-CP treated on Imatinib, presented with CML-BC within 2 months of treatment.

CBC: *Hb*: 11.8 g/dl, TLC: 39,100 cells/cu mm and Platelets: 2,10,000/cu mm.

DLC: Blasts: 76.0%, Myelocytes: 2.0%, Neutrophils: 16.0%, Lymphocytes: 4.0%, Eosinophils: 0.0%, Monocytes: 1.0%, Basophils: 1.0%.

Peripheral Smear and Bone Marrow
Smear showed leukocytosis ~76% blasts. Large cells with high N:C ratio, fine nuclear chromatin with inconspicuous nucleoli, and scant to moderate agranular cytoplasm. Bone Marrow showed similar blasts constituting 88% of total nucleated cells.

Peripheral blood and bone marrow findings were consistent with Chronic Myeloid Leukemia in blast phase.

Flow Cytometric Immunophenotyping
Blasts (~58% of acquired events); CD45 dim to moderate blast region with low side scatter.

CD19+, CD10+, CD22+, CD34+, CD9+, CD24+, CD20+, CD58+, CD38+, cCD79a+, CD81+, CD200+, CD13+/CD33+, CD66c+.

The flow cytometric features were consistent with B-lymphoid blast transformation in a diagnosed case of CML.

Cytogenetics
Conventional Karyotype
45,XY,dic(7;9)(p11;p11),t(9;22)(q34;q11.2)[12]/46,XY,t(9;22)(q34;q11.2)[4]/46,XY[4].

Cytogenetic analysis of 20 metaphases showed a neoplastic clone characterized by dicentric reciprocal translocation between chromosome 7 and chromosome 9. It also showed balanced translocation between the long arms of chromosome 9 (9q) and chromosome 22 (22q) in 12 metaphases. A separate clone in 4 metaphases showed only t(9;22).

M. Bhargava, *Hematologic Malignancies*, https://doi.org/10.1007/978-981-33-4799-1_7

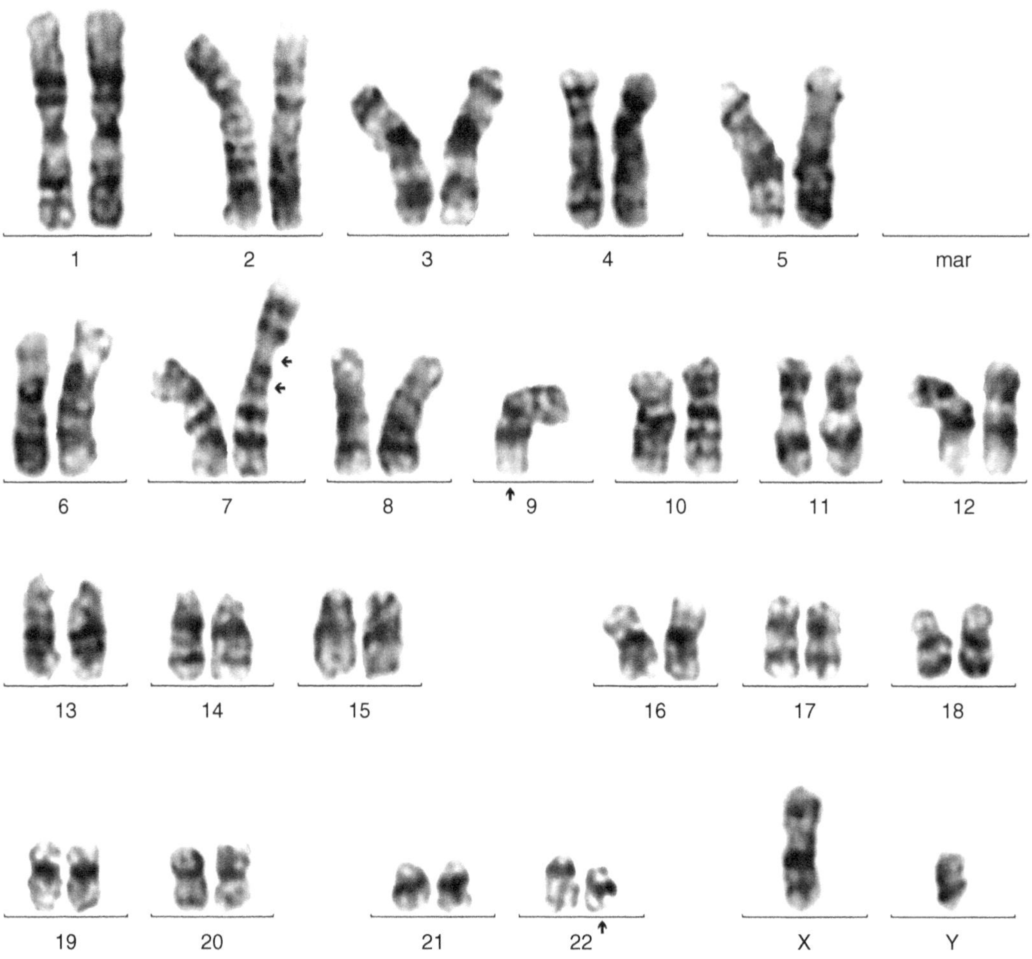

Case Study 7.1 Karyotype showing Philadelphia chromosome with dic(7;9)

Molecular Studies

The Real Time Quantitative PCR showed *BCR-ABL (IS) Positive (p210).*

Discussion

The frequency of additional chromosomal abnormalities in CMC-CP is 70% and increases to 40–70% in CML-BC. This patient first presented as CML-CP with karyotype 46,XY, t(9;22) (q34;q11.2) and within 2 months on Imatinib therapy showed CML in lymphoid blast crisis with dicentric (7;9) (p11;p11) retaining the clone t(9;22). Pan et al. [1] described dic (7;9) in 7 patients of ALL. In 30% of these BCR/ABL fusion product (p190) was seen unlike in this case who had p210. The outcome of patients with dic(7;9) together with t(9;22) is worse than with isolated dic (7;9). dic (7;9) is a rare but recurrent abnormality which in ALL, may represent a distinct cytogenetic subgroup.

7.2 Case Study 7.2: Chronic Myeloid Leukemia in Myeloid Blast Crisis with Hyperdiploid Complex Karyotype

Brief Clinical History

A 53-year-old male was diagnosed as CML in 2009 and put on Imatinib. Recently developed fever, weakness, and pancytopenia. Evaluation is done for suspected CML-BC.

CBC: *Hb*: 4.1 g/dl, TLC: 122,970/cu mm and Platelets: 20,000 /cu mm.

DLC: Neutrophils: 8.0%, Lymphocytes: 18.0%, Monocytes: 0.0%, Eosinophils: 1.0%, Basophils: 20%, Myelocyte: 1.0%, Blasts: 52.0%.

Peripheral Smear and Bone Marrow

Smear showed hyperleukocytosis with ~52% blasts and ~20% basophils. The blasts were medium to large in size with high N: C ratio, round nucleus, opened up chromatin, prominent nucleolus, and moderate amount of basophilic agranular cytoplasm. No Auer rods were seen.

Similar blasts (69%) replaced the bone marrow. Peripheral smear and bone marrow morphologic features were consistent with progression to blast crisis in a known case of CML on treatment.

Flow Cytometric Immunophenotyping

The picture was consistent with myeloid blast crisis in a known case of CML.

Immunophenotype

Blasts constitute ~66.7% of acquired events expressing dim to moderate CD45 and having moderate SSC. They were CD34+, CD13+, CD33+, CD11c+, CD11b+, HLA-DR+.

Cytogenetics

Conventional Karyotype

50,XY,+8,+8,+8,2ider(22)(q10)t(9;22)(q34;q11.2),i(17)(q10)[10]/49, XY, + 8, +8, 2ider (22) (q10) t(9;22)(q34;q11.2),i(17)(q10)[7]/47,XY,2ider(22)(q10)t(9;22)(q34;q11.2)[2]/46,XY[1].

Cytogenetic analysis of 20 metaphases showed a neoplastic clone characterized by a hyperdiploid karyotype with both structural and numerical abnormalities. There was tetrasomy 8 in 50% and pentasomy 8 in 35% of 20 metaphases analyzed. There was also gain of extra Philadelphia chromosome. Structural abnormalities showed two isochromosomes from the derivative chromosome 22 (Philadelphia isochromosome) and isochromosome 17 resulting in a loss of the short arm (17p) and duplication of the long arm (17q) leading to a single copy of 17p and three copies of 17q.

Case Study 7.2 (1) Hyperdiploid Complex Karyotype showing extra Philadelphia isochromosomes along with Tetrasomy 8 and isochromosome(17)

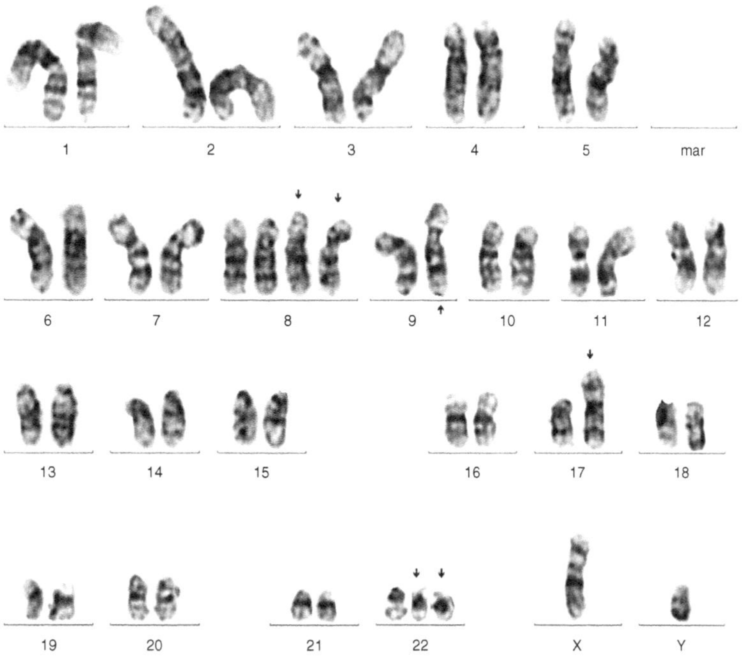

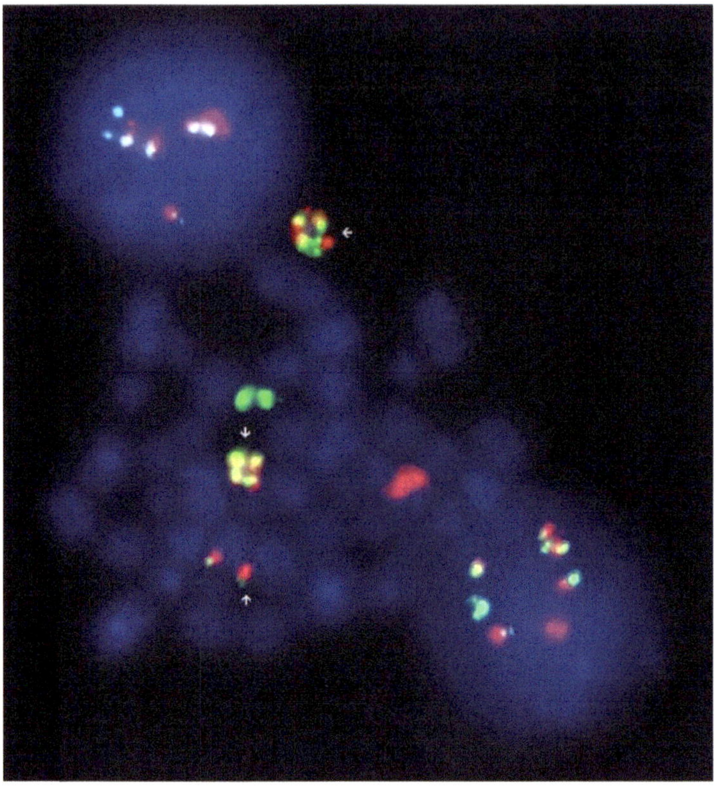

Case Study 7.2 (2)
FISH image showing 5F1R1G indicating two isochromosomes formed by Philadelphia chromosome

Molecular Studies

The Real Time Quantitative PCR showed *BCR-ABL(IS) positive (p210)*.

Discussion

This patient of CML-BC had two structural abnormalities of isochromosomes Philadelphia i(22) and isochromosome 17, i(17q). i(17q) is the most frequent abnormality seen in CML-BC [2]. In a study of p53 gene mutations in CML-BC, Nakai et al. [3], in 10 of 31 patients showed loss of q short arm of chromosome 17(17p) by forming i(17q) as in this case. Isochromosome 17q which results in loss of a short arm of chromosome 17 (17p) is well known to emerge in BC of CML suggesting the involvement of the p53 gene in the transition from CP-BC [4]. In an Indian study from Regional Cancer Centre, Trivandrum, Chandran et al. [5] in CML-BC reported additional chromosomal abnormalities in 50.9%. Numerical changes were seen in 15 and structural abnormalities in 11 patients. Multiple copies of Ph 1, hyperdiploid, high hyperdiploid triploidy, and tetraploidy were seen. Two to three copies or isoderivative chromosome 22(ider22) resulting in two- to sixfold amplification of Ph chromosome. These findings are very similar to those seen in this patient.

7.3 Case Study 7.3: Chronic Myeloid Leukemia in Myeloid Blast Crisis with t(9;22) and inv(16)

Brief Clinical History

A 48-year-old male, known case of CML on therapy. Evaluated for suspected CML-blast crisis.

CBC: *Hb*: 11.9 g/dl, TLC: 7480/cu mm and Platelets: 26,000/cu mm.

DLC: Blasts: 40.0%, Neutrophils: 14.0%, Lymphocytes: 24.0%, Eosinophils: 1.0%, Monocytes: 20.0%, Basophils: 1.0%.

Peripheral Smear and Bone Marrow

Smear showed 40% blasts. Bone marrow ~50.0% blasts were medium to large in size, having moderately condensed nuclear chromatin, inconspicuous nucleolus, scant to moderate amount of cytoplasm, and several showed coarse basophilic granules and cytoplasmic vacuolations.

The picture was consistent with CML-BC, morphologically myeloid.

Cytogenetics

Conventional Karyotype

46,XY,t(9;22)(q34;q11.2),inv(16)(p13q22)[17]/46,XY[3]

Cytogenetic analysis of 20 metaphases showed the presence of a neoplastic clone characterized by a balanced translocation between the long arms of chromosome 9 and chromosome 22 along with a clonal evolution with a pericentric inversion of chromosome 16 in 17 metaphases. The presence of additional cytogenetic abnormality, i.e., inv(16) is consistent with the accelerated/blast phase of chronic myeloid leukemia.

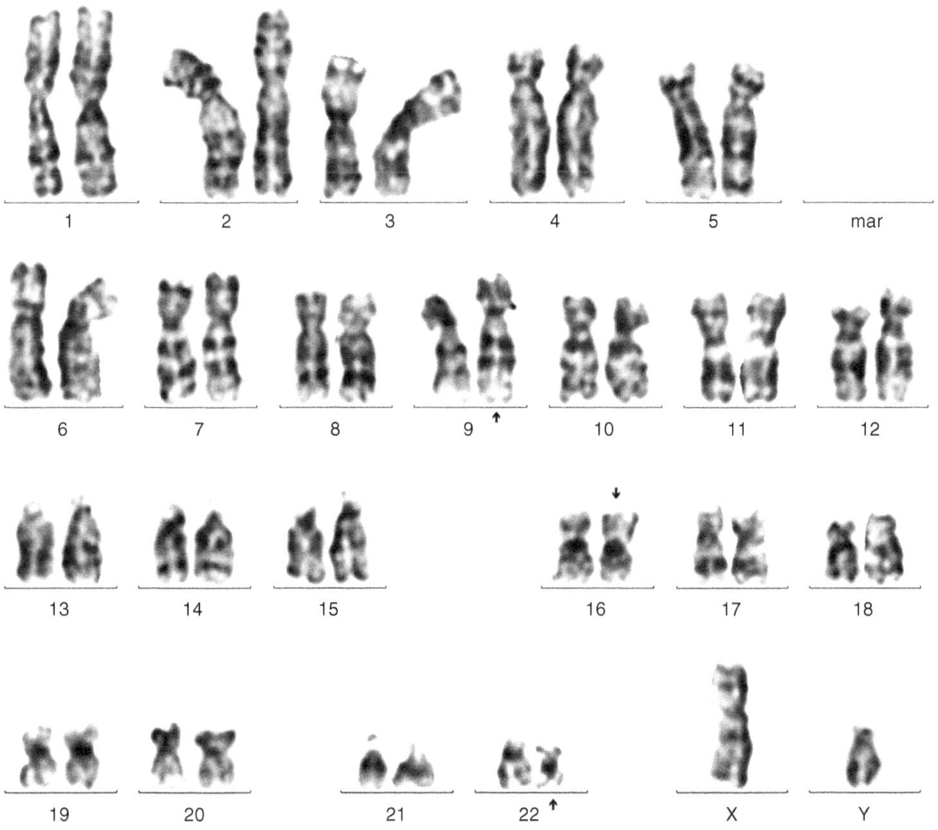

Case Study 7.3 Karyotype showing Philadelphia chromosome formed by t(9;22) and inv(16)

Molecular Findings
The Real Time Quantitative PCR showed *BCR-ABL1 (IS) positive (p210).*

Discussion
This patient of CML in blast crisis had been diagnosed as CML-CP at an outside facility and was on nilotinib. Karyotyping revealed t(9;22) with CBFB –MYH11 fusion resulting from inv(16)(p13q22). The co-occurrence of BCR-ABL1 and inv(16) is extremely rare with less than 20 cases described in literature and is seen in blast phase of CML. In a study of nine cases by Salem et al. [6]. The sequence of molecular alterations could be determined in all nine cases. BCR-ABL1 preceded CBFB rearrangement in seven CBFB preceded BCR-ABL1 in one (p210) and both alternations were discovered simultaneously in one patient (p190).

Additional cytogenetic alternations other than t(9;22) are seen in up to 80% cases of CML-BC. Most common are trisomy 8, the extra copy of Ph chromosome, 3q26 rearrangements, monosomy 7/del(7q),i(17)(q10), trisomy 21, 19, and -Y. Unlike de novo AML, inv(16) (p13q22) in CML-BC is not associated with a favorable prognosis. These patients can belong to two groups. In one, CML-CP progresses to CML-BC carrying p210 and acquire inv(16)(p13q22) or those with de novo AML who have both simultaneously (p210) and (p190). In rare cases, inv(16) precedes BCR-ABL1 [6].

7.4 Case Study 7.4: Chronic Myeloid Leukemia in Myeloid Blast Crisis with a Novel dic(16;17) and TP53 Gene Deletion

Brief Clinical History
A 48-year-old male was diagnosed in 2011 as CML-CP. Was on irregular treatment with Imatinib. Has now presented with breathlessness. Investigated as suspected CML in blast crisis.

CBC: *Hb*: 6.5 g/dl, TLC: 8590/cu mm and Platelets: 12,000/cu mm.

DLC: Blasts: 11.0%, Myelocytes: 3.0%, Neutrophils: 71.0%, Lymphocytes: 7.0%, Eosinophils: 0.0%, Monocytes: 3.0%, Basophils: 5.0%.

Peripheral Smear and Bone Marrow
Smear showed neutrophilic predominance with shift to the left with 11% blasts. Bone marrow imprint smears showed 24% blasts, myeloid in morphology.

Immunophenotype
Blasts (~21% of acquired events); present in the blast region.

Positive markers (on blasts): MPO+, CD13+ (moderate), CD33+ (moderate), HLA-DR+ (heterogenous), CD117+(moderate), CD34+(partial), CD38+.

The immunophenotypic features were suggestive of myeloid blast transformation in a diagnosed case of CML.

Cytogenetics
Conventional Karyotype

45,XY,t(9;22)(q34;11.2),+dic(16;17)(p11.2;p13.1),[12]/46,XY,t(9;22)(q34;q11.2)[8].

Cytogenetic analysis showed a neoplastic clone characterized by a balanced reciprocal translocation between long arms of chromosome 9 and 22 Philadelphia chromosome (Ph) in all the 20 metaphases analyzed. The clone also showed the formation of dicentric chromosome in 12 of the 20 metaphases by a translocation between short arm of chromosome 16 and short arm of chromosome 17 leading to the deletion of region p13 on chromosome 17, i.e., TP53 gene deletion which was also confirmed by FISH.

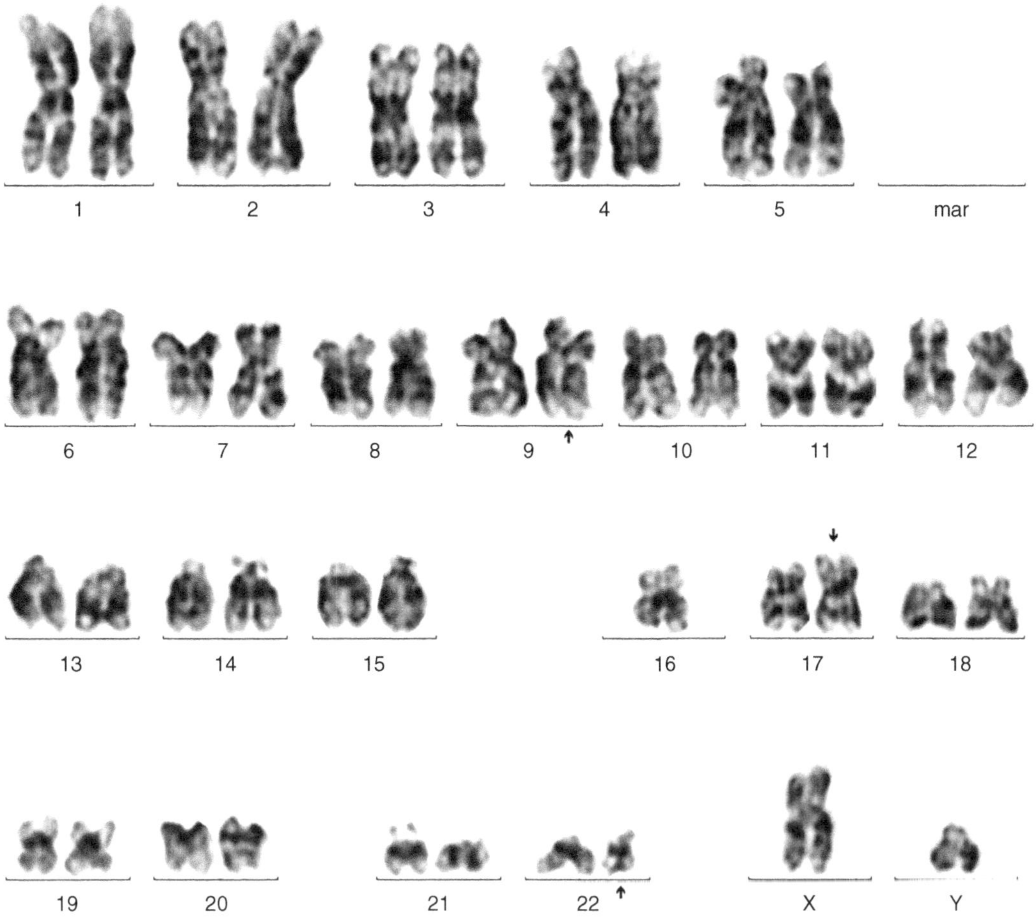

Case Study 7.4 Karyotype showing Philadelphisa chromosome formed by t(9;22) and dic(16;17) as a result of which TP53 gene is deleted

Molecular Studies

The Real Time Quantitative PCR showed *BCR-ABL (IS) positive (P210).*

Discussion

This patient of CML-BC had cytogenetic profile of t(9;22) but also the additional finding of dic (16;17) (p11.2:p13) leading to TP53 gene deletion confirmed on FISH. Nakai et al. [7] reported a close association between mutations of TP53 gene in CML myeloid blast crisis suggesting loss of 17p in the progression of the disease. Similarly, a frequent monoallelic deletion of the p53 gene supports the role of p53 gene alterations in CML-BC [8]. While i(17q) has been described in literature, dic(16;17) has not been described. However, dic (17;18) has been shown with loss of TP 53 and BCR-ABL rearrangement in an imatinib-resistant CML [9].

Annexure

- Case Study 7.1: Chronic Myeloid Leukemia with Lymphoid Blast crisis with rare dic (7;9)
- Case Study 7.2: Chronic Myeloid Leukemia in Myeloid Blast Crisis with Hyperdiploid Complex Karyotype
- Case Study 7.3: Chronic Myeloid Leukemia in Myeloid Blast Crisis with t(9;22) and inv(16)
- Case Study 7.4: Chronic Myeloid Leukemia in Myeloid Blast Crisis with a novel dic(16;17) and TP53 Gene Deletion

References

1. Pan J, Xue Y, Wu Y, Wang Y, Shen J, et al. Dicentric (7;9)(p11;p11) is a rare but recurrent abnormality in acute lymphoblastic leukemia: a study of 7 cases. Cancer Genet Cytogenet. 2006;169(2):159–63.
2. Fioretos T, Strombeck B, Sandberg T, et al. Isochromosome 17q in blast crisis of chronic myeloid leukemia and in other hematologic malignancies is the result of clustered breakpoints in 17p11 and is not associated with coding TP53 mutations. Blood. 1999;94(1):225–32.
3. Nakai H, Misawa S, Toguchida J, et al. Frequent p 53 gene mutations in blast crisis of chronic myelogenous leukemia, especially in myeloid crisis harboring loss of chromosome 17p. Cancer Res. 1992;52(23):6588–93.
4. Feinstein, et al. p53 gene in chronic myeloid leukemia in acute phase. Proc Nat Acad Sci USA. 1991;88:6293–7.
5. Chandran K, Geetha N, Murugesan K, et al. Impact of additional chromosomal aberrations on the disease progression of chronic myelogenous leukemia. Front Oncol. 2019. https://doi.org/10.3389/fonc.2019.00088
6. Salem A, Loghavi S, Tang G, et al. Myeloid neoplasms with concurrent BCR-ABL1 and CBFB rearrangements: a series of 10 cases of a clinically aggressive neoplasm. Am J Hematol. 2017;92(6):520–8.
7. Nakai H, Misawa S, Toguchida J, et al. Frequent p53 gene mutations in blast crisis of chronic myelogenous leukemia, especially in myeloid crisis harboring loss of a chromosome 17p. Cancer Res. 1992;52(23):6588–93.
8. Rovira A, Urbano-Ispizua A, Cervantes F, et al. P53 tumor suppressor gene in chronic myelogenous leukemia: a sequential study. Ann Hematol. 1995;70(3):129–33.
9. Al-achkar W, Wafa A, Othman M. A novel dic (17,18) (p13.1;q11.2) with loss of TP53 and BCR/ABL rearrangement with an imatinib resistant chronic myeloid leukemia. Mol Cytogenet. 2012;5:36.

8 Next Generation Sequencing for Diagnostics of Myeloid Malignancies for Routine Clinical Use: Pitfalls and Advantages

The use of Next Generation Sequencing is being increasingly sought to be included in routine clinical use for more differentiated disease classification, risk stratification and therapeutic decisions, particularly in AML, MDS, and MPNs. But unexpected challenges are being faced towards interpretation and implementation of the results obtained.

Commercially available panels are based on targeted sequencing and usually analyze 25–50 genes and generally add to the information routinely available on karyotyping, FISH, and Real Time PCR panels. In AML, more than one recurrent somatic mutation can be identified, in MDS in nearly 30% of cases (e.g., MDS with ring sideroblasts with SF3B1 mutation). In MPNs, however the significance of non-driver mutations, outside of driver mutations of JAK2, CALR and MPL are difficult to define. In triple negative MPNs, NGS may confirm the presence of a clonal disease. In spite of this, a large number of triple negative MPN patients show no detectable mutation. Another pitfall can be the discrimination of leukemia-initiating mutations from incidental passenger mutation or CHIP or germline mutations and requires extensive validation of the results in a large number of cases. Thus, the review of the reported genetic variants by experts with technical and clinical knowledge is highly relevant [1].

Shown here in the AML group are a representative number of four cases of AML, each different from the other with respect to cytogenetics and RT-PCR molecular findings. They were then analyzed by NGS. In these cases, it appears, that cytogenetics and RT-PCR investigation were adequate towards risk stratification and clinical decision making. While this may be true in many cases, it by no means diminishes the value of doing NGS to look for other mutations specially in AML and MDS, where nearly 50% of the patients are cytogenetically and by RT-PCR negative. This would be particularly true in the larger clinical context towards the use of targeted therapy.

Reference

1. Bacher U, Shumiov E, Flach J, et al. Challenges in the introduction of next generation sequencing (NGS) for diagnostic of myeloid malignancies into clinical routine use. Blood Cancer J. 2018;8:113. https://doi.org/10.1038/s41408-018-0148-6.

M. Bhargava, *Hematologic Malignancies*, https://doi.org/10.1007/978-981-33-4799-1_8

Part: II

Case Studies in Lymphoid Malignancies

9 Acute Lymphoblastic Leukemia

9.1 Case Study 9.1: Acute Lymphoblastic Leukemia with High Hyperdiploidy

Brief Clinical History
A 16-year-old female presented with history of fever off and on for 2 months. Diagnosed as Acute Leukemia at an outside facility. Admitted for further evaluation.

CBC: *Hb*: 9.8 g/dl, TLC: 8200 /cu mm and Platelets: 86,000 /cu mm.

DLC: Blasts: 71.0%, Neutrophils: 4.0%, Lymphocytes: 25.0%, Eosinophils: 0.0%, Monocytes: 0.0%, Basophils: 0.0%.

Peripheral Blood Smear and Bone Marrow
Smear showed Neutropenia along with ~71% blasts, large in size, having high N:C ratio with open chromatin, inconspicuous nucleoli, and agranular cytoplasm. Similar blasts (85.0%) replaced the entire bone marrow.

Peripheral blood and bone marrow picture were consistent with acute leukemia, morphologically lymphoid.

Flow Cytometric Immunophenotyping
Blasts (~72% of acquired events); CD45 dim to moderate blast region with low side scatter.

CD19+, CD10+, CD79a+, CD22+, CD34+, CD9+, CD24+, CD20+, CD58+, CD38+.

The picture was consistent with Precursor B cell Acute Lymphoblastic Leukemia, CALLA positive.

Cytogenetics
Conventional Karyotype

56,XX,+4,+6,+8,+10+11,+14,+17,+18,+21,+21[6]/46,XX[11]

Cytogenetic analysis of unstimulated bone marrow aspirate culture showed a neoplastic clone in 6 of the 17 metaphases analyzed, characterized by high hyperdiploidy (chromosome number: 56) showing gains of chromosomes 4, 6, 8, 10, 11, 14, 17 and 18 as trisomies and chromosome 21 as tetrasomies. No structural abnormality was seen.

M. Bhargava, *Hematologic Malignancies*, https://doi.org/10.1007/978-981-33-4799-1_9

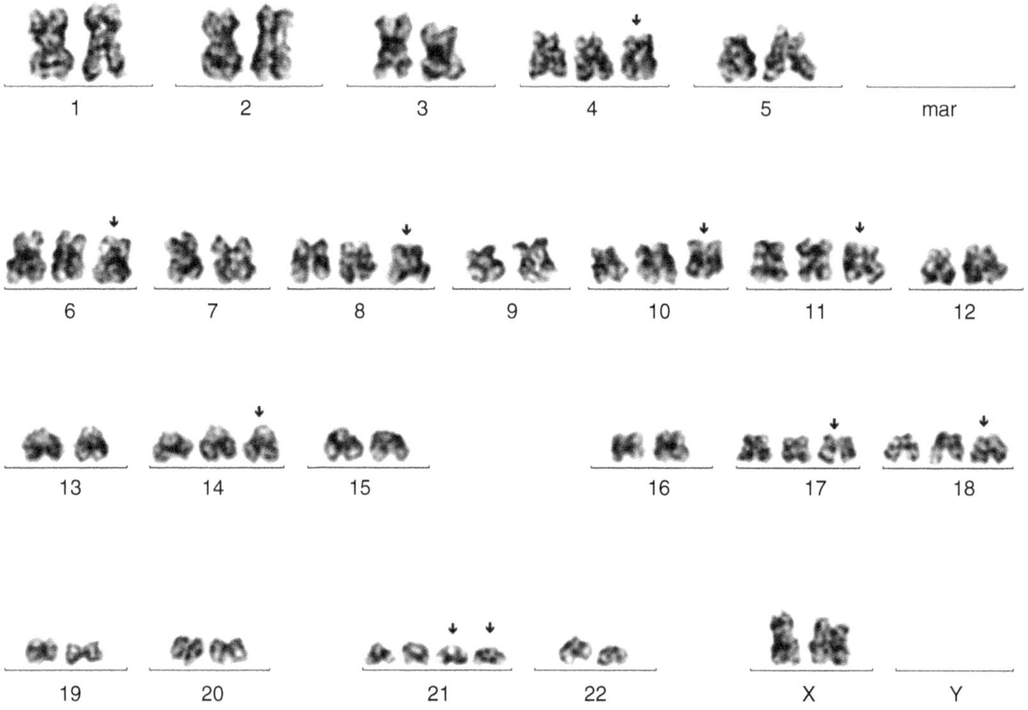

Case Study 9.1 Karyotype showing hyperdiploid (modal number 56) with gain of chromosome 4, 6, 8, 10, 11, 14, 17, 18, 21, and 21

Discussion

High hyperdiploidy is an important cytogenetic subtype of adolescent and adult ALL and is present in equal frequencies in Ph-positive and Ph-negative patients. This patient karyotypically had high hyperdiploidy without any structural abnormalities High hyperdiploid childhood ALL is characterized by multiple chromosome gains as in this patient [1]. Studies have shown that this subtype hardly harbors other genetic abnormalities. Multiple chromosome gains appear to arise early and are likely the main driver event. Structural chromosomal abnormalities may be present in half the cases. Dup (1q) is the most common. These leukemia are generally associated with good prognosis [2].

9.2 Case Study 9.2: Acute Lymphoblastic Leukemia with Complex Karyotype with t(12;21) (ETV6/RUNX1)

Brief Clinical History

A 6-year-old male presented with fever, pallor, and progressive myalgia for 1 month, investigated outside showed ALL with 70% blasts in peripheral blood. Received two units of blood transfusion.

CBC: *Hb*: 10.1 g/dl, TLC: 3180 /cu mm. Platelets: 22,000 /cu mm.

DLC: Blasts: 10.0%, Neutrophils: 5.0%, Lymphocytes: 81.0%, Eosinophils: 1.0%, Monocytes: 3.0%, Basophil: 0.0%.

Peripheral Smear and Bone Marrow
Peripheral blood smear showed leukopenia with marked neutropenia along with the presence of ~10% blasts. The Bone marrow showed blasts constituting ~93.0 of the total nucleated cells. Blasts were medium to large, having high N:C ratio, open chromatin, and inconspicuous nucleoli. No Auer rods were seen.

Peripheral blood and bone marrow findings showed acute leukemia, morphologically lymphoid.

Flow Cytometric Immunophenotyping
Blasts (~90% of acquired events); CD45 (negative to dim) in blast region with low side scatter.

CD19+, CD10+, CD22+, CD34+, CD9+, CD24+, CD58+, CD38+, cCD79a+

Picture showed B -Cell Acute Lymphoblastic Leukemia, CALLA positive.

Cytogenetics
Conventional Karyotype
46,XY,del(13)(q14)[4]/46,XY,del(6)(q21q25),del(13)(q14)[13]/46,XY[3]

Cytogenetic analysis of 20 metaphases showed a neoplastic clone characterized by a complex karyotype showing structural abnormalities such as deletion of the long arm of chromosome 6 (6q) and deletion of the long arm of chromosome 13(13q). FISH analysis of 200 interphase cells showed ETV/RUNX1 fusion positivity in 97% of cell the indicating the presence of t(12;21) which is a cryptic translocation not identified by karyotyping alone. Molecular panel for ALL also showed ETV6/RUNX1.

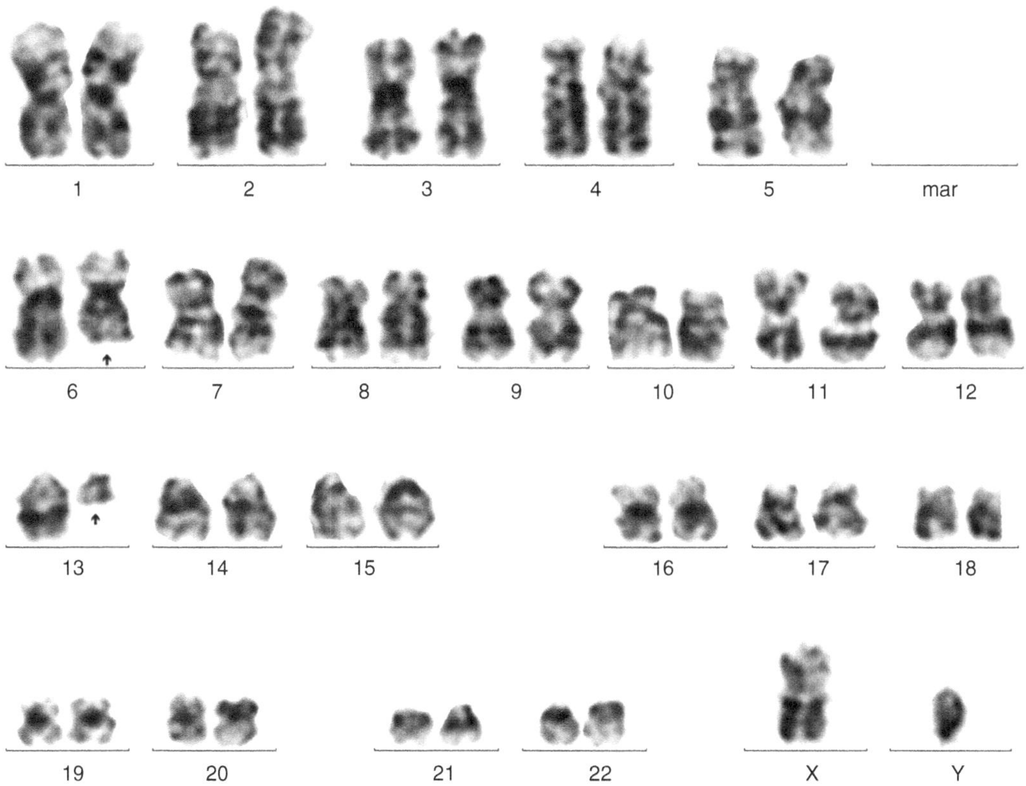

Case Study 9.2 (1) Karyotype showing deletion of the long arm of chromosome 6 and 13

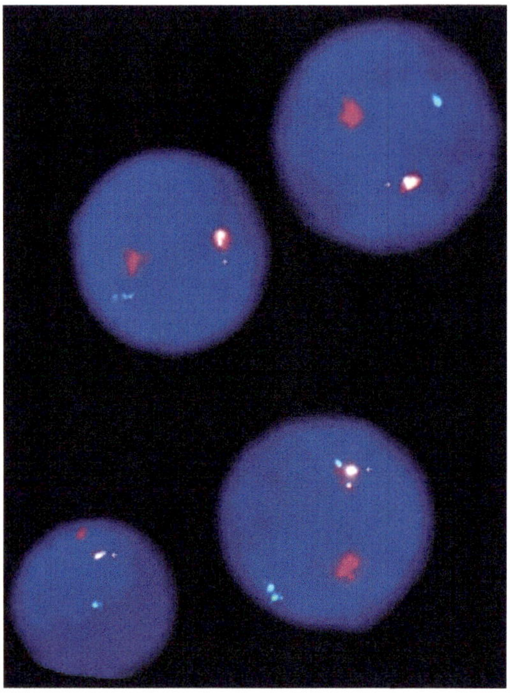

Case Study 9.2 (2) FISH image showing 1F1R1G aberrant signal pattern using *ETV6-RUNX1* LSI probe indicating *ETV6-RUNX1* fusion

Molecular Studies

The Real Time Quantitative PCR studies showed *ETV6-RUNX1 (TEL-AML1) Positive.*

Discussion

The t(12;21) is a cryptic abnormality seen in 25% children with B-Lineage ALL and is associated with a good prognosis. Raynaud et al. [3] looked for cytogenetic abnormalities associated with t(12;21) in 169 children. Only 13.6% had normal karyotype. Structural abnormalities were detected in 89.7% and numerical in 47%. Rearrangements of 12p were most frequent. Their results showed that t(12;21) is not associated with hyperdiploidy of 52 to 68 chromosomes or with prognostic t(1;19), t(4;11) or t (9;22).

This patient of CALLA positive precursor B cell acute lymphoblastic leukemia on FISH showed t(12;21) together with two other structural abnormalities of del(6q)and del (13q), a complex karyotype. It is a cryptic translocation not picked up on karyotyping and is a marker of good prognosis. t (12;21)(p13;q22) ETV6/RUNX1 is seen in 15 to 35% of childhood ALL, the most frequent translocation [4].

Del 13q in ALL may show higher risk for treatment failure but is not an independent marker for prognosis. Deletion of 6q is not a marker for adverse prognosis [5].

9.3 Case Study 9.3: Philadelphia Chromosome Positive Acute Lymphoblastic Leukemia with Hypodiploidy, Monosomy 7, and der 6 t(2;6)

Brief Clinical History

A 17-year-old male, known case of Philadelphia chromosome positive ALL, received CVAD. Has now relapsed. Bone marrow was done for evaluation of the disease.

CBC:*Hb*: 10.5 g/dl, TLC: 2060 /cu mm and Platelets: 7000 /cu mm.

DLC: Myelocytes: 1.0%, Blasts: 6.0%; Neutrophils: 17.0%, Lymphocytes: 70.0%, Eosinophils: 6.0%, Monocytes: 0.0%, Basophils: 0.0%.

Peripheral Blood and Bone Marrow

Smear showed leukopenia with neutropenia along with the presence of ~6% blasts. Bone marrow showed near total replacement by blasts which were large, with high N:C ratio, fine nuclear chromatin with inconspicuous nucleoli, and scant to moderate agranular cytoplasm.

The peripheral blood and bone marrow findings were consistent with relapsed Ph+ALL.

Flow Cytometric Immunophenotyping

Blasts (~80% of acquired events); CD45 (negative to dim) in blast region with low side scatter.

CD19+, CD34+, CD10+, CD79a+, CD22+, CD9+, CD24+, CD58+, CD38+, CD 13+33+

The immunophenotype was that of Precursor B Acute Lymphoblastic Leukemia, CALLA positive with aberrant expression of myeloid markers (CD13 and CD33), usually associated with BCR/ABL t(9;22) gene rearrangement.

Cytogenetics
Conventional Karyotype

45,XY,der(6)t(2;6)(q11.2;p25),-7,t(9;22)(q34;q11.2)[17]/46,XY[3]

Cytogenetic analysis of 20 metaphases showed the presence of a hypodiploid clone (modal number 45) with loss of chromosome 7 and translocation between the long arm of chromosome 2 and the short arm of chromosome 6 and balanced translocation between the long; arm of chromosome 9 and chromosome 22 in 17 metaphases.

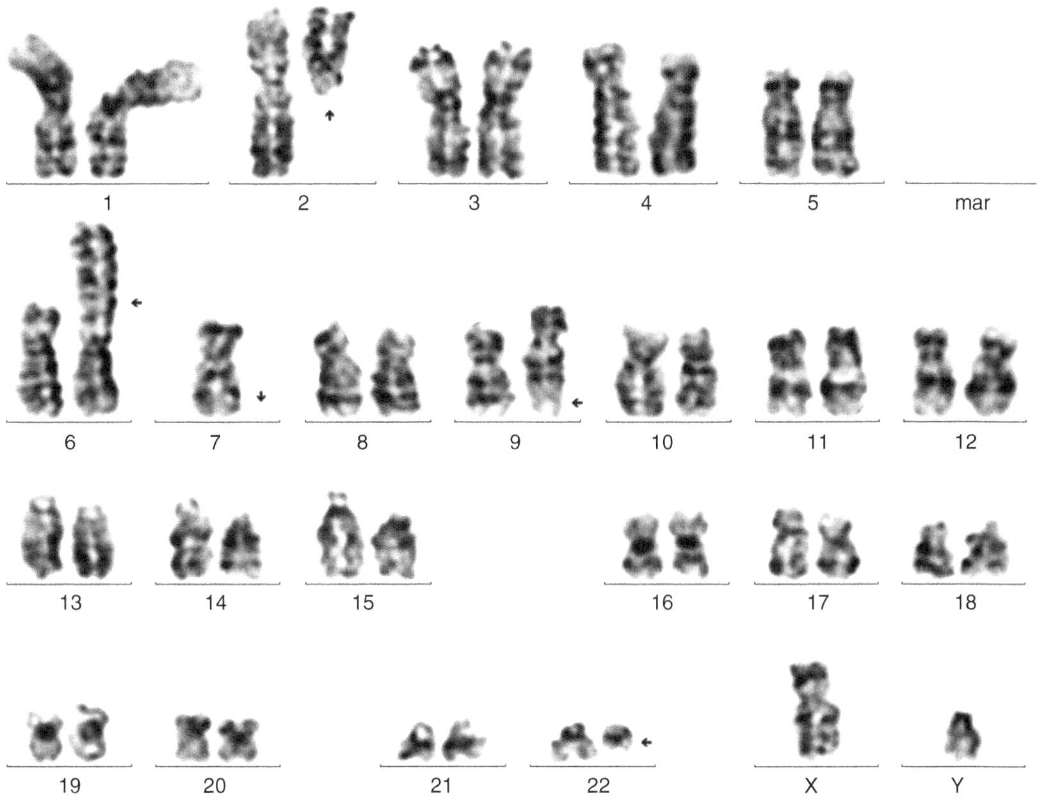

Case Study 9.3 Karyotype showing Philadelphia chromosome formed by t(9;22) as primary abnormality and monosomy 7 with t(2;6) as a secondary abnormality

Hypodiploidy with monosomy 7 and t(9;22) is associated with poor prognosis in precursor B cell acute lymphoblastic leukemia.

Molecular Findings
The Real Time Quantitative PCR showed *BCR-ABL positive (p210).*

Discussion
This patient of relapsed Ph-Positive Acute Lymphoblastic leukemia presented with secondary cytogenetic abnormalities characterized by high hypodiploidy with monosomy 7 and der(6) t(2;6) (q11.2;p25) [6]. In a report of 10 different study groups, Heerema et al. [7] reported secondary aberrations in 61% of the cases. Chromosomes 9, 22, 7, 14, and 8 were most frequently abnormal. 93% of karyotypes were unbalanced. The gain of second Ph chromosome, >50 chromosomes or loss of 7,7p and or 9p were seen. The loss group had the worst EFS. Hypodiploidy with monosomy 7and t(9;22) are associated with poor prognosis in precursor B cell ALL [8].

9.4 Case Study 9.4 Acute Lymphoblastic Leukemia with normal Karyotype with t(12;21) on FISH

Brief Clinical History:
A 3-year-old female child was admitted for bilateral lower limb pain, anemia, and thrombocytopenia. Bone marrow was done for evaluation of Acute Leukemia.

CBC: *Hb*: 7.5 g/dl, TLC: 7650 /cu mm and Platelets: 20,000 /cu mm.

DLC: Neutrophils: 1.0%, Lymphocytes: 71.0%, Eosinophils: 0.0%, Monocytes: 0.0%, Basophils: 0.0%, Blasts: 28.0%.

Peripheral Smear and Bone Marrow
Smear showed ~28% blasts, large having high N:C ratio, scanty cytoplasm, and inconspicuous nucleoli. Bone marrow showed similar blasts (90%) replacing the bone marrow. The peripheral blood and bone marrow picture were consistent with acute leukemia, morphologically lymphoid.

Flow Cytometric Immunophenotyping
Blasts (~72% of acquired events); CD45 dim to moderate blast region with low side scatter. CD19+, CD10+, CD22+, CD9+, CD24+, CD58+, CD38+, cCD79a+, CD81+, CD200+. Features were consistent with Precursor B cell Acute lymphoblastic Leukemia, CALLA positive.

Cytogenetics
Conventional Karyotype
46,XX[20]

Cytogenetic analysis of unstimulated bone marrow aspirate culture showed normal karyotype in all the analyzed 20 metaphases, none of the lymphoid neoplasia associated abnormalities were seen. FISH analysis of 200 interphase cells show ed t(12;21) along with the deletion of normal TEL allele in 97% of the cells. The t(12;21) is a cryptic translocation that cannot be identified by karyotyping alone.

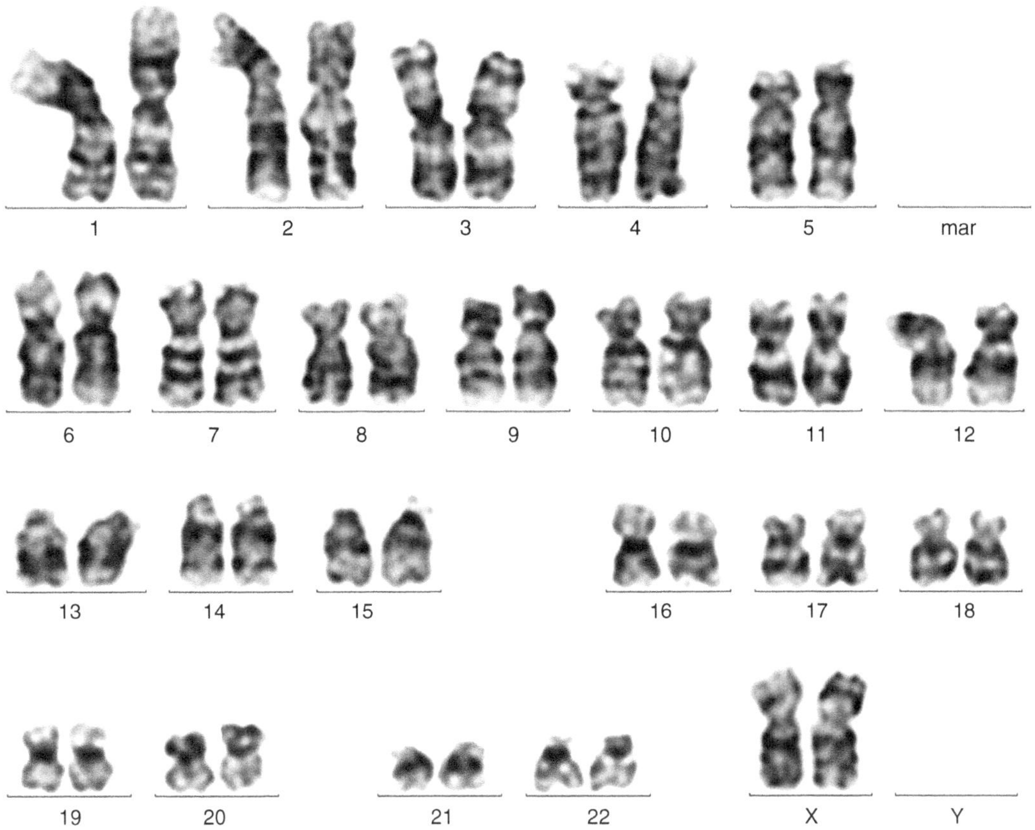

Case Study 9.4 (1) Karyotype showing normal female karyotype

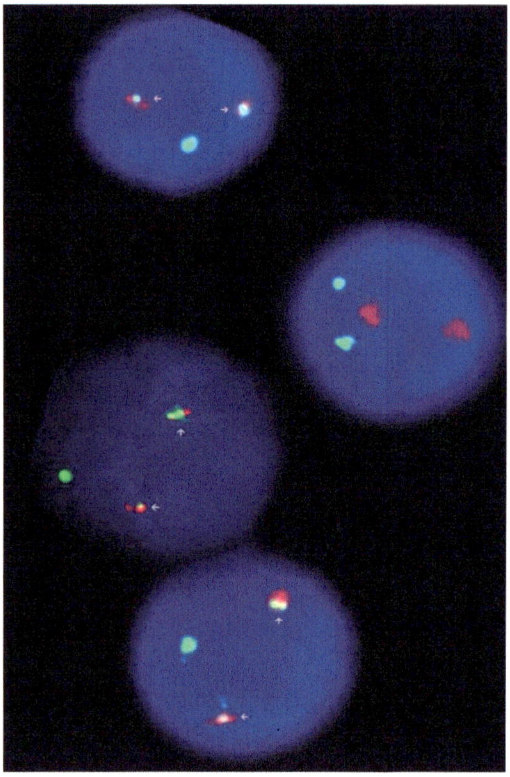

Case Study 9.4 (2) FISH image showing 2F1G aberrant signal pattern using *ETV6/RUNX1* LSI fusion probe indicating *ETV6/RUNX1* fusion

Molecular Studies

The Real Time Quantitative PCR was positive for *ETV6-RUNX1 (TEL-AML1) t(12;21) (p13;q22).*

Discussion

The t(12;21) is a cryptic abnormality seen in 25% children of B-lineage ALL, associated with a favorable prognosis. In a study of 169 patients Raynaud et al. [9] specific cytogenetic abnormalities were looked for. Only 13.6% had a normal karyotype. Rearrangements of 12p and trisomy 21 were the most frequent in those with abnormal karyotype. Further, t(12;21) was not associated with hyperdiploidy (52 to 68 chromosomes) or prognostic t(1;19),t(4;11) or t(9;22).

This patient showed t(12;21) involving TEL and deletion of the other TEL allele molecularly consistent with positive ETV6 RUNX1(TEL-AML1). This translocation fuses TEL and AML 1 genes. As reported by Raynaud et al. [10], the association between t(12;21) and deletion of the non-translocated allele of TEL is among the most frequent abnormalities in B-lineage ALLs (15–22%). It is suggested that the deletion of the normal TEL allele provides a proliferative advantage to leukemia cells.

9.5 Case Study 9.5: Acute Lymphoblastic Leukemia with t (1;19) (q23;p13.3) and i(1)(q10)

Brief Clinical History

A 15-year-old boy presented with fever, anemia and hepatosplenomegaly. Suspected Acute leukemia under evaluation.

CBC: *Hb*: 6.6 g/dl, TLC: 3160 /cu mm and Platelets: 82,000 /cu mm.

DLC: Blasts: 24%, Neutrophils: 45%, Lymphocytes: 27%, Monocytes: 5%, Reticulocyte count: 2.30%. Immature Platelet Fraction: 3.4%.

Peripheral Smear and Bone Marrow

Smear showed Leukopenia with neutropenia, ANC: 1160/uL with ~24% blasts. Bone marrow aspirate smears showed blasts ~88% of total nucleated cells. They were small to medium in size, having high N/C ratio, fine chromatin, and 1–2 prominent nucleoli with a scant amount of pale blue cytoplasm.

Peripheral smear and bone marrow morphology were consistent with acute leukemia.

Flow Cytometric Immunophenotyping

Blasts (~88% of acquired events); CD45 dim to moderate.

CD34−, CD19+, CD10+, CD22+, cCD79a+, CD20+, CD24+, CD9+, CD38+, CD58+, CD81+, CD200+, CD15+, CD34−.

The picture was of Precursor B cell Acute Lymphoblastic Leukemia, CALLA positive (Pre-B-ALL).

Cytogenetics
Conventional Karyotype

47,XY,t(1;19)(q23;p13.3),+i(1)(q10)[19]/46,XY[1]

Cytogenetic analysis of 20 metaphases showed an abnormal clone characterized by a balanced reciprocal translocation between the long arm of chromosome 1 and the short arm of chromosome 19. It also showed the formation of an isochromosome of the long arm of chromosome 1 (1q) in 19 of the 20 metaphases analyzed.

The t(1;19) is associated with intermediate prognosis and gain of isochromosome 1q is probably a secondary event.

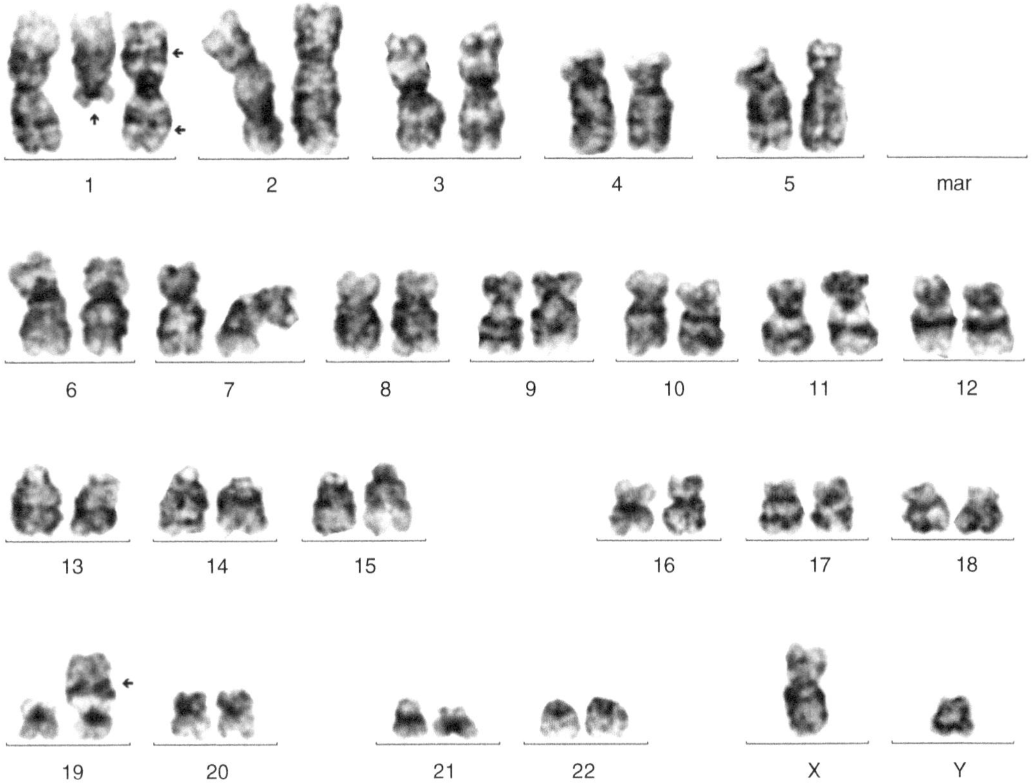

Case Study 9.5 Karyotype showing t(1;19) along with the gain of derivative chromosome 1

Molecular Studies
The Real time Quantitative PCR showed *ALL Panel positive for t(1:19)(q23;p13.3) TCF 3-PBX1 (E2A-PBX1).*

Discussion
This patient with Precursor B Cell Acute Lymphoblastic leukemia CALL positive (Pre-B-ALL) showed t(1:19) with i(1)(q10). The t(1,19) is one of the most frequent rearrangements in B cell ALL, both in adults and children at a frequency of 3–5% [11]. It results in the fusion of transcript factor 3(TCF3) located in 19p 13 with pre-B-cell leukemia homeobox 1 (PBX1) in 1q23 to form a chimeric gene whose protein causes cell differentiation to arrest [12]. In another study on 1322 children, Uckun et al. [13] found the non-random translocation t(1;19) associated with poor outcome in pediatric B-lineage ALL. These patients were more likely to be 10 years of age or greater and less likely to be hyperdiploid.

9.6 Case Study 9.6: Acute Lymphoblastic Leukemia with del(9p) and dup (1q)

Brief Clinical History
A 13-year-old presented with swelling of the ankle and elbow joints and pallor. Found to have lymphadenopathy, 30% blasts in the peripheral smear in an outside facility.

CBC: *Hb*: 9.8 g/dl, TLC: 11,820 /cu mm and Platelets: 20,000 /cu mm.

DLC: Neutrophils: 39.0%, Lymphocytes: 29.0%, Eosinophils: 1.0%, Monocytes: 0.0%, Basophils: 0.0%, Blasts: 31.0%.

Peripheral Smear and Bone Marrow
Smear showed ~31% blasts which had high N: C ratio with scant amount of basophilic cytoplasm, few with cytoplasmic vacuolation, large immature nucleus with inconspicuous nucleoli. Bone marrow showed similar blasts (95%) replacing the bone marrow. The picture was consistent with acute leukemia, morphologically lymphoid.

Flow Cytometric Immunophenotyping:
Blasts (~70.0% of acquired events); CD45 moderate blast region with low side scatter.

CD19+, CD10+, CD22+, cCD79a+, CD9+, CD24+, CD20+, CD58+, CD38+.

Picture was consistent with Precursor B Acute Lymphoblastic Leukemia, CALLA positive.

Cytogenetics
Conventional Karyotype

46,XY,dup(1)(q25q42),del(9)(p22)[18]/46,XY[2]

Cytogenetic analysis of unstimulated bone marrow aspirate culture showed neoplastic clone characterized by an interstitial duplication of the long arm of chromosome 1 and terminal deletion of the short arm of chromosome 9 in 18 of the 20 metaphases analyzed.

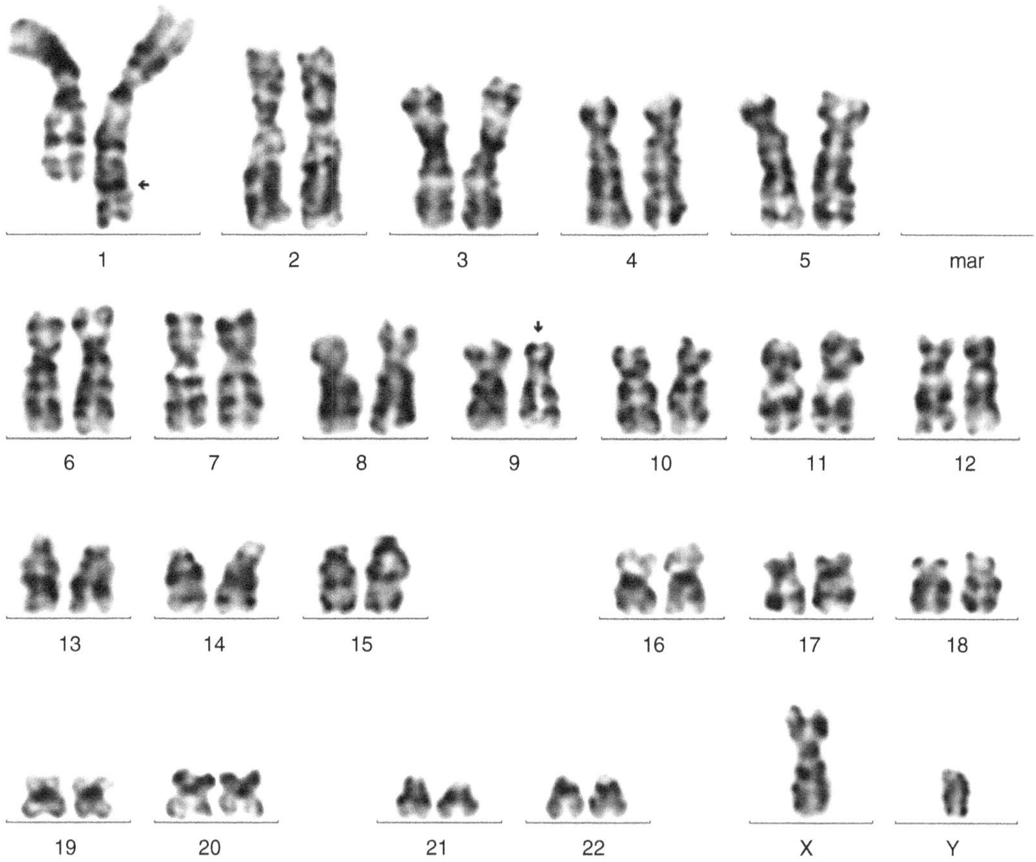

Case Study 9.6 Karyotype showing 1q duplication and deletion of the short arm of chromosome 9, i.e., del(9p)

Molecular Studies
The Real Time Quantitative PCR studies showed *ALL Molecular Panel Negative*.

Discussion
This patient of Precursor B Cell ALL-CALLA positive showed del(9p) together with dup1q. Del (9p) is a common recurring chromosomal aberration in ALL, both of B and T cell lineage (10%), and contains numerous cancer-associated genes [14]. It is the third most common abnormality after hyperdiploidy and t(12;21) in children. The gene involved is CDKN2A. The effect of this abnormality on prognosis is unclear, although some studies indicate adverse prognosis. The del of 9p is an unfavorable risk factor associated with a high ratio of relapses in B Precursor ALL in children [15]. Dup(1q) or gain of 1q is more often seen together with other abnormalities. It is also associated with t(1,19). Its presence has no effect on outcome [16].

9.7 Case Study 9.7: Relapsed Precursor B Acute Lymphoblastic Leukemia with t(12;17)(p13;q11.2)

Brief Clinical History
A 9 year female child diagnosed as ALL in April 2015, finished therapy in August 2017, presented with relapse in December 2019.

CBC: *Hb*: 12.6 g/dl, TLC: 6770 /cu mm and Platelets: 2,88,000 /cu mm.

DLC: Neutrophils: 30.0%, Lymphocytes: 54.0%, Eosinophils: 1.0%, Monocytes: 2.0%, Blasts: 13%.

Peripheral Smear and Bone Marrow

Smear showed ~13% blasts, medium in size with a high N: C ratio, scant agranular cytoplasm, and 1–2 inconspicuous nucleoli. Bone marrow showed 84% blasts. The picture was consistent with relapse in a treated case of ALL.

Flow Cytometric Immunophenotyping

Blasts ~85.22% of acquired events expressing dim to moderate CD45 and low SSC.

CD19+, CD10−, cCD79a+, CD34+, CD24+, CD22+, CD38+, CD58+, CD13+33+.

The picture was suggestive of Relapse of Precursor B Acute Lymphoblastic Leukemia, CALLA positive with aberrant CD13+33.

Cytogenetics

Conventional Karyotype

46,XX,t(12;17)(p13;q11.2)[5]/46,XX[15]

Cytogenetic analysis of unstimulated bone marrow aspirate culture showed neoplastic clone characterized by an balanced reciprocal translocation between short arm of chromosomes 12 and long arm of chromosome 17 in 5 of the 20 metaphases analyzed.

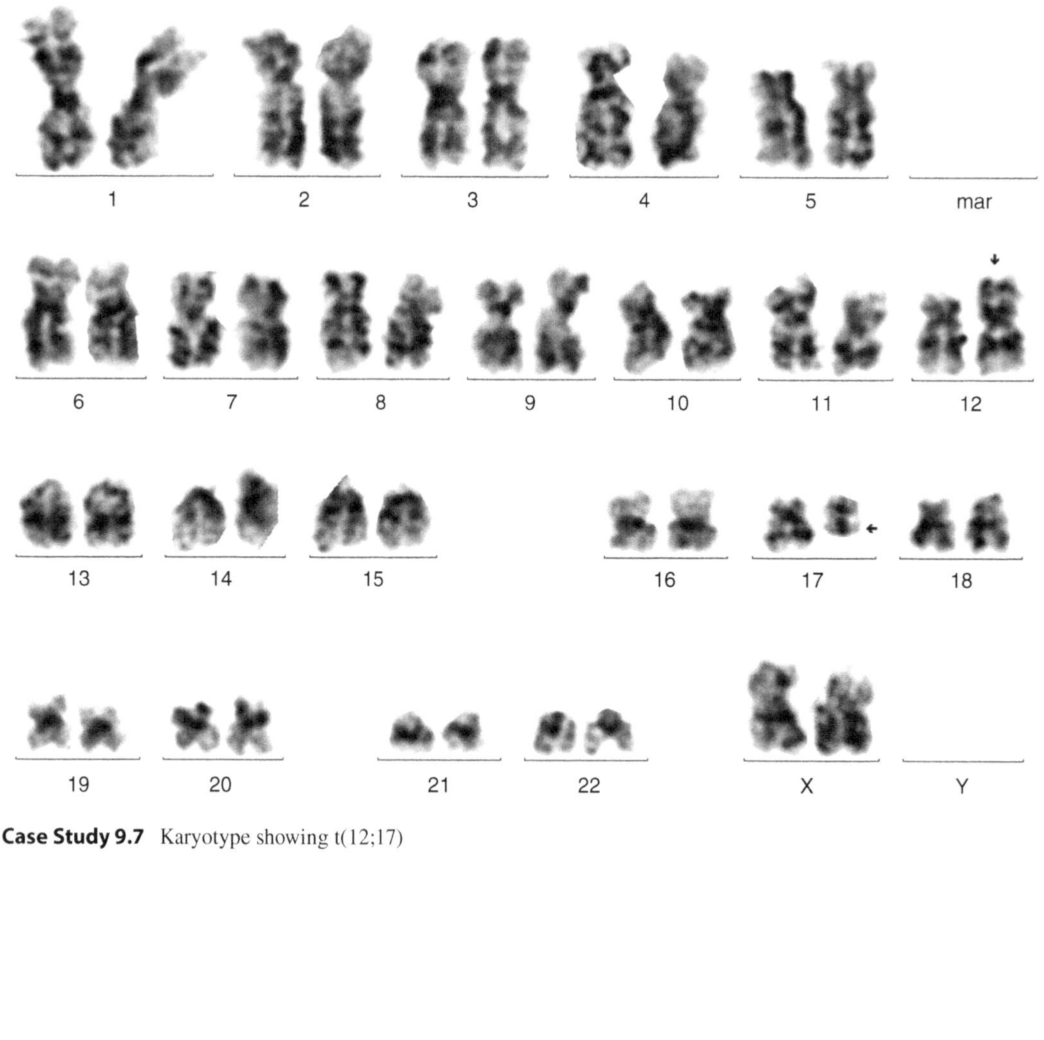

Case Study 9.7 Karyotype showing t(12;17)

Molecular Studies
The Real Time Quantitative PCR studies showed *ALL molecular Panel negative.*

Discussion
This patient presented with relapse of Precursor B CALLA positive ALL with expression of CD19+, CD10−, CD33+ and showed t(12;17)(p13;q12)/TAF15-ZNF384. In studies by Yammoto et al. [17], the t(12;17) translocation is a very rare but recurrent cytogenetic aberration observed in early pre-B-ALL with CD19+, CD−, CD33+ phenotype (as seen in this case). t(12;17) rearrangement is a rare chromosomal abnormality, reported both in lymphoid and myeloid leukemia and is involved in early differentiation of common progenitor [18]. This translocation forms a fusion gene between TAF15 at 17q12 and ZNF384 at 12p13. The clinical characteristics and prognostic impact of this translocation are not well characterized. Coexpression of myeloid markers and a lack of expression of CD10 are common immunophenotypic features as seen in this patient.

9.8 Case Study 9.8: Philadelphia Chromosome Positive Acute Lymphoblastic Leukemia with High Hypodiploidy with Monosomy 7

Brief Clinical History
A 32-year-old female presented with an outside diagnosis of ALL. Investigated for a complete evaluation.

CBC: *Hb*: 7.9 g/dl, TLC: 39,890 /cu mm and Platelets: 1,24,000 /cu mm.

DLC: Neutrophils: 1.0%, Lymphocytes: 15.0%, Eosinophils: 0.0%, Monocytes: 0.0%, Basophils: 0.0%, Myelocytes occasional, Blasts: 84.0%.

Peripheral Blood and Bone Marrow
Smear showed Leukocytosis with the presence of 84% blasts which were large with high N:C ratio, scant amount of agranular light blue cytoplasm, immature chromatin, and 1–2 prominent nucleoli.

Bone marrow imprint smear showed similar blasts replacing the bone marrow (90%).

Peripheral blood and bone marrow features were consistent with Acute Leukemia, morphologically acute lymphoblastic leukemia.

Flow Cytometric Immunophenotyping
Blasts (~90% of acquired events); CD45 dim to moderate blast region with low side scatter.

CD19+, CD10+, CD34+, CD9+, CD24+, CD20+, CD58+, CD38+ .

Picture was consistent with B cell Acute Lymphoblastic Leukemia, CALLA positive.

Cytogenetics
Conventional Karyotyping
45,XY,-7,t(9;22)(q34;q11.2)[20]

Cytogenetic analysis of unstimulated bone marrow aspirate culture showed the presence of a high hypodiploid clone (modal number 45) with monosomy 7 and balanced reciprocal translocation between the long arms of chromosome 9 and chromosome 22 in all the analyzed 20 metaphases. Hypodiploidy with monosomy 7 and t(9;22) are all associated with poor prognosis in precursor B cell acute lymphoblastic leukemia.

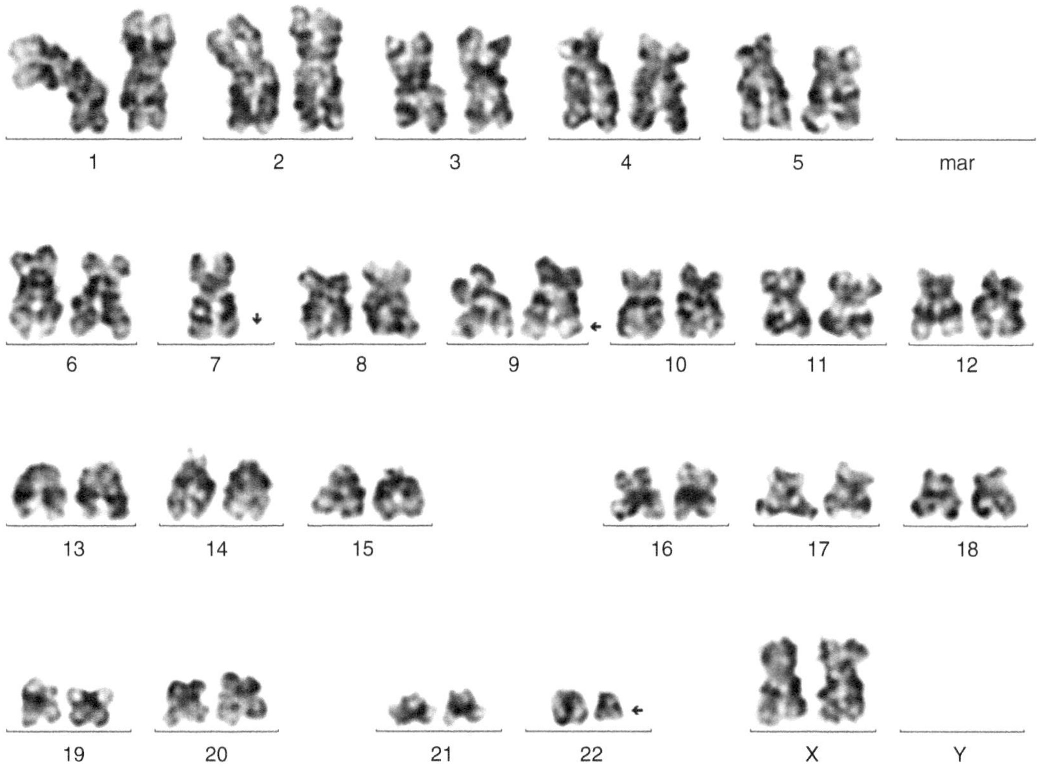

Case Study 9.8 Karyotype showing Philadelphia chromosome formed by t(9;22) along with monosomy 7 as a secondary abnormality

Discussion

Cytogenetically hypodiploid <46 chromosomes ALL may be divided into high hypodiploid (40–45 chromosomes), low hypodiploid, and near haploid ALL. High hypodiploid cases are cytogenetically heterogenous. It is believed that hypodiploidy may provide a selective advantage in tumorigenesis. Safavi et al. [19] found TP 53 mutations both in low hypodiploid and high hypodiploid cases of ALL and may harbor chromosomal instability. They are associated with a dismal prognosis.

The occurrence of additional chromosome abnormalities (ACA) is common in Philadelphia chromosome positive ALL (Ph+ALL). Among 130 adults with Ph+ALL, ACA was seen in 53% [20]. Monosomy 7 was the most common ACA (12 patients) (as also seen in this patient). The presence of ACA has significant effect on outcomes post HCT.

This patient morphologically and immunephenotypically on karyotyping belonged to Philadelphia positive ALL category with secondary abnormality of monosomy 7. Al-Achkar et al. [21] described secondary chromosomal abnormalities in Ph+ALL including that of monosomy 7, other abnormalities such as t(4;11), deletion of IKZF gene, etc. have also been described. They all confer poor prognosis in addition to BCR-ABL positivity in ALL, in itself a poor prognostic marker [22].

9.9 Case Study 9.9: Therapy Related Acute Lymphoblastic Leukemia with Complex Karyotype

Brief Clinical History

A 75-year-old male had radical prostatectomy and radiotherapy for carcinoma prostate in 2016. Now presented with history of on and off fever and low TLC. Suspected MDS. Bone marrow was done for evaluation.

CBC: *Hb*: 12.4 g/dl, TLC: 1620 /cu mm and Platelets: 1,20,000 /cu mm.

DLC: Neutrophils: 45.0%, Lymphocytes: 47.0%, Eosinophils: 4.0%, Monocytes: 2.0%, Basophils: 0.0%, Blasts: 2.0%.

Peripheral Smear and Bone Marrow
Smear showed leukopenia with neutropenia (ANC ~ 729 cells/cu mm) with ~2% blasts. Bone marrow showed trilineage dysplasia with 70% blasts. The picture was consistent with acute leukemia (therapy related).

Flow Cytometric/Immunophenotyping
Blasts (~49% of acquired events) CD45 moderate blast region with low side scatter.

CD19+, CD10−, CD20−, CD22+, CD34+, CD24+, CD58+, CD38+, CD81+, CD200+, HLA-DR+.

The immunophenotypic features were consistent with CALLA negative Pro-B acute lymphoblastic leukemia with aberrant CD13 expression.

Cytogenetics
Conventional Karyotype

77~82,XY,+X,1,+1,+2,+4,+5,+6,+6,+7,+7,+8,+8,+9,+10,+del(11)(q23)x2, +12, +13, +13, +14, +15, +16,+17,+17,+18,+19,+19,+20,+20, +21,+22,+22,+22,+2r,+mar1,+mar2[cp20]

Cytogenetic analysis of unstimulated bone marrow aspirate culture showed neoplastic clone characterized by a near triploidy to near tetraploidy (modal number 77~82) along with structural abnormality showing deletion of the long arm of chromosome 11 (11q23).There were gains of chromosomes 1, 2, 4, 5, 6, 7, 8, 9, 10, 11, 12, 13, 14, 15, 16, 17, 18, 19, 20, 21, 22, 2 copies of unidentified ring chromosomes and 2 marker chromosomes of unknown origin.

To confirm deletion of long arm of chromosome 11 metaphase FISH was done using MLL break apart probe. It showed MLL gene amplification resulting in 2 copies of ring chromosomes and marker chromosomes.

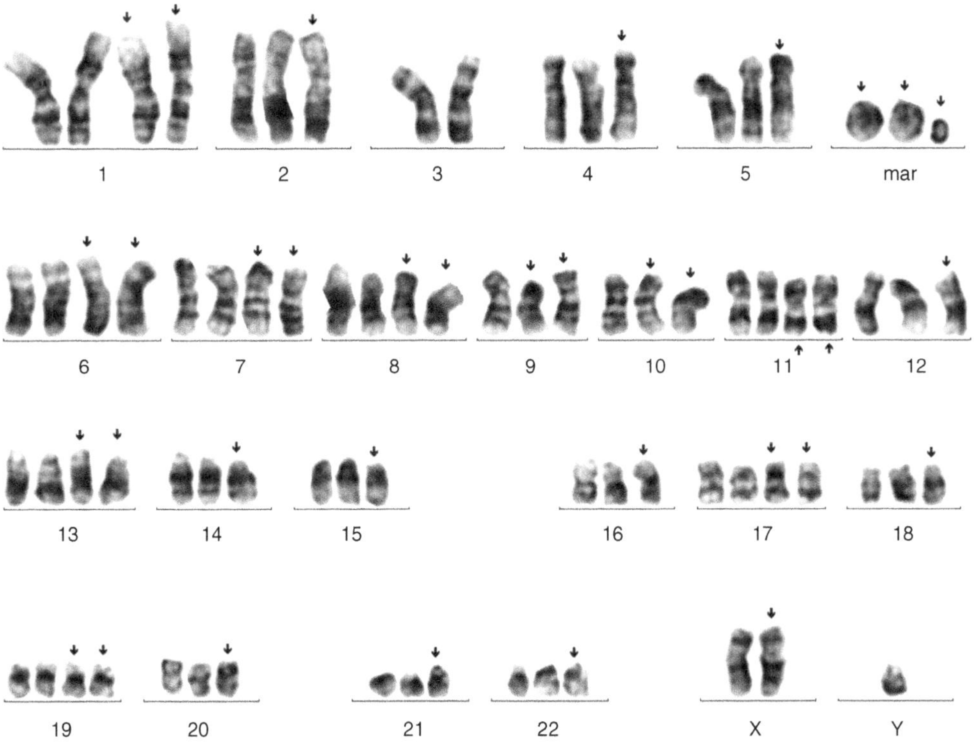

Case Study 9.9 (1) Hyperdiploid karyotype showing gain of different autosomes and two ring chromosome and one marker chromosome of unknown origin

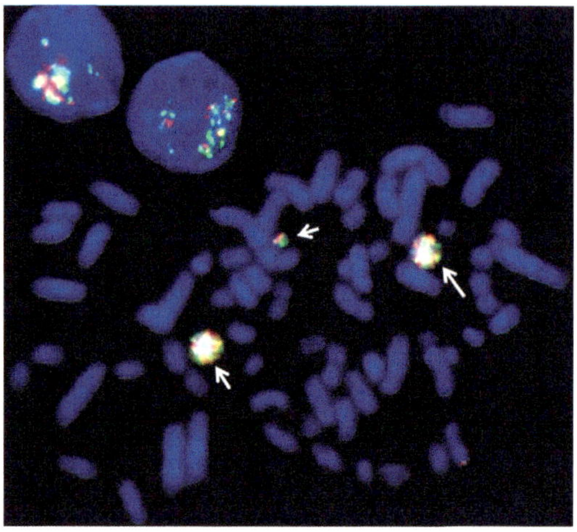

Case Study 9.9 (2) FISH image showing many fusion signal pattern using *MLL(=KMT2A)* BA probe indicating *MLL(=KMT2A)* amplification on ring chromosomes and one marker chromosome having part of long arm of chromosome 11 with *MLL(=KMT2A)* gene

Molecular Findings

The Real Time Quantitative PCR showed *Molecular ALL panel negative*.

Discussion

In a study of 1022 patients of ALL, Aldoss et al. [23], found that 9.1% were therapy related. The median latency for t-ALL was 6–8 years from original diagnosis and was shorter for patients carrying MLL gene rearrangement (as seen in this case). Compared to de novo ALL, t-ALL patients had more MLL gene rearrangements.

This patient developed ALL following chemotherapy and radiotherapy for prostate cancer. The cytogenetic abnormalities in these cases are similar to those seen in t-AML or t-MDS and the outcome of t-ALL patients is poorer than that of de novo ALL. The pathogenesis of t-ALL is attributed to genotoxic effects of cytogenetic drugs on hematopoietic progenitor cells. t-ALLs should be differentiated from s-ALL (secondary ALLs with the previous history of malignancy) because of an impact on morbidity and mortality. The cytogenetic features in these cases are mostly hypodiploid or complex karyotypes both with poor prognosis. MLL gene amplification in therapy related ALL, is known to be rare and indicates poor prognosis [24].

9.10 Case Study 9.10: Early T Cell Precursor Acute Lymphoblastic Leukemia (ETP-ALL) with Normal Karyotype and FLT3-ITD Mutation Positive

Brief Clinical History

A 12-year-old male diagnosed as a case of acute lymphoblastic leukemia in Iraq. Admitted for further evaluation.

CBC: *Hb*: 10.0 g/dl, TLC: 28,130 /cu mm and Platelets: 1,30,000 /cu mm.

DLC: Neutrophils: 11.0%, Myelocytes: 1.0%, Lymphocytes: 17.0%, Eosinophils: 0.0%, Monocytes: 4.0%, Basophils: 0.0%, Blasts: 67.0%.

Peripheral Blood and Bone Marrow

Smear showed Leukocytosis with ~67.0% blasts, small to medium sized with scant amount of basophilic cytoplasm, immature large nuclei with few showing prominent nucleoli and nuclear clefts. Bone marrow showed similar blasts (89%) replacing almost the entire bone marrow.

Flowmetric Immunotyping

Blasts (~85% of acquired events) in moderate positive CD45 blast region with low side scatter.

CD34+, cCD3+, CD7+, CD3+, CD99+ CD117+, CD2+, CD13+, CD33+, HLA-DR+ and CD1a−, CD8− and CD5−.

The immunophenotype was compatible with Early T cell precursor Acute Lymphoblastic Leukemia (ETP-ALL).

Cytogenetics

Conventional Karyotype

46,XY[20]

Cytogenetic analysis of unstimulated bone marrow aspirate culture showed normal karyotype in all the analyzed 20 metaphases, none of the lymphoid neoplasia associated abnormalities were seen.

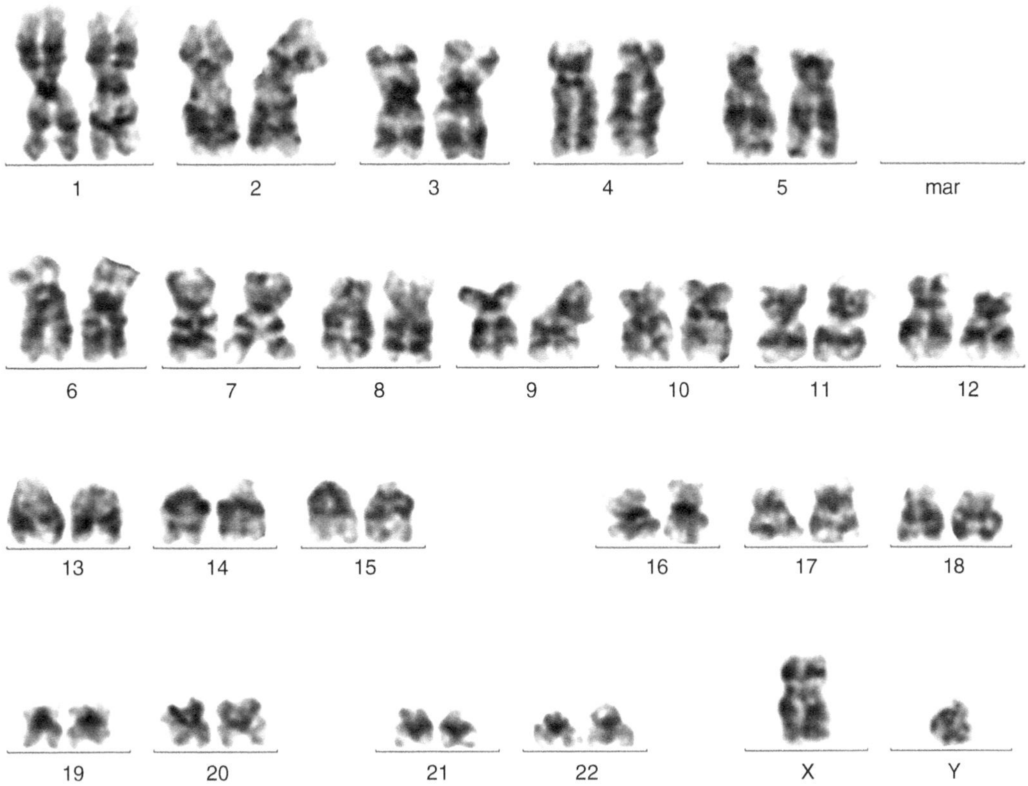

Case Study 9.10 Karyotype showing normal male karyotype

Molecular Studies

The Real Time Quantitative PCR showed FLT3-*ITD Positive with Allelic ratio 46.28% and FLT3 –TKD Negative.*

Discussion

In a study of 178 patients of ETP-ALL, Neumann et al. found that 32% were classified as ETP-ALL based on immunophenotype. FLT3 mutations, rare in overall cohort of T-ALL were very frequent and nearly exclusively found in ETP-ALL. This is a particularly high-risk subgroup of ETP-ALL.

This patient had ETP-ALL together with FLT3-ITD mutation. He had a typical immunophenotypic profile of T cell markers with negative CD1a, CD8, CD5 and expression of stem cell or myeloid markers CD13, CD33, CD117. ETP-ALL is a distinct molecular subgroup of T-ALL and has a high frequency of FLT3 mutation as seen in this patient [25]. Although FLT 3 mutations are frequent in AML and have prognostic implications, they are rare in T-ALL. They are mostly found in ETP T-ALL and are more common in adults. Patients with FLT3 mutation show a distinct phenotype such as CD117, CD34, and CD13 positivity which was also seen in this patient. The treatment of choice is allogeneic SCT since the disease has poor response to treatment [26].

9.11 Case Study 9.11: Acute Lymphoblastic Leukemia with del (9p)

Brief Clinical History

An 18-year-old male presented with anemia, thrombocytopenia, and hyperleukocytosis with splenomegaly. Investigated for acute leukemia.

CBC: *Hb*: 8.1 g/dl, TLC: 1,70,340 /cu mm and Platelets: 20,000 /cu mm.

DLC: Neutrophils: 1%, Lymphocytes: 5%, Blasts: 94%.

Peripheral Smear and Bone Marrow
Smear showed hyperleukocytosis with 94% blasts medium to large sized with high N:C ratio, opened up chromatin,1–3 nucleoli, and scant to moderate pale agranular cytoplasm. Some of the blasts showed deep nuclear clefts. Similar blasts showed near total replacement of the bone marrow. The picture was consistent with acute leukemia.

Flow Cytometric Immunophenotyping
Blasts ~86.6% of acquired events expressing moderate CD45 and low SSC.

CD19+, CD10+, CD34+, CD24+, CD22+, CD20+, CD38+, CD58+.

Picture was suggestive of Precursor B Acute Lymphoblastic Leukemia, CALLA positive.

Cytogenetics
Conventional Karyotype

46,XY,del(9)(p22)[20]

Cytogenetic analysis of 20 metaphases showed the presence of an abnormal clone characterized by a terminal deletion of the short arm of chromosome 9 in all 20 metaphases.

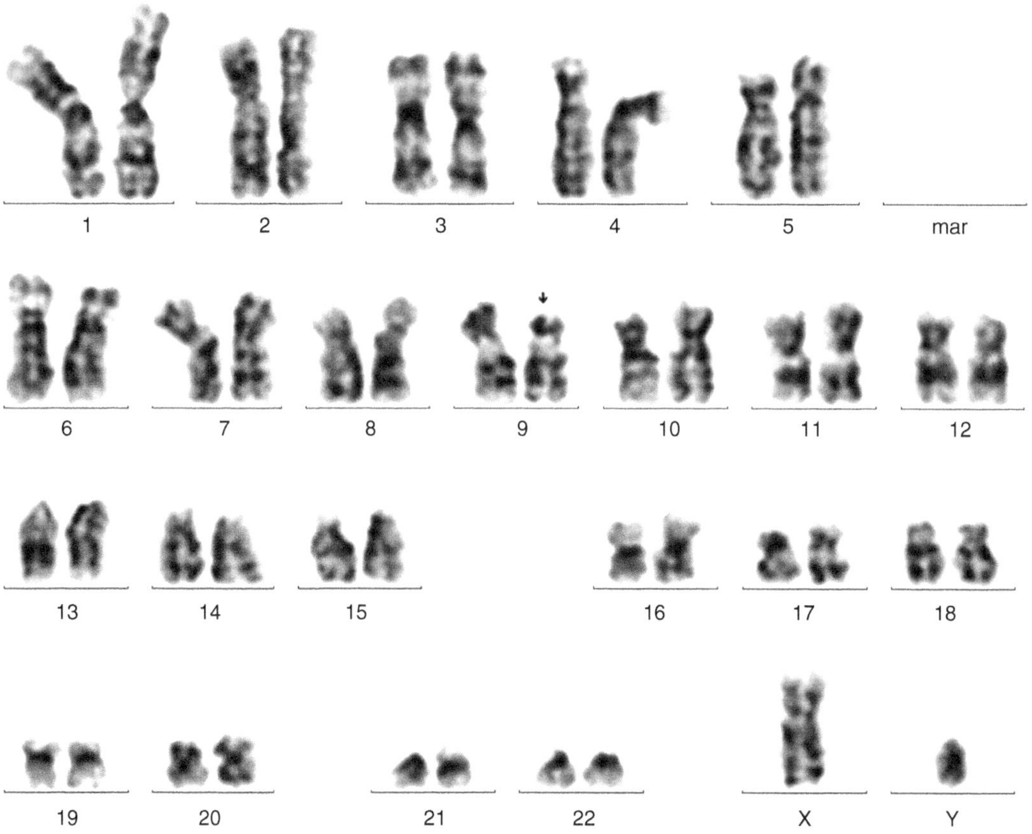

Case Study 9.11 Karyotype showing deletion of short arm of chromosome 9, i.e., del(9p)

Molecular Studies
The Real Time Quantitative PCR for *ALL Panel Negative*.

Discussion
This patient of precursor B-ALL (CALLA+) showed an isolated deletion of (9p). del(9p)is a common recurring chromosomal aberration both in B and T cell lineage ALLs. The 9p region contains several cancer-associated genes such as JAK2, PDL1, PDL2 at 9p24.1 and others, many of which are implicated in leukemogenesis of ALL. The loss of 9p is the second most frequent after Ph1 chromosome in adult ALL and the third after high hyerdiploidy and t(12;21)in pediatric ALL. An abnormal 9p usually occurs as part of a more complex karyotype; it occurs as a sole abnormality (as in this case) in less than 10% of cases with an abnormal 9p. Both pediatric and adults ALL with 9p deletion have overall poor survival and an increased risk of relapse [27].

9.12 Case Study 9.12: Acute Lymphoblastic Leukemia with High Hyperdiploidy with Complex Karyotype and TP53 Gene Deletion

Brief Clinical History
A 6-year-old female was admitted with joint pains and fever for investigations for acute leukemia.

CBC: *Hb*: 9.6 g/dl, TLC: 3930 /cu mm and Platelets: 3,07,000 /cu mm.

DLC: Blasts: 4%, Myelocytes: 1%, Neutrophils: 33%, Lymphocytes: 59%, Eosinophils: 1%, Monocytes: 2%.

Peripheral Smear and Bone Marrow
Smear showed ~4% blasts and neutropenia, ANC: 1296/uL.

Bone marrow showed blasts constituting ~86% of total nucleated cells. They were medium to large in size with scant to moderate amount of pale blue cytoplasm, high N/C ratio, fine chromatin, and 1–2 prominent nucleoli. The picture was consistent with acute leukemia.

Flow Cytometric Immunophenotyping
Blasts (~82% of acquired events); CD45 negative to dim with low side scatter.

CD19+, CD10+, CD34+, cCD79a+, CD22+, CD9+, CD24+, CD20+, CD58+, CD38+.

Picture was consistent with Precursor B cell ALL, CALLA positive.

Cytogenetics
Conventional Karyotype
63,XX,+X,+4,+5,+6,del(6)(q21),+7,del(7)(p13q11.2),+8,+10,+11,+12,+14,+14,+17,del(17)(p13),+18,+21,+21,+mar1,+mar2[5]/46,XX[15]

Cytogenetic analysis of 20 metaphases showed a neoplastic clone characterized by a high hyperdiploid complex karyotype showing numerical and structural abnormalities. Numerical abnormalities showed (Chromosome number: 63) gains of chromosomes 4, 5, 6, 7, 8, 10, 11, 12, 17, and 18 as trisomies and chromosome 14 and 21 as tetrasomies. Also, there was gain of sex chromosome X and 2 marker chromosomes of unknown origin. Structural abnormalities showed deletion of long arm of chromosome 6, interstitial deletion of chromosome 7 and deletion of short arm of chromosome 17, i.e., TP53 deletion which was also confirmed by FISH.

The above cytogenetic alterations indicate poor prognosis.

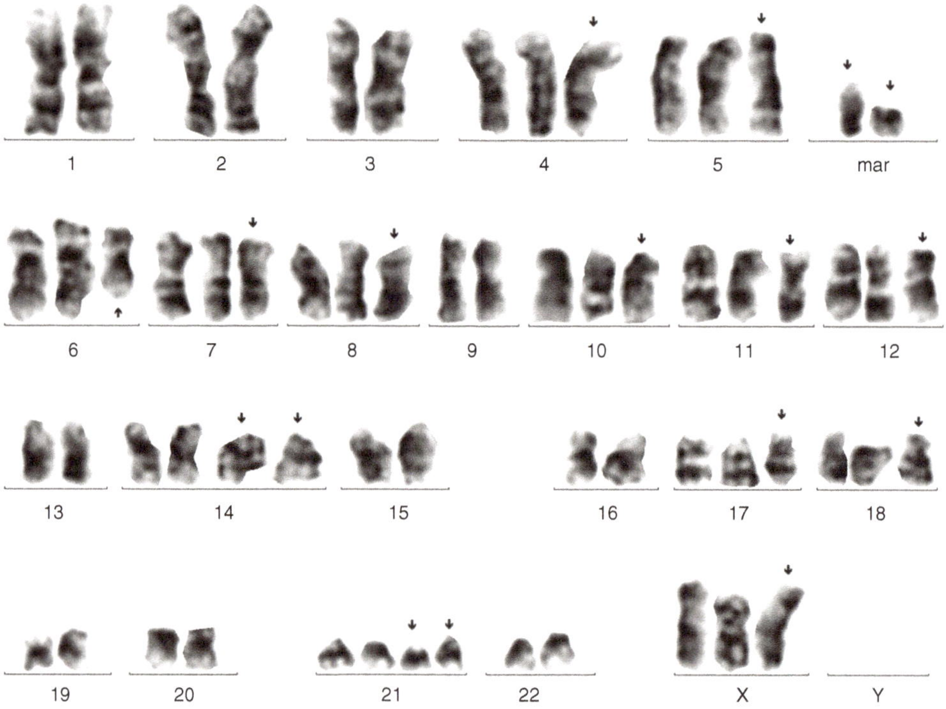

Case Study 9.12 (1) Karyotype showing hyperdiploidy in the form of gain of chromosome X, 4, 5, 7, 8, 10, 11, 12, 14, 14, 17, 18, 21, 21 along with deletion of long arm of chromosome 6. The clone also has two different marker chromosomes of unknown origin

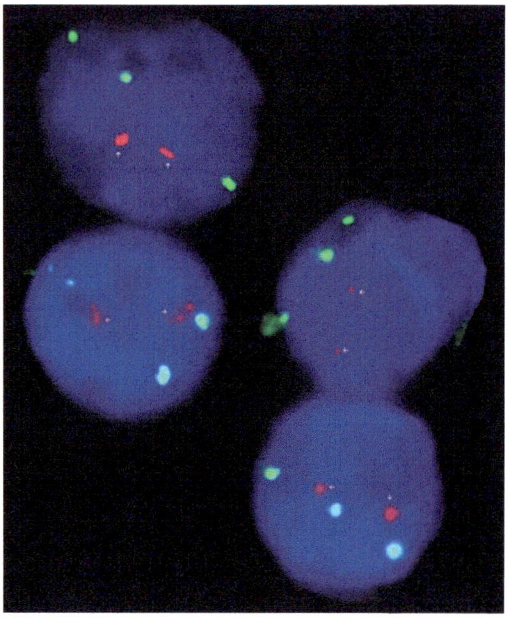

Case Study 9.12 (2) FISH image showing 3G2R signal pattern using *TP53* DC deletion probe indicating *TP53* deletion along with trisomy 17

Molecular Studies

The Real Time Quantitative PCR for *Acute Lymphoblastic Leukemia panel negative*.

Discussion

This patient presented with high hyperdiploidy (chromosome number 63) and numerous numerical and structural abnormalities, notably deletion of TP53. High hyperdiploidy (chromosome number 51–67) is seen in almost 30% of all pediatric B Cell precursor B-ALL. The genetic features are mostly trisomies and tetrasomies. More than 70% cases harbor+X,+4,+6,+10, +14,+17, or +21 as seen in this patient too [28]. Hyperdiploidy appears to be an epiphenomenon without any leukemogenic effect and no cryptic fusion gene has been found on whole genome sequencing. Putative driver event identified genes involved are RAS(KRAS, NRAS), FLT3, and PTPN 11 as suggested by Paulson et al. [29]. Hyperdiploidy is associated with favorable prognosis with overall survival rates >90% on current therapy protocols. Alterations of the TP53 gene are generally associated with resistance to therapy or relapse. In this patient, its presence may negate the favorable impact of hyperdiploidy and may be a predictor of early relapse [30].

9.13 Case Study 9.13: Acute Lymphoblastic Leukemia with Complex Karyotype with t(4;11) (q21;q23)

Brief Clinical History
Four-month-old infant with fever, leukocytosis, anemia, and thrombocytopenia admitted for the exclusion of acute leukemia. There was hepatosplenomegaly.

CBC: *Hb*: 8.3 g/dl, TLC: 2250 /cu mm and Platelets: 12,000 /cu mm.

DLC: Blasts: 42.0%, Neutrophils: 1.0%, Lymphocytes: 57.0%, Eosinophils: 0.0%, Monocytes: 0.0%, Basophils: 0.0%.

Peripheral Smear and Bone Marrow
Smear post therapy of IT MTX and prednisolone showed leukopenia with marked neutropenia along with the presence of ~42% blasts. Blasts were medium to large, with high N:C ratio, open chromatin, agranular cytoplasm, and inconspicuous to single nucleolus.

Bone marrow showed similar blasts (75%) of the total nucleated cells.

The picture was consistent with acute leukemia morphologically lymphoid.

Flow Cytometric Immunotyping
Blasts (~73% of acquired events); CD45 moderate blast region with low side scatter.

CD19+, CD10−, CD22+, CD34+, CD9+, CD58+, CD38+, and CD15+,05.

The picture was consistent with Precursor-B acute lymphoblastic leukemia, CALLA negative with the expression of CD15 which is commonly associated with KMT2A (MLL) gene rearrangement.

Cytogenetics
Conventional Karyotype

46,XY,t(4;11)(q21;q23)[12]/46,XY[4]

Cytogenetic analysis of unstimulated peripheral blood culture showed a neoplastic clone characterized by balanced reciprocal translocation between long arms of chromosome 4 and chromosome 11 in 12 of the 16 analyzed metaphases. t(4;11) in infant ALL is associated with poor risk.

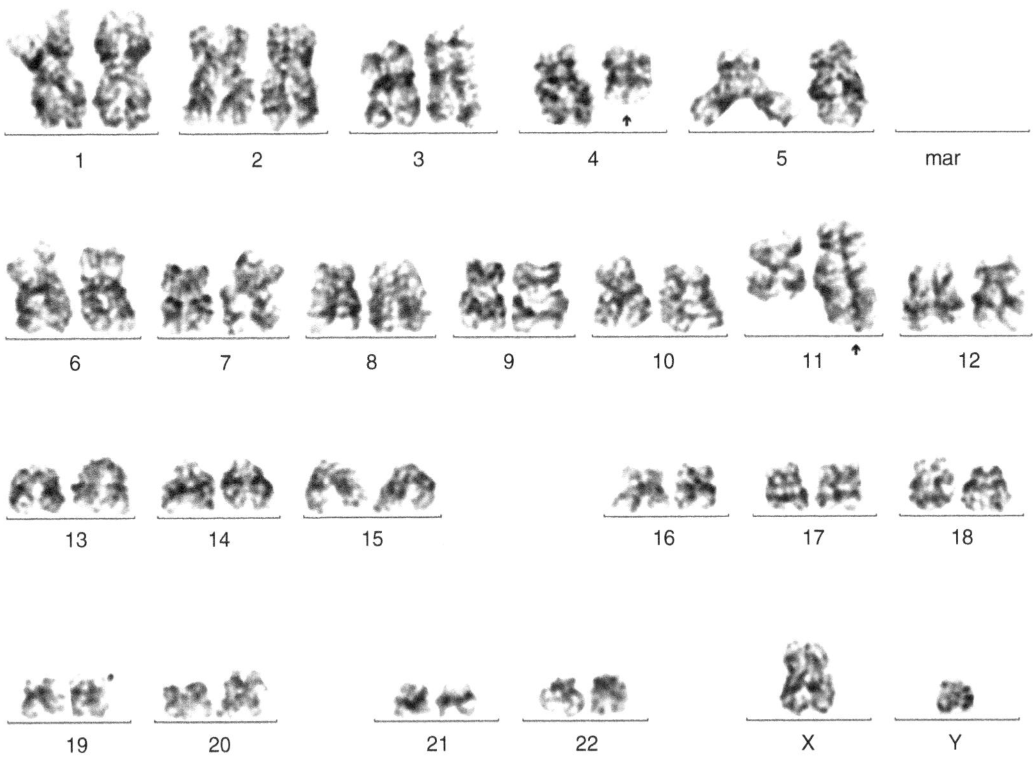

Case Study 9.13 Karyotype showing t(4;11)

Molecular Findings

The Real Time Quantitative PCR was *Positive for t (4;11):(MLL;AF4).*

Discussion

Infant Leukemia with mixed lineage leukemia (MLL) gene rearrangement might arise without any cooperating genetic lesions. There is a considerable subclonal diversity indicating existing of multiple clones. Most ALL relapses arise from clones already existing at diagnosis. The MLL translocation (with the AF4 partner gene) most frequently is believed to be initiating event occurring in utero [32]. They typically display a mixed lineage phenotype and such infants have a dismal prognosis resulting from drug resistance and relapses.

This 4-month-old male had Precursor B Acute Lymphoblastic Leukemia immunophenotypically CALLA negative with CD15 expression commonly associated with KMT2A (MLL) gene rearrangement which indeed it was both on karyotype and molecular ALL panel, t,(4;11) (MLL;AF4). Infant B-ALL accounts for 10% of childhood ALL and has a particularly poor prognosis [31]. There is compelling evidence of a single cell of prenatal origin as the target for MA4 fusion which explains brief leukemia latency [32]. Infant B-ALL is mediated by mixed lineage leukemia rearrangement (MLL-r). It functions as perhaps the sole driving oncogenic event dysregulating the epigenetic events [33]. Many partners have been identified in MLL-r leukemia, the most common being AF4 (AFF1), AF9, and ENL [34].

9.14 Case Study 9.14: Acute Lymphoblastic Leukemia with both Hyperdiploid and Hypodiploid Clones

Brief Clinical History

A 47-year-old female with no known comorbidities presented with fever and epistaxis off and on for 15 days. Bone marrow was done for evaluation.

CBC: *Hb*: 6.6 g/dl, TLC: 5590 /cu mm and Platelets: 10,000 /cu mm.

DLC: Neutrophils: 10%, Lymphocytes: 23%, Eosinophils: 1%, Metamyelocytes: 1%, Blasts: 65%, 4nRBC/100 WBC seen.

Peripheral Blood Smear and Bone Marrow

Smear showed ~65% blasts, medium to large cells with high N:C ratio, fine nuclear chromatin with 1–2 prominent nucleoli, and moderate amount of agranular basophilic cytoplasm with fine vacuolations. Similar blasts (92%) infiltrated the bone marrow.

Peripheral blood and bone marrow were consistent with acute leukemia morphologically lymphoid.

Flow Cytometric Immunophenotyping

Blasts ~61% of acquired events expressing moderate CD45 and low to moderate SSC on CD45 versus SSC plot. CD19+, cCD79a+, CD10+, nTDT+, CD24+, CD22+, HLA-DR+, CD38+, CD58+, CD81+, CD20+ partial.

The picture of Immunophenotype was consistent with Precursor B Acute Lymphoblastic Leukemia, CALLA positive.

Cytogenetics

Conventional Karyotype

31~33,X,-X,-2,-3,-4,-4,-7,-10,-11,-12,-14,-14,-15,-16,-17,-20[cp7]/59~63,XX,+8, +8, +9, +11, +14, +14, +16,+19,+20,+21,+21,+22,+22[cp5]/46,XX[8].

Cytogenetic analysis of 20 metaphases showed the presence of two clones. A neoplastic clone showing composite karyotype characterized by low hypodiploidy showing loss of chromosomes 2, 3, 4, 7, 10, 11, 12, 14, 15, 16, 17, 20 and X in 7 metaphases (modal number 31 to 33). Another clone showed high hyperdiploidy showing gain of chromosomes 8, 9, 11, 14, 16, 19, 20, 21, and 22 in 5 metaphases (modal number 59–63).

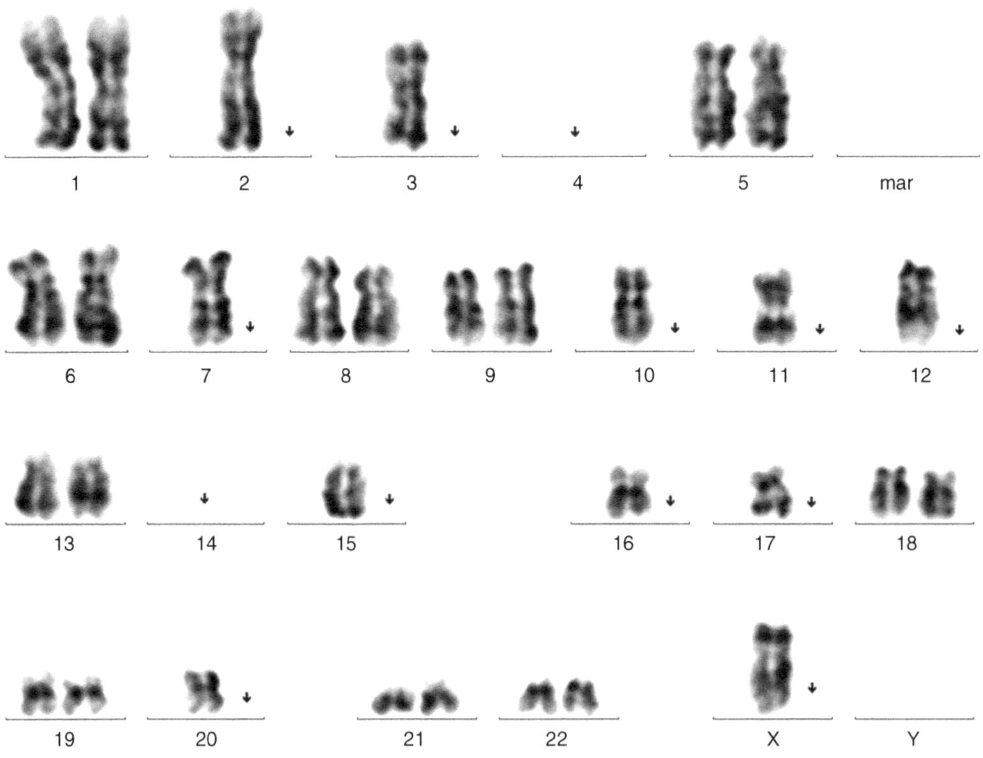

Case Study 9.14 (1) Hypodiploid karyotype showing loss of chromosome X,2,3,4,4,7,10,11,12,14,14,15,16,17,20

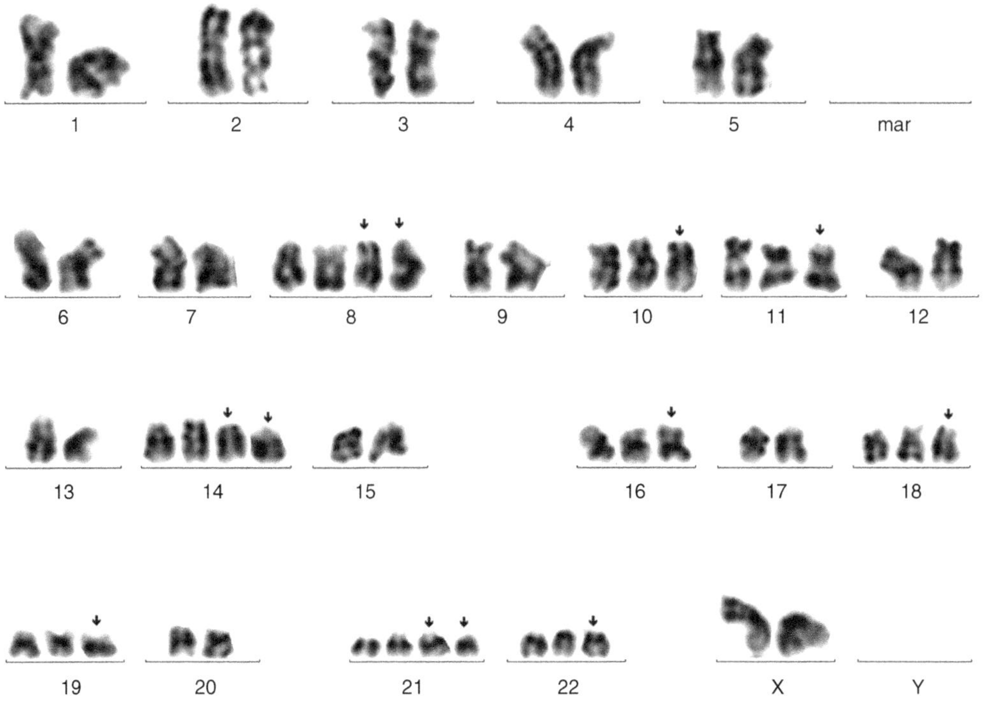

Case Study 9.14 (2) Hyperdiploid karyotype showing gain of chromosomes 8,8,9,11,14,14,16,19,20,21,21,22,22

Molecular Studies
The Real Time Quantitative PCR showed *Molecular ALL Panel negative*.

Discussion
This adult patient of ALL showed a composite karyotype with both a hypodiploid and hyperdiploid clone. Safavi et al. [35] reported hypodiploidy (<40 chromosomes) as an uncommon genetic feature in both children and adults. (This patient had high hypodiploidy with 31 chromosomes). Hypodiploidy is divided into two groups, near haploid (24–30 chromosomes) and low hypodiploid (31–39 chromosomes). Both groups have a different mutational profile and both have poor prognosis and could benefit from novel therapies. A frequent phenomenon in hypodiploid ALL is the doubling of chromosomes resulting in clones with 50–78 chromosomes (in this case 59–63) showing hyperdiploidy. The two clones in this case could be due to endo-reduplication. The prognosis remains poor. The role of hypodiploidy in leukemogenesis remains unclear.

9.15 Case Study 9.15: Philadelphia Chromosome Positive Acute Lymphoblastic Leukemia with additional Cytogenetic Abnormalities

Brief Clinical History
A 16-year-old male presented to the emergency with complaints of headache since 10–15 days high grade fever for last 7 days and weakness. Peripheral smear done outside showed blasts. Patient has been admitted for evaluation in view of suspicion of Acute Leukemia and appropriate management.

CBC: *Hb*: 7.7 g/dl, TLC: 6300 /cu mm and Platelets: 90,000 /cu mm.

DLC: Blasts: 28.0%, Myelocytes: 6.0%; Neutrophils: 26.0%, Lymphocytes: 37.0%, Eosinophils: 0.0%, Monocytes: 3.0%, Basophil: 0.0%.

Peripheral Smear and Bone Marrow
Smear showed 28% blasts. Blasts were large, having high N:C ratio, scanty granular basophilic cytoplasm, moderately condensed chromatin, and inconspicuous nucleoli. Similar blasts showed near total replacement of the bone marrow (94.0%).

Peripheral blood and bone marrow picture were consistent with acute leukemia, morphologically Lymphoid.

Flow Cytometric Immunophenotyping
Blasts ~33.6% of acquired events expressing negative to moderate CD45 and low SSC.

CD34+, CD19+, cCD79a+, CD10+, CD20+, CD24+, CD22+, CD38+, CD58+, CD200+, CD81+, CD13+33+, CD66c+, CD73+.

Picture was consistent with Precursor B cell ALL, CALLA+,13+,33+

Cytogenetics
Conventional Karyotype

46,XY,t(9;22)(q34;q11.2),del(9)(p22),-20,+mar[11]/47, XY,t(9;22)(q34;q11.2),+10[5]/46,XY, t(9;22) (q34;q11.2)[2]/46,XY,[2]

Cytogenetic analysis showed a neoplastic clone characterized by a balanced reciprocal translocation between long arms of 9 and 22 forming the Philadelphia chromosome (Ph) in 18 of 20 metaphases analyzed. This clone also showed terminal deletion of the short arm of chromosome 9, Monosomy of chromosome 20, and gain of a marker chromosome of unknown origin in 11 of the 18 metaphases. Another cell clone showed gain of chromosome 10 as trisomy in 5 of the metaphases analyzed.

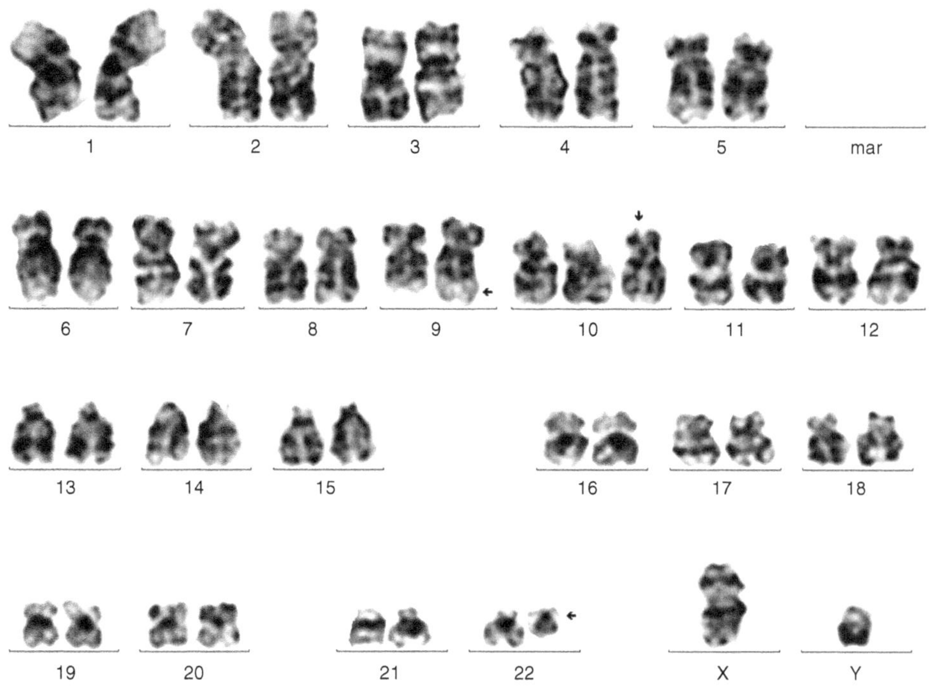

Case Study 9.15 (1) Karyotype showing Philadelphia chromosome formed by t(9;22) along with trisomy 10 as secondary abnormality

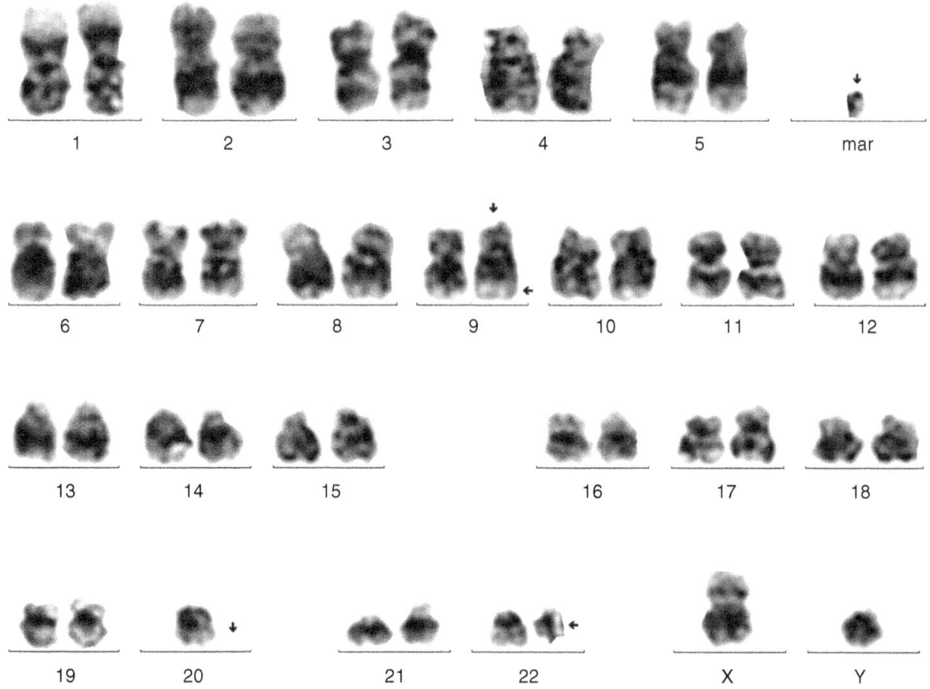

Case Study 9.15 (2) Karyotype showing Philadelphia chromosome formed by t(9;22) along with monosomy 20 and an marker chromosome of unknown origin

Molecular Studies
The Real Time Quantitative PCR showed *BCR-ABL positive in ALL panel.*

Discussion
This patient morphologically an ALL, on immunophenotyping showed a CALLA positive precursor B cell ALL with the aberrant impression of CD13 and CD33 suggesting a BCR-ABL positive ALL. On karyotyping, it showed t(9;22) together with secondary cytogenetic abnormalities. In a study of 111 newly diagnosed adults with ALL and t(9;22)(q34;q11.2) or its variants, secondary alterations were seen in 68% of the patients [36]. They were in descending order of frequency +der (22) t(9;22),+21, abnormalities of 9p, high hyperdiploidy,+8,-7,+X. Deletion of 9p as well as monosomy 20 and trisomy 10 are of unusual interest in this case. It was found that additional chromosomal abnormalities (ACA) have a significant deleterious effect on outcomes post HCT. In children, Heerema et al. [7], the presence of ACAs showed heterogeneous outcomes. ACAs were found in 61% of them and included gain of the second Ph and/or >50 chromosomes or abnormalities of chromosomes 9,22,7,14 and 8 or loss of 7,7p and/or 9p. Patients with loss of 7,7p and/or 9p have an inferior outcome compared to those with gain of a second Ph and/or >50 chromosomes.

t(9;22) is associated with poor prognosis in precursor B Cell acute lymphoblastic leukemia. The additional cytogenetic abnormalities (deletion 9p, Monosomy 20, gain of marker chromosome, and Trisomy 10) are likely due to a secondary event. The clinical features in B-ALL cases showing additional cytogenetic abnormalities are governed by the presence of the t(9;22) [37].

9.16 Case Study 9.16: Acute Lymphoblastic Leukemia with Masked t(9;22)

Brief Clinical History
A 60-year-old female presented with complaints of generalized weakness and abdominal pain. She was found to have leukocytosis and was referred for evaluation of suspected acute leukemia.

CBC: *Hb*: 6.3 g/dl, TLC: 22,210 /cu mm and Platelets: 16,000 /cu mm.

DLC: Neutrophils: 10.0%, Lymphocytes: 10.0%, Eosinophils: 0.0%, Monocytes: 1.0%, Basophils: 0.0%, Blasts: 79.0%.

Peripheral Blood Smear and Bone Marrow
Smear showed Leukocytosis with the presence of ~79.0% blasts. The blasts were variably sized, medium to large in size with high N: C ratio, scant cytoplasm, homogeneous, fine nuclear chromatin with prominent 1–2 nucleoli.

Bone marrow showed similar blasts replacing the marrow (94%). Peripheral blood and bone marrow showed features consistent with Acute Leukemia morphologically lymphoid.

Flow Cytometric Immunophenotyping
Blasts (~86.0% of acquired events) showed low side scatter with negative to dim positive CD45.

CD19+, CD10+, CD34+, CD22+, CD24+, CD9+, cCD79+, CD58+, CD38+.

The picture was of precursor B-ALL-CALLA positive.

Cytogenetics
Conventional Karyotyping

46,XX[16]

Cytogenetic analysis of unstimulated bone marrow aspirate culture showed normal karyotype in all the analyzed 16 metaphases, none of the lymphoid neoplasia associated abnormalities were seen.

Since in house RT-PCR showed BCR-ABL to be positive with normal karyotype, theFISH analysis was done which showed a positive test in this case, indicating the presence of a cryptic translocation of t(9;22).

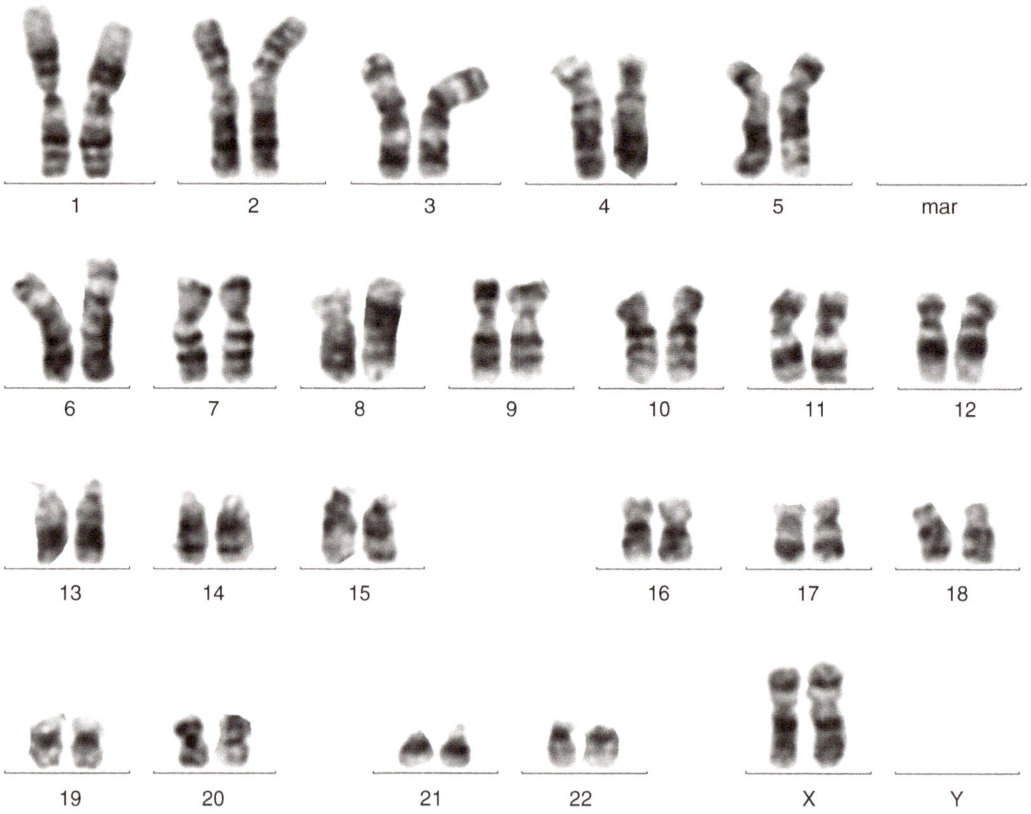

Case Study 9.16 (1) Karyotype showing normal female karyotype

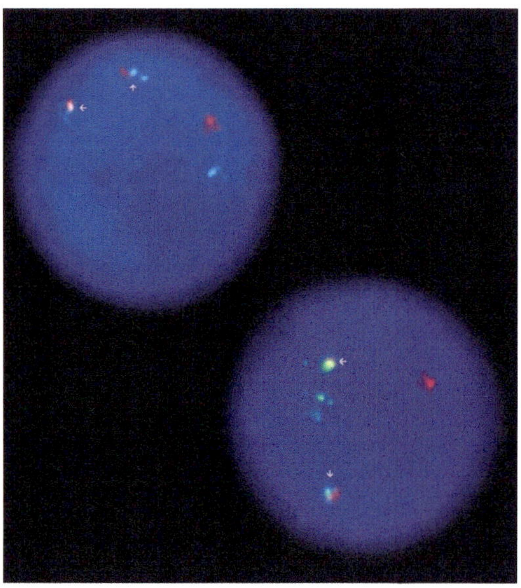

Case Study 9.16 (2) FISH image showing 2F1G1R signal pattern using *BCR/ABL1* DC DF probe indicating *BCR/ABL1* fusion probe

Molecular Studies

The Real Time Quantitative PCR showed *ALL panel positive for BCR-ABL gene rearrangement: Positive (P210).*

Discussion

This case identified morphologically and immunophenotypically as ALL showed a normal karyotype but was positive for BCR-ABL1 on molecular ALL panel on RT-PCR FISH also revealed the presence of t(9;22), a situation of a masked Philadelphia chromosome in ALL. Such cases have been described Kawahara et al. [38] described one such case. Such cases are considered to be due to the cryptic insertion of 3′ABL within a chromosome 22. These patients often have a high TLC and frequent CNS involvement and are mostly CD10+ as in this case. Prognosis in such cases is very poor [39].

9.17 Case Study 9.17: Philadelphia Chromosome Positive Acute Lymphoblastic Leukemia

Brief Clinical History

A 43-year-old male presented with fever. Peripheral blood and bone marrow showed acute leukemia. Admitted for further investigations.

CBC: *Hb*: 6.7 g/dl, TLC: 5170 /cu mm and Platelets: 11,000 /cu mm.

DLC: Blasts: 32.0%, Myelocytes: 4.0%, Neutrophils: 8.0%, Lymphocytes: 52.0%.

Peripheral Smear and Bone Marrow

Smear showed total count within normal limits with ~32.0% blasts. The blasts were large in size with high N:C ratio, immature nuclear chromatin with inconspicuous nucleoli and scant to moderate agranular cytoplasm.

Bone marrow showed similar blasts 77% of the total nucleated cells.

Flow Cytometric Immunophenotyping

Blasts (~37% of acquired events); CD45 moderate blast region with low side scatter.

CD19+, CD34+, CD10+, CD79a+, CD22+, CD9+, CD24+, CD20+, CD58+, CD38+.

The immunophenotype was that of Precursor B cell Acute Lymphoblastic Leukemia, CALLA positive (B cell-ALL).

Cytogenetics

Conventional Karyotype

46,XY,t(9;22)(q34;q11.2)[11]/46,XY[9]

Cytogenetic analysis of unstimulated PB culture showed Philadelphia chromosome, t(9;22) (q34;q11.2) in 11 of the 20 metaphases analyzed. t(9;22) is associated with poor prognosis in precursor B cell acute lymphoblastic leukemia.

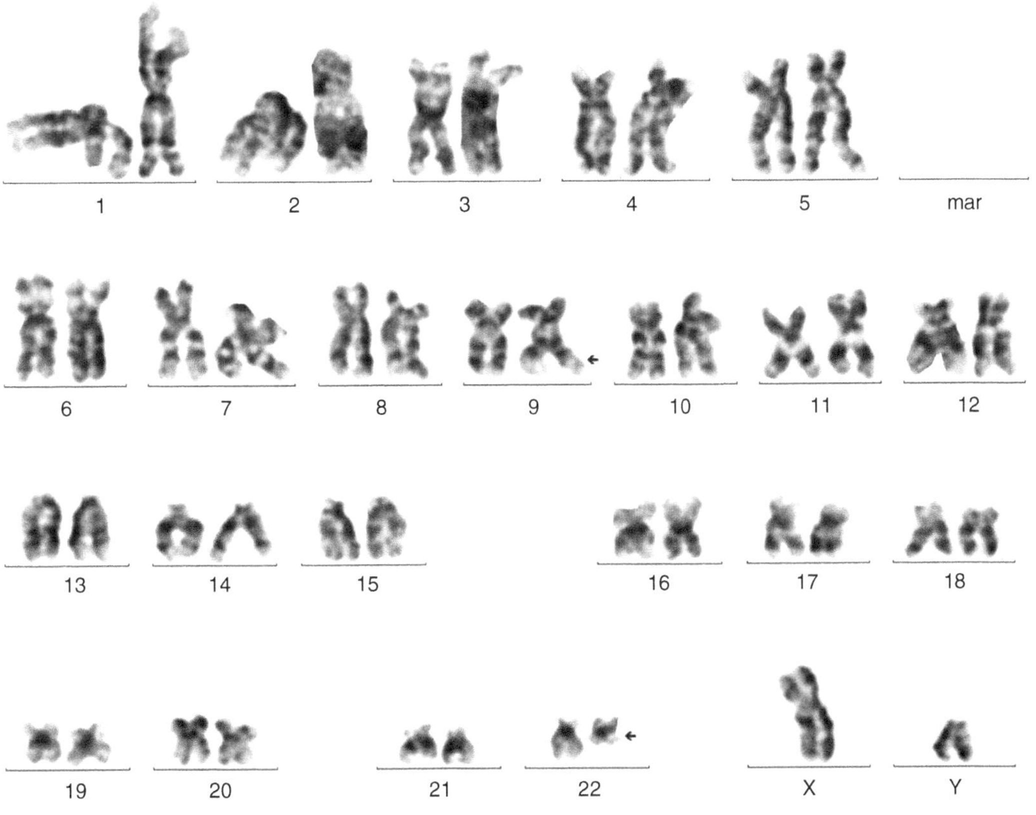

Case Study 9.17 Karyotype showing Philadelphia chromosome formed by t(9;22)

Molecular Findings

The Real Time Quantitative PCR showed *BCR-ABL gene rearrangement: Positive (P210).*

Discussion

Philadelphia positive ALL is seen in nearly 25–30% cases of adult ALLs. It typically presents with aggressive clinical course, poor response to standard therapy, and a high risk of relapse. The TKIs has completely changed the therapeutic approach, Imatinib was the first to be incorporated in the treatment followed by nilotinib, dasatinib, bosutinib, and ponatinib. Novel drugs such as antibody-drug conjugates, bispecific monoclonal antibodies, and CAR-T cell therapy are being developed [40].

This patient morphologically and immunophenotypically was B cell precursor ALL-CALLA positive.

On karyotyping and molecularly he was Ph chromosome positive t(9;22). RT-PCR studies showed BCR-ABL1(p210). Ph+ALL is the subtype with the worst prognosis. In spite of the use of TKI inhibitors, allogeneic stem cell transplantation (aSCT) remains the most curative option following post-remission therapy in adults [41]. It appears to be less critical in children indicating a clinical and biological heterogeneity within the subgroup of Ph+ALL. The Ph chromosome is the most frequent cytogenetic abnormality in adult patients with ALL seen in 20% to 30% adults but only 5% of children. Based on the location of breakpoint within the BCR gene, P190 protein is seen in 2/3 of patients and p210 in one-third of the children [42].

9.18 Case Study 9.18: Acute Lymphoblastic Leukemia with der 19 t(1;19) (q23;p13)

Brief Clinical History

A 15-year-old female was admitted for investigations of acute febrile illness, severe anemia, and bleeding per vaginum.

CBC: *Hb*: 7.5 g/dl, TLC: 2460 /cu mm and Platelets: 30,000 /cu mm.

DLC: Neutrophils: 15.0%, Lymphocytes: 76.0%, Eosinophils: 0.0%, Monocytes: 1.0%, Basophils: 0.0%, Blasts: 6.0%, Myelocytes: 2.0%.

Peripheral Blood Smear and Bone Marrow

Smear showed Leukopenia with neutropenia with ~6% blasts and 2% myelocytes. Blasts showed total replacement of the bone marrow with (~95.5%) which showed the scant amount of basophilic cytoplasm, few showing cytoplasmic vacuoles, round, and convoluted nuclei with condensed nuclear chromatin and indistinct nucleoli. The features were consistent with acute leukemia, morphologically lymphoid.

Flow Cytometric Immunophenotyping

Blasts (~53.0% of acquired events) in moderate positive blast region with low side scatter. cCD79+, CD19+, CD22+, CD24+, CD9+, CD58+, CD24+, CD20+, CD58+, CD10+, CD38+, CD20+, CD200+, CD81+ consistent with precursor B cell Acute Lymphoblastic Leukemia, CALLA positive.

Cytogenetics

Conventional Karyotype

46,XX,der(19)t(1;19)(q23;p13)[11]/46,XX[9]

Cytogenetic analysis of 20 metaphases showed the presence of an abnormal clone characterized by an unbalanced translocation between the long arm of chromosome 1 and the short arm of chromosome 19 with the formation of derivative chromosome 19 in 11 metaphases.

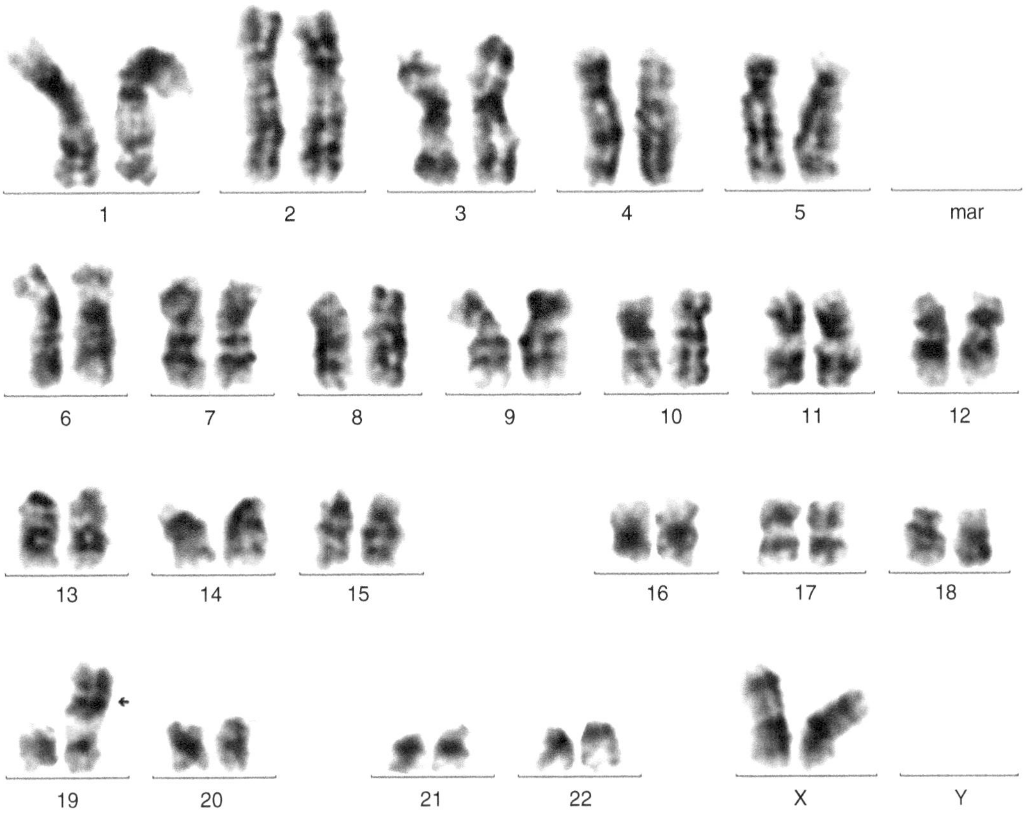

Case Study 9.18 Karyotype showing reciprocal unbalanced translocation(1;19)

Molecular Findings

The Real Time Quantitative PCR was *Positive for t(1;19) (TCF3-PBX1) in ALL panel.*

Discussion

This patient of Precursor B cell ALL presented with the sole abnormality of t(1;19) (q23;p13) with formation of derivative chromosome 19. It is one of the most frequent translocations in ALL both in children and adults accounting for 6%. This translocation can occur as balanced t(1;19) or unbalanced der (19)t(1;19)(q23:p13), the latter being the case in this patient [43]. There is a fusion of transcription factor TCF3 found at 19p13 and PBX1 (Pre-B cell leukemia homeobox 1) found at 1 q23 to form a Chimeric gene whose protein alters cell differentiation arrest. The t(1;19) has intermediate prognosis [44].

9.19 Case Study 9.19: Acute Lymphoblastic Leukemia with Philadelphia Chromosome t(9;22) (q34;q11.2) with i(8)(q10) and der (5) t (5;8)

Brief Clinical History

A 42-year-old male with past history of chronic pancreatitis and diabetes presented with perianal pain and pancytopenia. Investigated for suspected acute leukemia.

CBC: *Hb*: 7.2 g/dl, TLC: 1270 /cu mm and Platelets: 2000 /cu mm.

DLC: Blasts: 13.0%, Neutrophils: 11.0%, Lymphocytes: 70.0%, Eosinophils: 0.0%, Monocytes: 6.0%, Basophils: 0.0%.

Peripheral Smear and Bone Marrow

Smear showed leukopenia with neutropenia along with the presence of ~13% blasts. Blasts were large, having high N:C ratio, open chromatin, and inconspicuous nucleoli. Bone marrow aspirate smears showed replacement by blasts constituting ~77.0% of the total nucleated cells with similar morphological features as in peripheral smear. Peripheral blood and bone marrow findings were consistent with acute leukemia.

Flow Cytometric Immunophenotyping

Blasts (~84% of acquired events); dim to CD45 negative region with low side scatter.

CD19+, CD79a+, CD10+, CD22+, CD34+, CD9+, CD24+, CD58+, CD38+, CD81+, CD200+, CD13+, CD33+, CD66c+.

The immunophenotype was that of Precursor B Acute Lymphoblastic Leukemia with aberrant expression of myeloid markers (CD13 and CD33) and CD66c, which is usually associated with BCR/ABL t(9;22) gene rearrangement.

Cytogenetics

Conventional Karyotype

46,XY,der(5)t(5;8)(q35;q13),t(9;22)(q34;q11.2)[12]/46,XY,i(8)(q10),t(9;22)(q34;q11.2)[2]/46,XY,t(9;22)(q34;q11.2)[4]/46,XY[2]

Cytogenetic analysis showed a neoplastic clone characterized by a balanced reciprocal translocation between long arms of chromosome 9 and 22 leading to the formation of a Philadelphia chromosome (Ph) in 18 of 20 metaphases analyzed. The clone also shows the formation of an isochromosome of the long arm of chromosome 8 (8q) in 2 of the 18 metaphases and another cell clone shows derivative chromosome 5 formed by an unbalanced translocation between the long arm of chromosome 5 and the long arm of chromosome 8 in 12 of the 18 metaphases analyzed. t(9;22) is associated with poor prognosis in precursor B cell acute lymphoblastic leukemia and isochromosome 8q and der(5)t(5;8) is probably a secondary event.

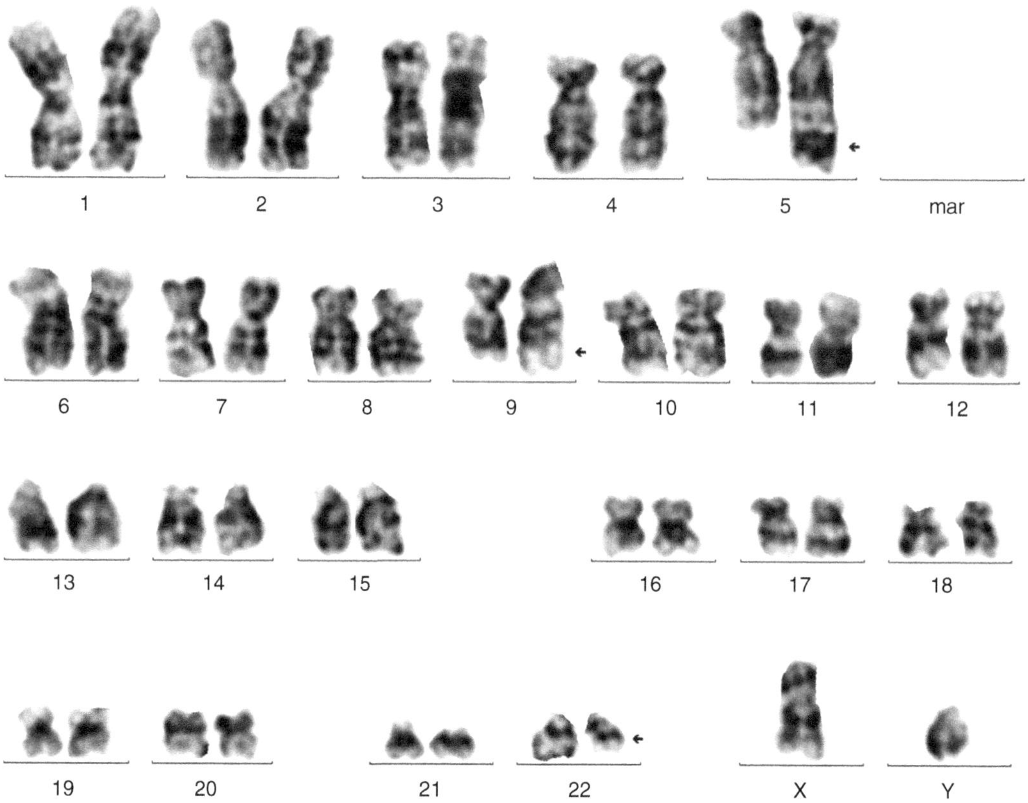

Case Study 9.19 (1) Karyotype showing Philadelphia chromosome formed by t(9;22) along with t(5;8)

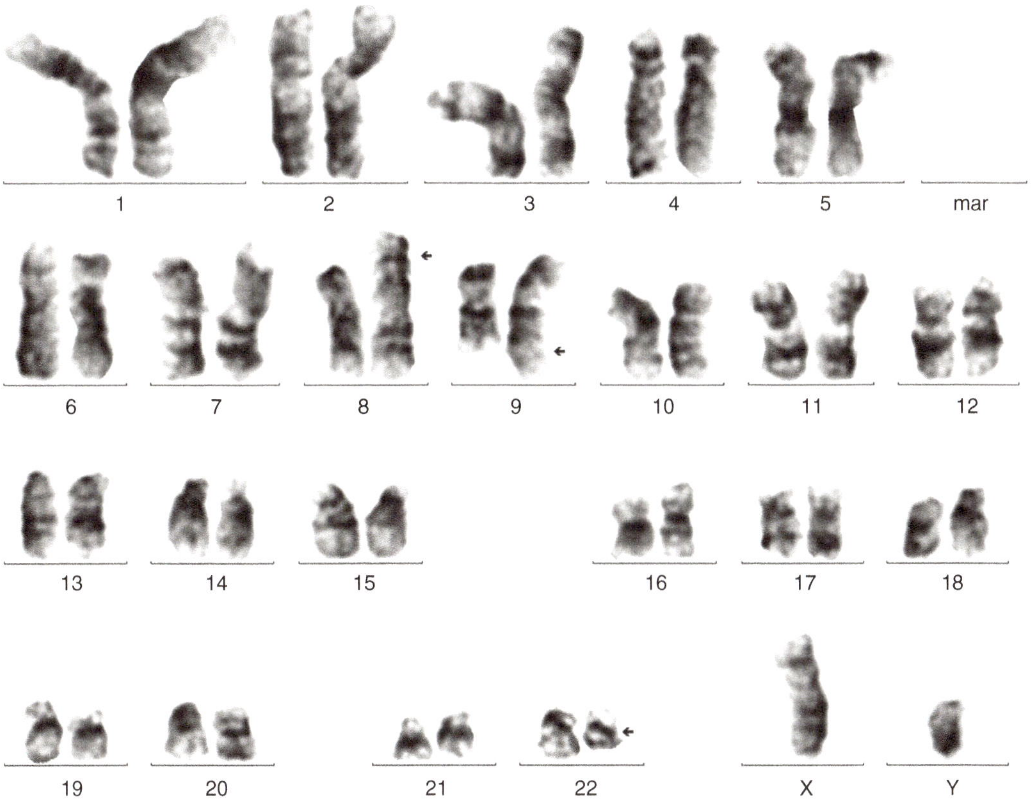

Case Study 9.19 (2) Karyotype showing Philadelphia chromosome formed by t(9;22) along with isochromosme 8

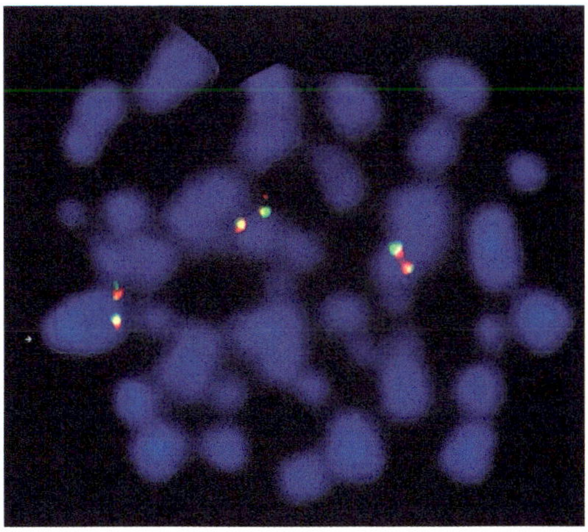

Case Study 9.19 (3) FISH image showing 3F signal pattern using *CMYC* BA probe indicating t(5;8)

Molecular Studies

The Real Time Quantitative PCR on *ALL panel showed BCR-ABL1 Positive (P210).*

Discussion

The Ph1 chromosome is the most frequent cytogenetic abnormality in adult patients seen in 20% to 30% adults and 5% of children. Based on the location of break point within the BCR gene, P190 protein is seen in 2/3 of patients and P210 in one-third. This patient is primarily a case of Philadelphia chromosome positive ALL with secondary abnormalities of i(8) (q10) and a der (5) t(5:8) seen in another clone. Such abnormalities are known in Ph1+ALL. Al-Achkar reported a case of B cell precursor ALL with multiple secondary cytogenetic aberrations using GTG bandings, FISH, and microarray. The findings revealed a complex structural rearrangement, der (2) along with t(9;22) (q34;q11), t(12;14)(q12:p12), and monosomy of chromosome 7 [45]. *Isochromosomes are nonrandom chromosomal anomalies.* Very few cases have been described with i(8) (q10) and are mostly a part of complex karyotypes suggesting that they are secondary in nature associated with clonal evolution [46]. t (9;22) is associated with poor prognosis in precursor B ceil acute lymphoblastic leukemia [47].

9.20 Case Study 9.20: Acute Lymphoblastic Leukemia/ Non-Hodgkin's Lymphoma with a Variant t(8;22), MYC/ IGL

Brief Clinical History

A 44-year-old male presented with complaints of fever, generalized weakness, headache for the last 1 month. There was abdominal, mediastinal and right supraclavicular lymphadenopathy with FNAC favoring NHL. Bone marrow was done for evaluation.

CBC: *Hb*: 7.0 g/dl, TLC: 7470 /cu mm and Platelets: 1,30,000 /cu mm.

DLC: Neutrophils: 35.5%, Myelocytes and metamyelocytes: 10.5%, Blasts: 16.5%, Lymphocytes: 30.0%.

Peripheral Smear and Bone Marrow

Smear showed left shift with the presence of ~16.5% blasts. The blasts were intermediate in size with high N:C ratio, immature nuclear chromatin with inconspicuous nucleoli, and scant basophilic cytoplasm. Bone marrow showed similar cells replacing the bone marrow (92%).

Peripheral blood and bone marrow aspirate features were suggestive of Acute leukemia, morphologically lymphoid.

Flow Cytometric Immunophenotyping

Blasts (~88.0% of acquired events) in moderate positive (CD45/SSc) blast region with low side scatter. cCD79+, CD19+, CD22+, CD24 +, CD9+, CD58+, CD20+, CD38+, CD10+, CD81+, HLA-DR+.

Picture was consistent with B cell Acute lymphoblastic Leukemia, CALLA positive.

Cytogenetics

Cytogenetic Karyotype

46,XY,t(8;22)(q24;q11)[7]/46,XY[13]

Cytogenetic analysis of 20 metaphases showed the presence of an abnormal clone characterized by a balanced translocation between the long arm of chromosome 8 and the long arm of chromosome 22 t(8;22) (q24;q11)in 7 metaphases analyzed (MYC/IGL).

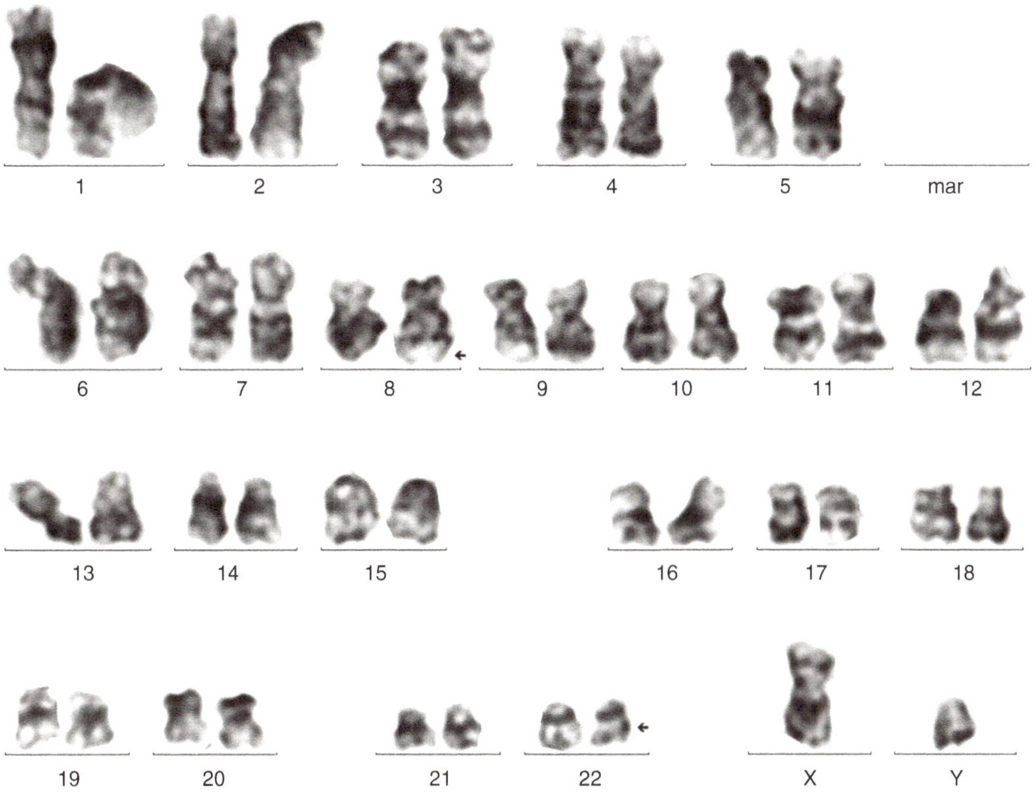

Case Study 9.20 (1) Karyotype showing t(8;22)

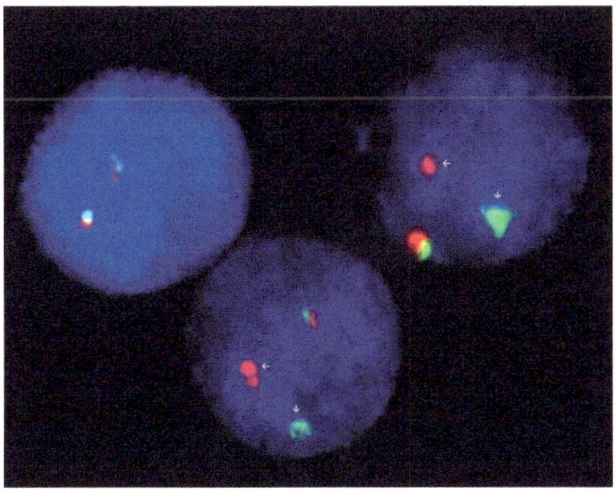

Case Study 9.20 (2) FISH image showing 1F1G1R signal pattern using *CMYC* BA probe indicating *CMYC* rearrangement

Discussion

This patient presenting with generalized lymphadenopathy showed morphologically a picture of a Lymphoproliferative disease Leukemia/ Lymphoma with B cell immunophenotype. On karyotyping, there was t(8:22)(q24:q11). This profile can be seen both in B cell acute lymphoblastic leukemia as well as non-Hodgkin's Lymphoma (NHL) particularly in Burkitt Lymphoma and DLBCL [48]. t(8;22) is seen in nearly 10% of these cases and is a variant of t(8;14) just like t(2;8). In a case study Abe et al. [49] a variant t (8;22) was seen. The survival of patients with t(8;22)-type BL and ALL-L3 is short and comparable to the more common t(8;14). Since the supraclavicular lymph node biopsy showed DLBCL, this patient was treated on the lymphoma protocol.

9.21 Case Study 9.21: Relapsed Philadelphia Chromosome Positive Acute Lymphoblastic Leukemia with Composite Karyotype and del(7p)

Brief Clinical History
A 20 year old male diagnosed as Ph Chromosome positive ALL 10 years back treated on Hyper CVAD2, cranial RT, and Dasatinib. Has now presented with relapse with 84% blasts in PBS.

CBC: *Hb*: 11.7 g/dl; TLC: 53,060 /cu mm and Platelets: 68,000 /cu mm.

DLC: Myelocytes: 1.0%, Blasts: 90.0%; Neutrophils: 3.0%, Lymphocytes: 6.0%, Eosinophils: 0.0%, Monocytes: 0.0%, Basophils: 0.0%.

Peripheral Blood and Bone Marrow
Smear showed Leukocytosis with ~90% blasts which were large with high N:C ratio, fine nuclear chromatin with inconspicuous nucleoli, and scant to moderate agranular cytoplasm. Bone marrow showed similar blasts (92%) replacing the bone marrow.

Flow Cytometric Immunophenotype
Blasts (~88% of acquired events); extending from CD45 negative region to moderate CD45 with low side scatter. CD19+, CD79a+, CD10+, CD22+, CD34+, CD9+, CD24+, CD20+, CD58+, CD38+, CD13+, CD33+. The immunophenotype was that of Precursor B Acute Lymphoblastic Leukemia with aberrant expression of myeloid markers (CD13 and CD33) and CD66c, which is usually associated with BCR/ABL t(9;22) gene rearrangement.

Cytogenetics
Conventional Karyotype
51~52,XY,+X,+2,+4,+6,del(7)(p15),t(9;22)(q34;q11.2),+der(22)t(9;22)(q34;q11.2),+14[cp20]

Cytogenetic analysis of 20 metaphases showed a neoplastic clone characterized by a high hyperdiploidy karyotype (chromosome number 51~52) along with the presence of structural and numerical abnormalities. Structural abnormalities were characterized by a deletion of the short arm of chromosome 7, i.e., del(7p) and a balanced reciprocal translocation between the long arms of chromosome 9 and chromosome 22 (Philadelphia chromosome). This clone also showed a gain of an extra Philadelphia chromosome 22, i.e., der(22)t(9;22). It was a composite karyotype showing gains of chromosomes 2, 4, 6, and 14 as trisomies along with gain of sex chromosome X. The t(9;22), extra Philadelphia chromosome and del(7p) is present in all the 20 metaphases analyzed and overrides the good prognostic effect of high hyperdiploidy, conferring a poor prognosis.

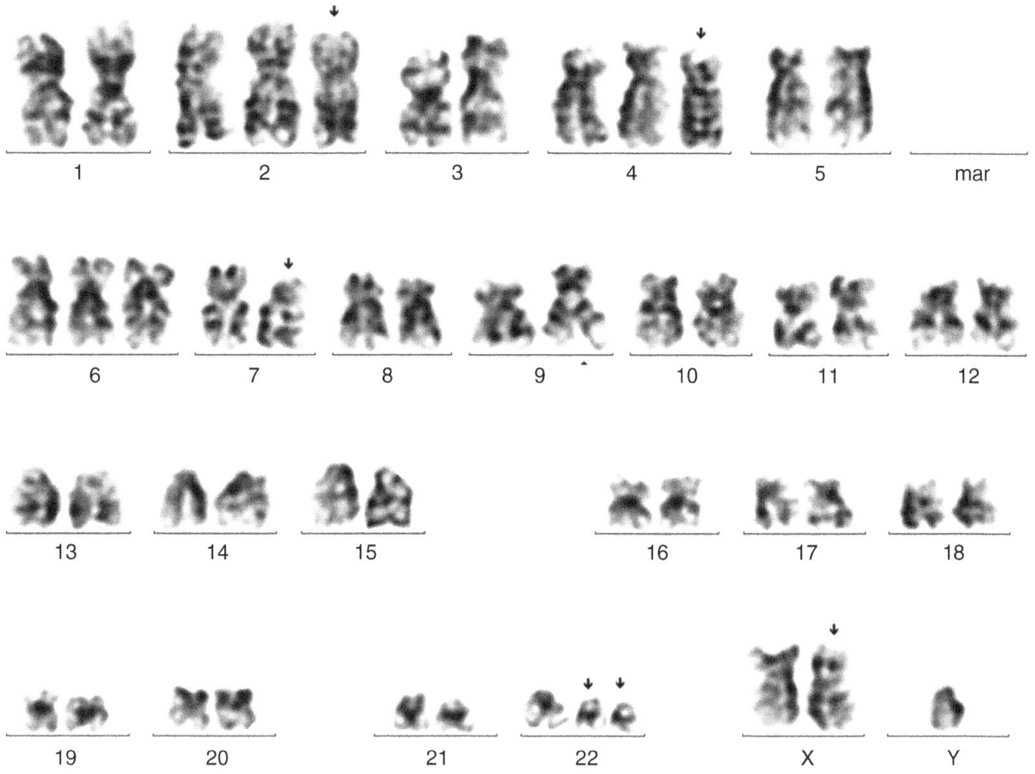

Case Study 9.21 Hyperdiploid Karyotype showing Philadelphia chromosome along with gain of chromosome X,2,4,6,deletion of short arm of chromosome 7, i.e., del(7)(p15), +14, an extra der(22)t(9;22)(q34;q11.2)

Molecular Studies

The Real Time Quantitative PCR was *Positive for BCR-ABL1 (P190, Minor transcript).*

Discussion

This patient of treated Ph 1 chromosome positive ALL presented in relapse with multiple cytogenetic abnormalities comprising composite karyotype with double Ph chromosome and del (7p) together with hyperdiploidy. Additional chromosomal abnormalities can be seen both at presentation as well as relapse [50]. In a study of 30 patients [51], such abnormalities were found in addition to Ph 1 chromosome in 60%. Four had double Ph chromosome (22%), 3 had monosomy 7 or 7 q deletion (17%) (as seen in this patient), one had trisomy 8, and one trisomy 19. Only one was found to have hyperdiploidy (>50 chromosomes). In another study by Bacher et al. [52], 154 patients were studied who had Ph+ALL at diagnosis or relapse and 174 patients with different phases of CML. Ph1+All at diagnosis showed a high frequency of numerical and structural abnormalities. At relapse, there was a higher number of abnormalities but the pattern was not very different from that at diagnosis.

9.22 Case Study 9.22: Acute Lymphoblastic Leukemia with Complex Karyotype

Brief Clinical History

A 15-year-old female from Turkmenistan presented with a history of pain in hands and feet since past 2 months, pallor since past 2 weeks. On examination hepatosplenomegaly was present. Peripheral smear showed ~43% blasts. Bone marrow was done for evaluation.

CBC: *Hb*: 8.2 g/dl, TLC: 10,160 /cu mm and Platelets: 144,000 /cu mm.

DLC: Neutrophils: 35.0%, Lymphocytes: 23.0%, Eosinophils: 4.0%, Monocytes: 2.0%, Basophils: 0.0%, Blasts: 36.0%.

Peripheral Smear and Bone Marrow

Smear showed mild leukocytosis with ~36% blasts small to moderate in size, with scant amount of basophilic cytoplasm, immature nuclei. Similar blasts infiltrated the bone marrow (85%). The features were consistent with acute leukemia, morphologically lymphoid.

Flow Cytometric Immunophenotype

Blasts ~85% of acquired events expressing negative to moderate CD45 and low SSC. Two distinct populations of lymphoblasts are seen on CD34 versus CD19 scatter plot. One population was CD34 negative (48% of total blasts). CD19+, cCD79a+, CD10+, CD20+, CD24+, CD22+, CD38+, CD58+, CD9+, CD200+, CD81+, CD13+33+. A second population was CD34+ (36% of total blasts) CD19+, cCD79a+, CD10+, CD24+, CD22+, CD38+, CD58+, CD9+, CD200+, CD81+.

The features were consistent with Precursor B cell acute lymphoblastic leukemia, CALLA positive.

Cytogenetics

Conventional Karyotype

45~47,del(1)(q32),del(2)(p13),del(6)(q13),del(9)(q13),add(9)(p24),del(10) (p11.2), −10, +19 [cp13]/46, XX[7]

Cytogenetic analysis of 20 metaphases showed a neoplastic clone in 13 metaphases characterized by composite (complex karyotype) showing numerical (45–47 modal number) and several structural abnormalities. Structural abnormalities showed deletions of long arms of chromosomes 1(1q), 6(6q) and 9(9q) and deletions of short arms of chromosomes 2(2p) and 10(10p). Also, there was addition of unknown material on short arm of chromosome 9. Numerical abnormalities were Trisomy of chromosome 19 and Monosomy of chromosome 10.

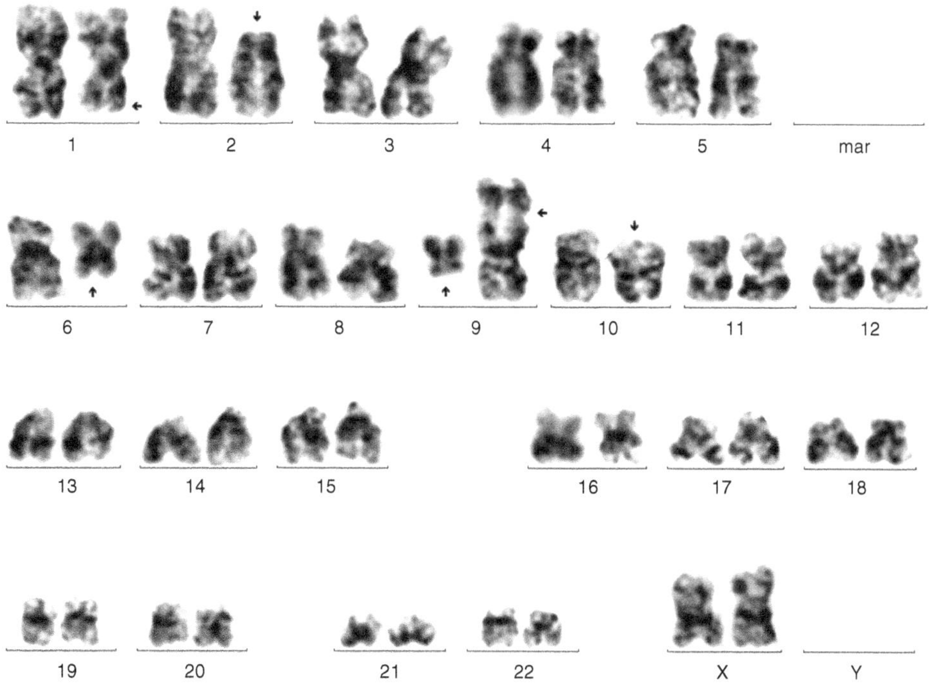

Case Study 9.22 (1) Complex Karyotype showing deletion of long arm of 1, short arm of chromosome 2, long arm of chromosome 6, addition of short arm of chromosome 9 and deletion of short arm of chromosome 10

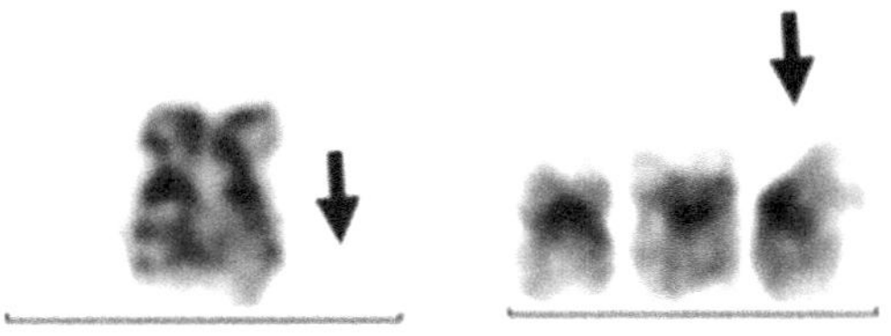

Case Study 9.22 (2) Partial karyotype showing monosomy 10 and trisomy 19

Molecular Studies
The Real Time Quantitative PCR was *Negative for ALL Leukemia Panel.*

Discussion
This patient of ALL showed complex karyotype with multiple numerical and structural abnormalities. In a study of 411 patients in a multicentre study by the Groupe Francais De Cytogenetique Hematologique [53], clonal abnormality was seen in 71%. In numerical abnormalities, pseudodiploidy was seen in 28.2%, hyperdiploidy in 26.8%, hypodiploidy in 4.9%. Fifty percent patients with hyperdiploidy also showed structural abnormality such as t(9;9),t(7;9), t(7;12), t(4;12), and t(1;17).

In a cytogenetic analysis of 79 childhood acute lymphoblastic leukemia, Jarosova et al. [54] showed chromosomal abnormalities in 96%. Complex karyotype was identified in 26%, 11 of these had recurrent chromosomal abnormalities, such as t(12;21) in 7 cases, t(9;22) in 2 and 1 case each with t(2,1,19) and 11q 23. In 10 patients, there were miscellaneous abnormalities such as five patients showed hyperdiploidy (45–57 chromosome), 3 complex karyotype with del(9p), one with (2;4;12;13) and t(11;20), and one with dic (12;21).

9.23 Case Study 9.23: Acute Lymphoblastic Leukemia with Eosinophilia

Brief Clinical History
A 15-year-old female presented with complaints of fever since 3 weeks and chills, fatigue, and lower backache since 1 week. Admitted for further evaluation and management.

CBC: Hb: 10.2 g/dl, TLC: 44,490 /cu mm and Platelets: 25,400 /cu mm.

Peripheral Blood Smear and Bone Marrow
Smear showed leukocytosis with marked eosinophilia, mostly mature forms (AEC- 30,698 cells/cu mm). Eosinophilic precursors were also seen. Occasional blast was present.

Bone marrow showed 52% blasts with high nuclear cytoplasmic ratio, regular nuclear contour, fine chromatin, inconspicuous to prominent nucleoli surrounded by scant cytoplasm. Normal hematopoietic precursors were markedly reduced. Picture was suggestive of Acute Leukemia, likely acute lymphoblastic leukemia with eosinophilia.

Immunophenotyping
Blasts (~15% of acquired events); CD45 dim to moderate. Granulocytic cells including eosinophils, lymphocytes & monocytes were also present. Positive markers: CD34+, CD19+, CD10+, cCD79a+, CD20+, CD22+, CD24+, CD9+, CD66c+, CD38+, CD58+, CD81+, CD200+, CD13/33+, HLA-DR+.

Immunophenotype showed the presence of Precursor B-lineage blasts. Together with bone marrow morphology, findings were consistent with Precursor B cell Acute Lymphoblastic Leukemia, CALLA positive.

Cytogenetics
Conventional Karyotype
46,XX,t(5;14)(q31;q32)[6]/46,XX[14]

Cytogenetic analysis of 20 metaphases showed the presence of an abnormal cell clone characterized by balanced reciprocal translocation between the long arm of chromosome 5 (5q) and the long arm of chromosome 14 (14q) in six metaphases analyzed.

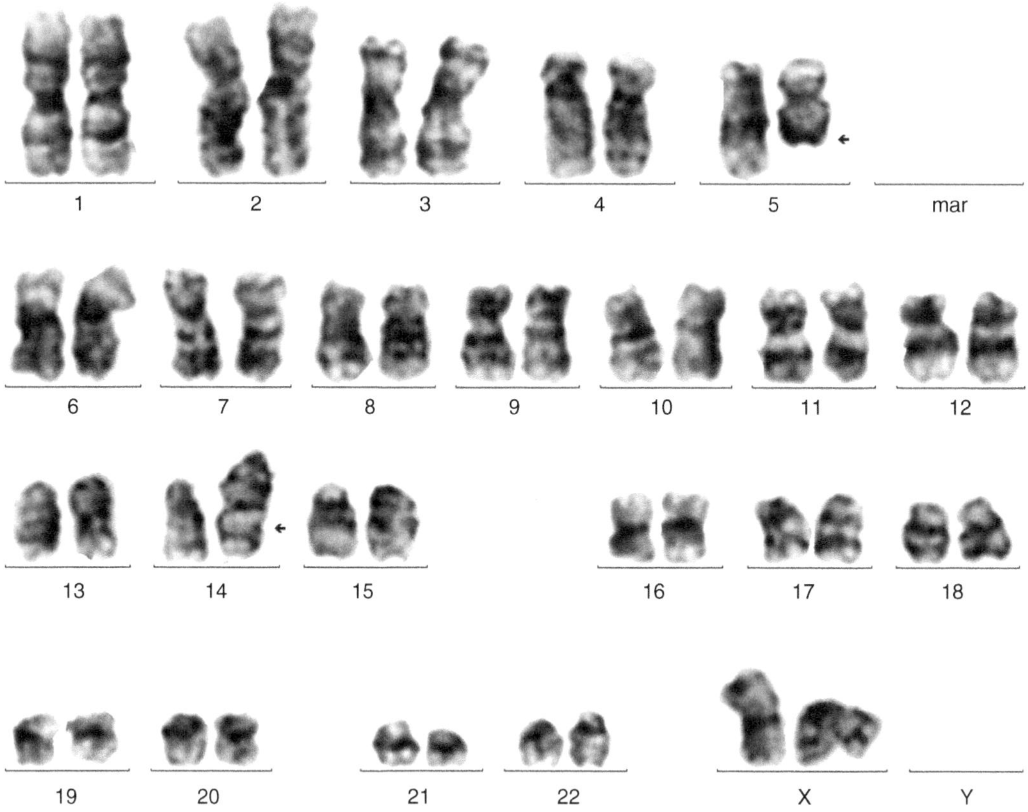

Case Study 9.23 Karyotype showing t(5;14)

Molecular Findings

The Real time Quantitative PCR showed *ALL comprehensive panel negative*.

Discussion

This patient with immunophenotype of precursor B-ALL, CALLA positive showed on karyotyping t(5;14), a rare translocation for which clinical, biological data are extremely scarce in literature. It is an exceptional cause for eosinophilia in ALL as seen in this patient in whom hypereosinophilia was seen both in the peripheral blood and bone marrow. The IHC on bone biopsy was non-contributory [55]. t(5;14)(q31;q32)c IGH-IL3 shows IGH enhancer on 14q32 juxtaposed to the IL3 gene on 5q31 leading to interleukin 3 overproduction and release of mature eosinophils in the blood [56]. It has been classified as a distinct entity in WHO classification. It has an intermediate prognosis and clinical features related to eosinophilia [57].

Annexure

- Case Study 9.1 Acute Lymphoblastic Leukemia with High Hyperdiploidy
- Case Study 9.2 Acute Lymphoblastic Leukemia with complex karyotype with t(12;21) (ETV6/RUNX1)
- Case Study 9.3 Philadelphia Chromosome Positive Acute Lymphoblastic Leukemia with Hypodiploidy, Monosomy 7, and der 6 t(2;6)
- Case Study 9.4 Acute Lymphoblastic Leukemia with normal Karyotype with t(12;21) on FISH
- Case Study 9.5 Acute Lymphoblastic Leukemia with t (1;19) (q23;p13.3) and i(1) (q10)
- Case Study 9.6 Acute Lymphoblastic Leukemia with del(9p) and dup (1q)
- Case Study 9.7 Relapsed Precursor B Acute Lymphoblastic Leukemia with t(12;17) (p13;q11.2)

- Case Study 9.8 Philadelphia Chromosome positive Acute Lymphoblastic Leukemia with High Hypodiploidy with Monosomy 7
- Case Study 9.9 Therapy Related Acute Lymphoblastic Leukemia with Complex Karyotype
- Case Study 9.10 Early T Cell Precursor Acute Lymphoblastic Leukemia (ETP-ALL) with Normal Karyotype and FLT3-ITD Mutation Positive
- Case Study 9.11 Acute Lymphoblastic Leukemia with del (9p)
- Case Study 9.12 Acute Lymphoblastic Leukemia with High Hyperdiploidy with Complex Karyotype and TP53 Gene Deletion
- Case Study 9.13 Acute Lymphoblastic Leukemia with Complex Karyotype with t(4;11) (q21;q23)
- Case Study 9.14 Acute Lymphoblastic Leukemia with both Hyperdiploid and Hypodiploid Clones
- Case Study 9.15 Philadelphia Chromosome positive acute Lymphoblastic Leukemia with additional Cytogenetic Abnormalities
- Case Study 9.16 Acute Lymphoblastic Leukemia with masked t(9;22)
- Case Study 9.17 Philadelphia Chromosome Positive Acute Lymphoblastic Leukemia
- Case Study 9.18 Acute Lymphoblastic Leukemia with der 19 t(1;19) (q23;p13)
- Case Study 9.19 Acute Lymphoblastic Leukemia with Philadelphia Chromosome t(9;22) (q34;q11.2) with i(8)(q10)and der (5) t (5;8)
- Case Study 9.20 Acute Lymphoblastic Leukemia / non Hodgkin's Lymphoma with a Variant t(8;22), MYC/IGL
- Case Study 9.21 Relapsed Philadelphia Chromosome Positive Acute Lymphoblastic Leukemia with Composite Karyotype and del(7p)
- Case Study 9.22 Acute Lymphoblastic Leukemia with Complex Karyotype
- Case Study 9.23 Acute Lymphoblastic Leukemia with Eosinophilia

References

1. Gibbons B. High hyperdiploid acute lymphoblastic leukemia. Atlas Genet Cytogenet Oncol Haematol. 1999;3(3):145–6.
2. Paulsson K. High hyerdiploid childhood acute lymphoblastic leukemia: chromosomal gains as the main driver event. Mol Cell Oncol. 2016;3(1):e1064555.
3. Raynaud, et al. Cytogenetic abnormalities associated with the t(12;21): a collaborative study of 169 children with t(12;21)-positive acute lymphoblastic leukemia. Leukemia. 1999;13:1325–30.
4. Romana SP, Poirel H, Leconiat M, et al. High frequency of t(12;21) in childhood B-lineage acute lymphoblastic leukemia. Blood. 1995;86(11):4263–9.
5. Huret JL, Bernheim A. t(12;21)(p13;q22), Atlas Genet Cytogenet Oncol Haematol. 1997;1(1).
6. Yadav A, Bhargava M. High hypodiploidy karyotype showing t(9;22) (q34,q11.2) along with monosomy 7 in adult ALL. ASH Image Bank Published date 08/07/2018.
7. Heerema NA, Harbott J, Galimberti S, et al. Secondary cytogenetic aberrations in childhood, Philadelphia chromosome positive acute lymphoblastic leukemia are non random and may be associated with outcome. Leukemia. 2004;18(4):693–702.
8. Achkar W, Wafa A, Othman M. An adult B cell precursor acute lymphoblastic leukemia with multiple secondary cytogenetic aberrations. Mol Cytogenet Published online. 2014. https://doi.org/10.1186/s13039-014-0060-0
9. Raynaud SD, Cave H, Baens M, et al. The 12;21 translocation involving TEL and deletion of the other TEL allele: two frequently associated alterations found in childhood acute lymphoblastic leukemia. Blood. 1996 May;87(7):2891–9.
10. Raynaud SD, Dastugue N, Zoccola D, et al. Cytogenetic abnormalities associated with the t(12;21): a collaborative study of 169 children with t(12;21)-positive acute lymphoblastic leukemia. Leukemia. 1999;13:1325–30.
11. Wafa A, Assad M, Liehr T. Childhood pre-B cell acute lymphoblastic leukemia with translocation t(1;19)(q23;p13.3)and two additional chromosome aberrations involving chromosomes 1,6,13: a case report. J Med Case Rep. 2017;11:94.
12. Borowitz MJ. B-lymphoblastic leukaemia/lymphoma with recurrent genetic abnormalities: WHO classification of Tumours of Haematopoietic and lymphoid tissues (revised 4th ed). IARC: Lyon; 2017.
13. Uckun FM, Sensel MG, Sather HN. Clinical significance of translocation t(1;19) in childhood acute lymphoblastic leukemia in the context of contemporary therapies: a report from the Children's Cancer group. 1998;16(2)527–35.

14. Mohamed A. N: del (9p) in acute lymphoblastic leukemia. Atlas Genet Cytogenet Oncol Haematol. 2017;21(7):252–5.
15. Iacobucci I, Papayannidis C, Lonetti A. Cytogenetic and molecular predictors of outcome in acute lymphocytic leukemia: recent developments. Current Haematol Malig Rep. 2012;7:133–43.
16. Mohamed AN. dup(1q) in ALL. Atlas Genet cytogenet Oncol Hemato. 2018;22(6):255–9.
17. Yamamoto K, Kawamoto S, Mizutani Y, et al. Mixed phenotype acute leukemia with t(12;17) (p13;q21) TAF15-ZNF384 and other chromosome abnormalities. Cytogenet Genome Res. 2016;149(3):165–70.
18. Kim J, Kim HS, Choi JR. t(12;17) (p13;q12) /TAF15-ZNF 384 in acute lymphoblastic leukemia. Ann Lab Med. 2016;36(4):396–8.
19. Safavi S, et al. Genetic and epigenetic characterisation of hypodiploid acute lymphoblastic leukemia. Oncotarget. 2015;6:42793–802.
20. Aldoss I, Stiller T, Cao TM, et al. The impact of additional cytogenetic abnormalities in adult with Philadelphia Chromosome-positive acute lymphoblastic leukemia undergoing allergenic hematopoietic cell transplantation. Brol Blood Marrow Transplant. 2015;21(7):1326–1329.
21. Achkar W, Wafa A, Othman M. An adult B cell precursor acute lymphoblastic leukemia with multiple secondary cytogenetic aberrations. Mol Cytogenet. 2014;7:60. https://doi.org/10.1186/s13039-014-0060-0.
22. Hoelzer D, Gokbuget N. New approaches to acute lymphoblastic leukemia in adults: where do we go? Seminar Oncol. 2000;27:540–59.
23. Aldoss I, Stiller T, Tsai NC, et al. Therapy- related acute lymphoblastic leukemia has distinct clinical and cytogenetic features compared to de-novo acute lymphoblastic leukemia but outcomes are comparable in transplanted patients. Haematologica. 2018;103(10):1662–8.
24. Ribera JM. Therapy related acute lymphoblastic leukemia haematologica. 2018;103(10):1581–3.
25. Neumann M, Heesch S, Gokbuget N, et al. Clinical and molecular risk subgroup in adult Characterisation of early T-cell precursor leukemia: a high of FLT3 mutations. Blood Cancer J. 2012;2:e55.
26. Gilliland DG, Groffin JD. The roles of FLT3 in hematopoiesis and leukemia. Blood. 2002;100:1532–42.
27. Heerema NA, Sather HN, Sensel MG, et al. Association of chromosome 9p abnormalities with adverse risk in childhood acute lymphoblastic leukemia: a report from the Children's Cancer Group. Blood. 1999;94(5):1537–44.
28. Paulson K, Lilljebjor H, Biloglav A, et al. The genomic landscape of high hyerdiploid childhood acute lymphoblastic leukemia. Nat Genet. 2015;47:672–6.
29. Paulson K. High hyperdiploid childhood acute lymphoblastic leukemia: chromosomal gains as the main driver event. Mol Cell Oncol. 2016;3(1):e1064555.
30. Hof J, Groeneveld-Krentz S, Schewick C, et al. Mutations and deletions of the TP53 gene predict non response to treatment and poor outcome in first relapse of childhood acute lymphoblastic leukemia. J Clin Oncol. 2011; 29(23):3185–93.
31. Sanjuan PA, Bueno C, Prieto C, et al. Revisiting the biology of infant t(4;11)/MLL-AF4^{+}B-cell acute lymphoblastic leukemia. Blood. 2015;126(25):2676–85.
32. Greaves M. When one mutation is all it takes. Cancer Cell. 2015;27(4):433–4.
33. Bardini M, Woll PS, Corral L, et al. Clonal variegation and dynamic competition of leukemia initiating cells in infant acute lymphoblastic leukemia with MLL rearrangements. Leukemia. 2015;29(1).
34. Raimondi SC. 11q23 rearrangements in childhood acute lymphoblastic leukemia. Atlas Genet Cytogenet Oncol Haematol. 2004;8(2):97–102.
35. Safavi S, Paulson K. Near-haploid and low-hypodiploid acute lymphoblastic leukemia: two distinct subtypes with consistently poor prognosis. Blood. 2017;129(4):420–3.
36. Wetzler M, Dodge R, Morozek K, et al. Additional cytogenetic abnormalities in adults with Philadelphia chromosome-positive acute lymphoblastic leukaemia: a study of the Cancer and Leukaemia. Group B Br J Haematol. 2004;124(3):275–88.
37. Swerdlow SH, Campo E, Harris NL, Jaffe ES, Pileri SA, Stein H, Thiele J (Eds). WHO classification of tumours of haematopoietic and lymphoid tissues (revised 4th ed). IARC: Lyon; 2017.
38. Kawahara Y, Iioka F, Fukutsuka K, et al. Masked Philadelphia chromosome in acute lymphoblastic leukemia. Tenri Med Bull. 2015;18(1):36–9.
39. Huret JL. t(9,22) (q34;q11) in ALL. Atlas Gunet Cytogenet Oncol Haematol. 1997;1(1):26–8.
40. Yilmaz M, Kantarjian H, et al. Philadelphia chromosome positive acute lymphoblastic leukemia in adults: current treatments and future perspectives. Clin Adv Hematol Oncol. 2018;16(3).
41. Ravandi F, Kebriael P. Phildelphia chromosome positive acute lymphoblastic leukemia. Hematol Oncol Clin North Am. 2009;23(5).
42. Kantarjian HM, Talpraz M, Dhingra K. Significance of the P210 versus P 190. Molecular abnormalities in adults with Philadelphia chromosome positive acute leukemia. Blood. 1991;78:2411–8.
43. Barber KE, Harrison CJ, Broadfield ZJ, et al. Molecular cytogenetic characterization of TCF3 (E2A/19P13.3) rearrangements B cell precursor acute lymphoblastic leukemia. Genes Chromosomes Cancer. 2007;46(5):478–86.
44. Tirado CA, Shabsovich D, Yeh L, et al. A t(1:29) translocation involving TCF3-PBx1 fusion within the context of a hyperdiploid karyotype in adult B-ALL: a case report and review of literature. Biomark Res. https://doi.org/10.1186/s40364-015-0029-0
45. AL-Achkar W, Wafa A, Liehr T. An adult B-cell precursor acute lymphoblastic leukemia with multiple secondary cytogenetic aberrations. Mol Cytogenet. 2014;7:60.
46. Zamecnikova A. i(8)(q10) on ALL. Atlas Genet Cytogenet Oncol Haematol. 2018;22(8):336–40.

47. Ravandi F, Kebriaei P. Philadelphia chromosome positive acute lymphoblastic leukemia. Hematol Oncol Clin North Am. 2009;23(5):1043–63.
48. Van Der Berg E. Marian Stevens-Kroef: t(8:14) (q23:q32) / t(2;8) (p12;q24)/t(8;22) q(24;q11). Atlas Genet Cytogenes Oncol Haematol. 2017;21(2):56–9.
49. Abe R, Tebbi CK, Yasuda H, et al. North American Burkitt type ALL with a variant translocation t(8;22). Cancer Genet. 1982;7(3):185–95.
50. Ko Bor S, Tang JL, Lee FY. Additional chromosomal abnormalities and variability of BCR breakpoints in Philadelphia chromosome / BCR-ABL positive acute lymphoblastic leukemia in Taiwan. Am J Hemat. 2002;71:291–9.
51. Walters R, Kantarjian HM, Keating MJ. The importance of cytogenetic studies in adult acute lymphoblastic leukemia. Am J Med. 1990;89:579–87.
52. Bacher U, Haferlach T, Hiddemann W. Additional clonal abnormalities in Ph+ALL and CML demonstrate a different cytogenetic pattern at diagnosis and fellow at different pathways at progression. Cancer Genet. 2005;157(1):53–61.
53. Collaborative study of karyotype in childhood acute lymphoblastic leukemia. Groupe Francais de cytogenetique Hematologique. Leukemia. 1993;7(1):10–9.
54. Jarosova M, et al. Complex karyotypes in childhood acute lymphoblastic leukemia: cytogenetic and molecular cytogenetic study of 21 cases. Cancer Genet Cytogenetics. 2003;145(2):161–8. https://doi.org/10.1016/S0165-4608(03)00099-2.
55. Toboso DG, Campos CB. Peripheral eosinophilia as the first manifestation of B-cell acute lymphoblastic leukemia with t(5;14) (q31;q32). Blood. 2017;130:380. https://doi.org/10.1182/blood-2016-12-754812.
56. Grimaldi JC, Meeker TC. The t(5;14) chromosomal translocation in a case of acute lymphocytic leukemia joins the interleukin- 3 gene to the immunoglobulin heavy chain gene. Blood. 1989;73:2081–5. https://doi.org/10.1182/blood.V73.8.2081.2081.
57. Arber DA, Orazi A, Hasserjian R, et al. The 2016 revision to the World Health Organisation classification of myeloid neoplasms and acute leukemia. Blood. 2016;127(20):2391–405.

Chronic Lymphocytic Leukemia and Non-Hodgkin's Lymphomas

10

10.1 Case Study 10.1: Chronic Lymphocytic Leukemia with ATM Gene (11q22) Gene Deletion and Trisomy 12

Brief Clinical History
A 70-year-old male diagnosed as CLL in January 2019 (FISH 17p-negative) has now presented (April 2020) with fever, low hemoglobin, and rising TLC. Investigated further.

CBC : *Hb*: 8.10 g/dl, TLC: 96,660/cu mm and Platelets: 1,25,000/cu mm.

DLC: Neutrophils: 3.60%, Lymphocyte: 89.50%, Monocytes: 6.40%, Eosinophil: 0.40%, Basophil: 0.10%.

Peripheral Smear and Bone Marrow
Peripheral blood smear showed hyperleukocytosis, lymphocytosis with absolute lymphocytosis (89,500/mm^3).

Bone marrow not done.
Immunophenotyping not done.

FISH-CLL/Lymphoma Panel on Peripheral Blood
FISH results showed ATM (11q22) gene deletion in 92% cells and Trisomy 12 in 90% in 200 interphase cells analyzed. There was no evidence of TP53 gene deletion, RB1 gene deletion, or IgH gene rearrangement.

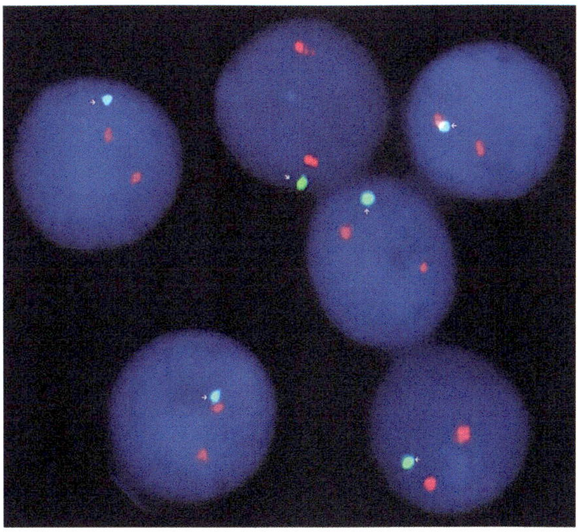

Case Study 10.1 (1) Interphase nuclei showing 2R1G signal pattern using *TP53/ATM* dual-color probe indicating *ATM* gene deletion

M. Bhargava, *Hematologic Malignancies*, https://doi.org/10.1007/978-981-33-4799-1_10

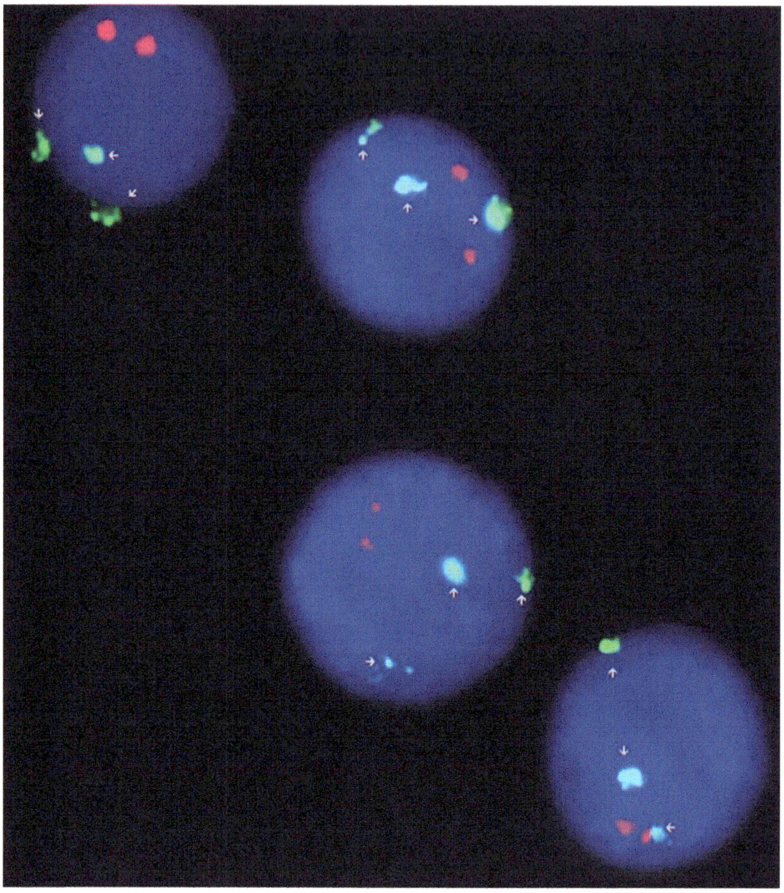

Case Study 10.1 (2) Interphase nuclei showing 2R3G signal pattern using DLEU/LAMP1/CEN 12 probe indicating trisomy 12

Discussion

Trisomy 12 is the third most frequent chromosomal aberration seen in CLL (10–20% cases). As a sole abnormality, it is seen in 40–60% cases but can be associated with other alterations such as +18, +19, or recurrent CLL deletions (e.g., 14q, 13q, 11q, or 17p) and IGH translocations. It is considered as a driver mutation occurring early in CLL and facilitates the appearance of secondary mutations in genes such as NOTCH1, TP53, and FBX7. Association of 11q deletion with other chromosomal abnormalities (as in this case) shows increased copy number indicating genomic instability. The presence of 11q implies clinically progressive disease with shorter TTFT, shorter remission duration, and shorter OS [1].

This patient showed coexistence of two cytogenetic abnormalities of ATM (11q22) gene deletion and trisomy 12 in 90% of the cells analyzed. Deletion (11q) can be found in 25% of patients with advanced-stage disease and 10% of patients with early stage. They typically show bulky lymphadenopathy, rapid progression, and reduced survival [2]. There is a possibility that this patient had trisomy 12 as a primary abnormality and ATM (11q22) gene deletion has occurred with the progression of the disease. Trisomy 12 by itself has an intermediate prognosis. ATM (11q22) gene deletion, however, is associated with poor prognosis. In a study of 238 B-CLL patients by FISH, Kadam Amare [3] reported genetic alterations such as 13q deletion (63%) trisomy 12 (28%), 11q deletion (18%) 6q deletion (11%) with a comparatively higher frequency of TP53 deletion (22%). del(13q) was the most frequent sole abnormality. In a group with coexistence of ≥2 aberrations13 q was a major clone indicating del(13q) as a primary event followed by 11q deletion, TP53 deletion, trisomy 12, and 6q deletion. Trisomy 12 group with coexistence of more than ≥2 aberrations was associated with poor prognosis such as hyperleukocytosis (as in this case) advanced stage and lymphadenopathy.

10.2 Case Study 10.2: Chronic Lymphocytic Leukemia with Isolated Trisomy 12

Brief Clinical History

A 66-year-old male presented with marked lymphocytosis and cervical lymphadenopathy and moderate splenomegaly, diagnosed outside as CLL.

CBC: *Hb*: 7.8 g/dl; TLC: 3,71,270/cu mm and Platelets: 1,60,000/cu mm.

DLC: Neutrophils: 1.2%; Lymphocytes: 90.8%; Eosinophils: 0.0%; Monocytes: 5.0%; Basophil: 0.0%.

Peripheral Smear and Bone Marrow

Smear showed Hyperleukocytosis with absolute lymphocytosis and smudge cells.

Flow Cytometric-CLL/Lymphoma Panel: Immunophenotyping

B lymphoid cells (gated on CD19-SSC) (~96% of the acquired events).

CD19+, CD5+, CD23+, CD22+, CD200+, CD20+, CD11c+, Kappa+, CD38+, lambda−. The immunophenotype was that of a B cell, Chronic Lymphoproliferative Disorder, compatible with B-Chronic Lymphocytic Leukemia (CLL) kappa light chain restricted. CD200 positivity helps differentiate CLL from other CLPDs.

FISH-CLL/Lymphoma Panel on Peripheral Blood

FISH results showed trisomy of chromosome 12 in 90% of interphase cells analyzed. There was an absence of TP53 gene deletion, ATM gene deletion, RB1 gene deletion as well as of IgH gene rearrangement.

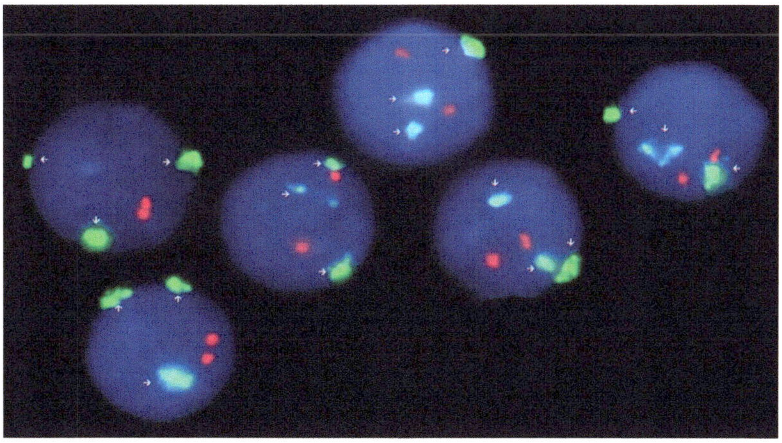

Case Study 10.2 Interphase nuclei showing 2R3G signal pattern using RB1/Trisomy 12 probe indicating trisomy 12

Discussion

Trisomy 12 is considered as a driver mutation occurring early in CLL and facilitates the appearance of secondary mutations in genes such as NOTCH1, TP53, and PBXW7. A higher frequency of IGHV unmutated genes has been described with trisomy 12 [1].

This patient of CLL had a classical morphologic and immunophenotypic profile of CLL including positivity for CD 200. As per FISH, he showed a sole abnormality of trisomy 12. It is the second most frequent aberration at the time of diagnosis in CLL (10–15%) and confers an intermediate prognosis with a median overall survival of 114 months [4]. Patients with trisomy 12 generally have unmutated IGHV, ZAP 70 positivity. They rarely show TP53 mutations but frequently have NOTCH 1 mutations, which can be seen in 40% of those with rapidly progressive clinical course. Concomitant abnormalities include +12 sole abnormality (70%), associated with del 13q, del11q, or del 17p in 18%, 8%, and 4% of +12 cases, respectively. It can also be associated with trisomy 18, trisomy 19, del 14 q, or t(14;19). +12 is more frequent in small lymphocytic lymphoma/leukemia than in those with CLL (28–36% vs 13–16%) which may well be the case in this patient who had cervical lymphadenopathy and mild splenomegaly. +12 is also more frequent in patients who develop Richters syndrome (50%). Importantly in patients with adverse cytogenetic abnormalities such as del17p or del11q, the co-occurrence of a favorable clone such as +12 portends an improved prognosis indicating relevance of intratumoral cytogenetic heterogeneity. +12 as a driver mutation can facilitate the appearance of NOTCH 1 and FBXW7 mutations, disruption of which can lead to cell proliferation and evasion from apoptosis [5]. CLL patients may have a higher incidence of autoimmune cytopenias. An atypical immunephenotype and morphology is more frequent in +12 and del17p subgroups of CLL. An atypical immunophenotype can be defined by knowing Matute's score (positive expression of CD5 and CD23, negative expression of FMC7, and weak expression of slg and CD22). The score is atypical if it is less than 4.

10.3 Case Study 10.3: Chronic Lymphocytic Leukemia with TP53 (17p13) and RB 1 (13q14) Gene Deletions

Brief Clinical History

A 54-year-old male was found to have leukocytosis in 2013 diagnosed as CLL Rai stage 0, now having multiple infections. He was treated on FCR, 6 cycles, and is being investigated further for evaluation of the disease.

CBC: *Hb*: 12.6 g/dl; TLC: 41,910/cu mm and Platelets: 1,31,000/cu mm.

DLC: Neutrophils: 13.3%; Lymphocytes: 74.5; Eosinophils: 0.5%; Monocytes: 11.2 0%; Basophil: 0.5%.

Peripheral Blood Smear and Bone Marrow

Smear showed Neutropenia (ANC~560 cells/cu mm). There is lymphocytic predominance. Most of the lymphoid cells are small and mature. Three percent prolymphocytes are seen.

Bone marrow was not done.

Flow Cytometric-CLL/Lymphoma Panel: Immunophenotyping

Lymphoid cells: B cells (gated on CD19-SSC) (~40% of the acquired events), CD45 (moderate).

CD19+, CD5+, CD23+, CD22+, CD200+, CD20+, CD11c+, CD38+ (dim/partial), kappa+, lambda−.

Peripheral blood showed the presence of monoclonal B lymphocytosis compatible with B-Chronic Lymphocytic, Leukemia phenotype. Kappa light chain restricted.

Expression of CD200 (MRC OX-2) helps in differentiating CLL from non-CLL CLPDs. CD38 expression is a prognostic marker.

FISH-CLL/Lymphoma Panel on Peripheral Blood

FISH results showed gene deletions of TP53 gene deletion in 50% cells and RB1 (13q14) gene deletion in 70% cells, respectively, among 200 interphase cells analyzed. There was no evidence of ATM gene deletion, IgH gene rearrangement, or Trisomy 12.

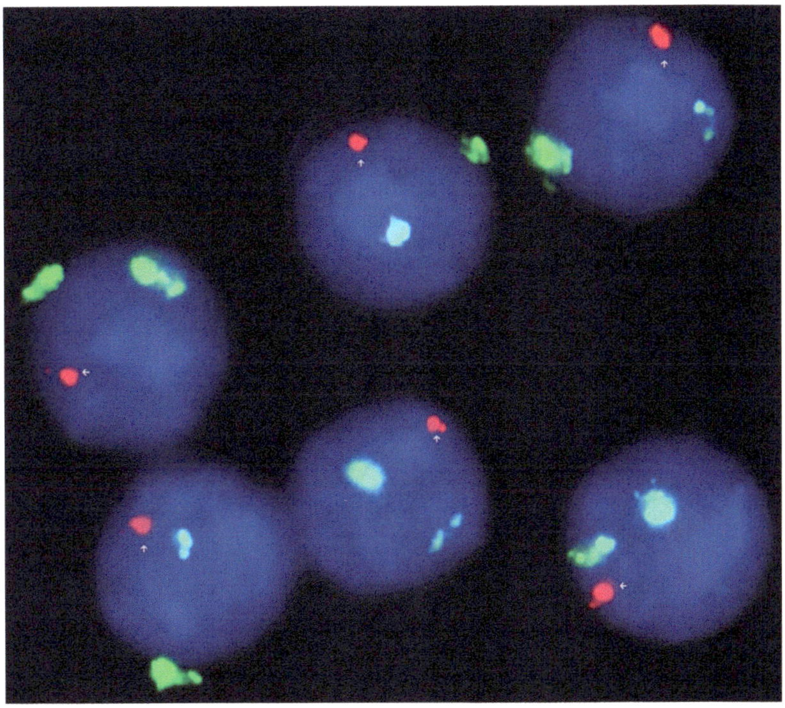

Case Study 10.3 (1) Interphase nuclei showing 1R2G signal pattern using DLEU/ LAMP1/CEN 12 tricolor probe indicating *RB1* gene deletion

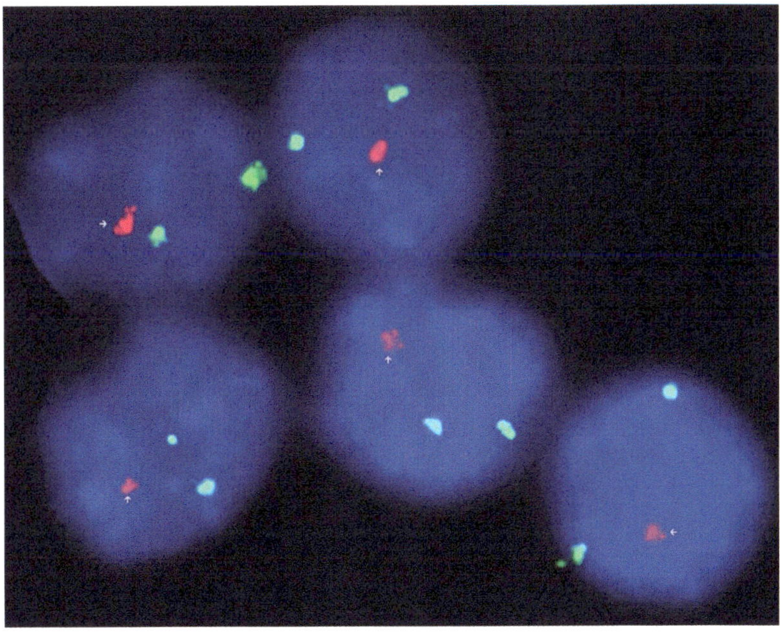

Case Study 10.3 (2) Interphase nuclei showing 2G1R signal pattern using *TP53/ATM* dual-color probe indicating deletion of *TP53* gene

Discussion

Deletion of 17p is found in nearly 3–8% of CLL patients at diagnosis but can account for up to 30% following chemotherapy. Patients with 17p deletion have been included in the highest risk prognostic category with shortest OS and PFS. Deletion of 17p can be detected in TP53 mutated patients of CLL and vice versa and can significantly affect OS and PFS; on the other hand, 13q deletion is found in more than 50% patients of CLL and is associated with good prognosis. Biologically, underlying 13q deletions, miR-15a and miR16-1 located in the minimally deleted region (MDR) have been shown to have a tumor suppressor function. In FISH, Patients with isolated 13q can be risk-stratified by the percentage of altered nuclei [1].

This patient in 2013 was diagnosed to have CLL Rai stage 0 with a molecular abnormality of 13q14 in 18% cells. In 2019, he showed del 17p13 (TP53) in 50% and del 13q14 (RB1) in 70% of the cells, indicating disease progression. Using FISH, by interphase cytogenetics, the most frequent abnormalities are del 13q14, followed by del 11q22, trisomy 12, del 6q21–q23, and del/mutation of TP53 tumor suppressor gene at 17p13 [6]. These aberrations provide the basis for more accurate correlations with clinical characteristics and outcome. Deletions/mutations of the TP53 gene are associated with resistance to treatment and are an independent marker for poor survival. 11q deletions (ATM gene) are associated with extensive nodal investment, rapid disease progression, and short survival times [7].

The understanding of these lesions leads to "precision medicine" through targeted therapy. Two to five percent of CLL patients transform to DLBCL due to: (1) TP53 disruption, CDKN2A loss, and MYC activation in 30% of cases; (2) trisomy 12 and NOTCH 1 mutations in 30%; and (3) a group with no obvious lesions [8].

10.4 Case Study 10.4: Chronic Lymphocytic Leukemia with ATM (11q22) and RB 1 (13q14) Gene Deletions

Brief Clinical History

A 60-year-old male presented with generalized lymphadenopathy. CECT and PET-CT showed lymph nodes both above and below the diaphragm. Evaluated for lymphoma.

CBC: *Hb*: 12.3 g/dl; TLC: 21,520/cu mm and Platelets: 1,30,000/cu mm.

DLC: Neutrophils: 20.0%; Lymphocytes: 77.0%; Eosinophils: 0.0%; Basophil: 0.0%.

Peripheral Blood Smear and Bone Marrow

Smear showed Leukocytosis with absolute lymphocytosis. Most of the lymphocytes were small and mature. Peripheral smear findings were suggestive of a chronic lymphoproliferative disorder, Chronic lymphocytic leukemia. Bone marrow not done.

Flow Cytometric-CLL/Lymphoma Panel: Immunophenotyping

Lymphoid cells (gated on CD19-SSC) (~53% of the acquired events).

CD19+, CD5+, CD23+, CD22+, CD200+, CD20+, CD11c+, CD38+, IgM+, kappa +, lambda−.

The immunophenotype was that of a B cell, Chronic Lymphoproliferative Disorder, compatible with B--Chronic Lymphocytic Leukemia. Kappa light chain restricted.

Expression of CD200 (MRC OX-2) helps in differentiating CLL from non-CLL CLPDs. CD38 expression is positive which is a prognostic marker.

FISH-CLL Panel on Peripheral Blood

FISH results showed ATM (11q22) gene deletion in 83% cells and RB1 (13q14) gene deletion in 84% interphase cells analyzed. There was no evidence of TP53 gene deletion, IgH gene rearrangement, and Trisomy 12.

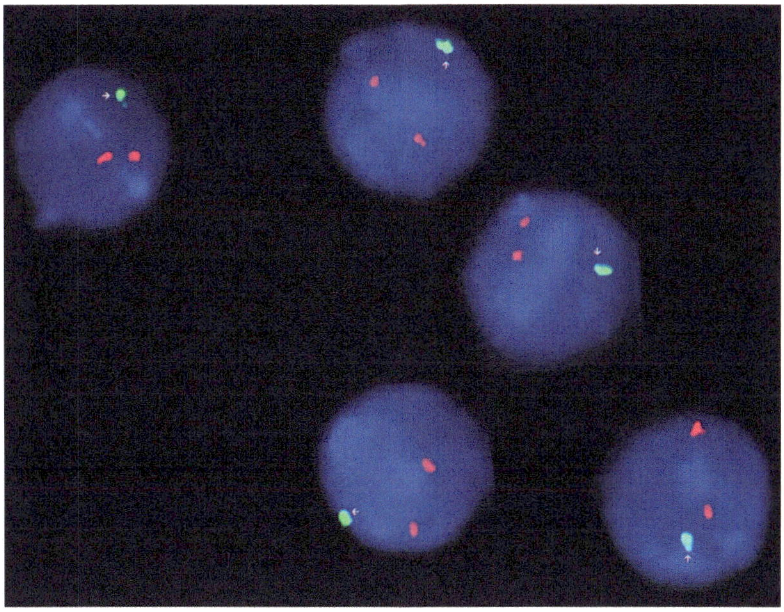

Case Study 10.4 (1) Interphase nuclei showing 2R1G signal pattern using *TP53/ATM* dual-color probe indicating *ATM* gene deletion

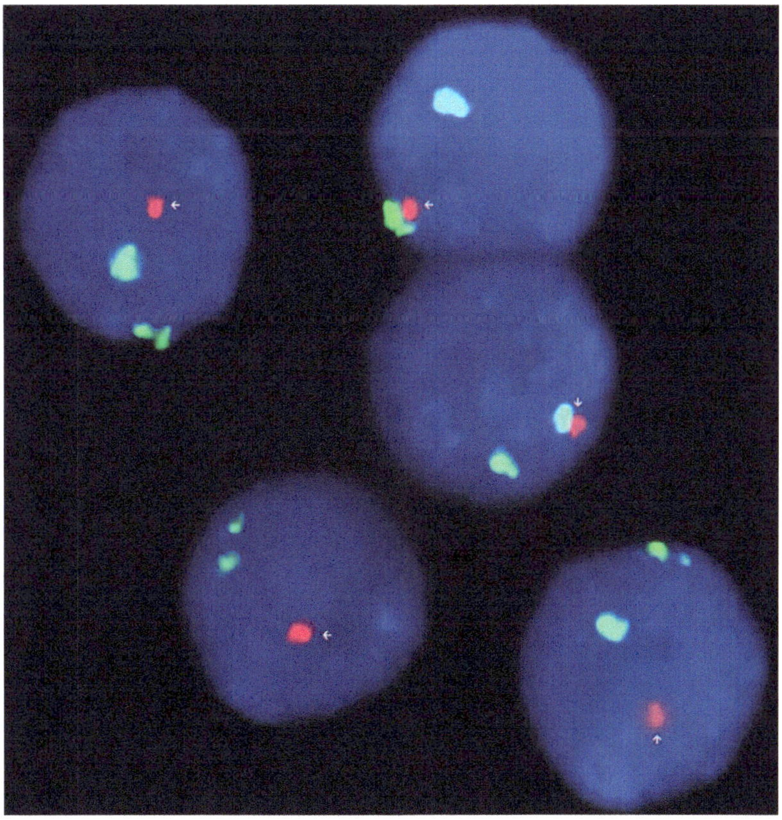

Case Study 10.4 (2) Interphase nuclei showing 1R2G signal pattern using DLEU/LAMP1/CEN 12 probe indicating *RB1* gene deletion

Discussion

Deletion of the long arm of chromosome 11 is detected in 5–20% of CLL patients. The deletions are highly variable in size. No homozygous deletions have been described so far. The MDR includes 11q23.3-q 23.1 chromosome bands which harbors the ATM gene in most cases.

13q deletion is found in more than 50% of patients of CLL and is associated with a good prognosis. Biologically underlying 13q deletions, miR-15a, and miR-16-1 located in the MDR region have been shown to have a tumor suppressor function.

This patient presented with prominent lymphadenopathy and absolute lymphocytosis identified by Flow cytometry immunophenotype as CLL. FISH identified the presence of both 11q22 gene (ATM) deletion and del 13q14 (RB1). Patients with ATM gene deletion generally show large and multiple lymphadenopathies as was also seen in this case [9]. They have a poor prognosis and relatively aggressive disease. del 13q14, on the other hand, has favorable prognosis. del 11q22 and del 17p13 are independent prognostic markers in multivariate analysis identifying subgroups of patients with rapid disease progression and short survival in patients of CLL [10].

10.5 Case Study 10.5: Chronic Lymphocytic Leukemia with IgH and Gene Rearrangement and Trisomy 12

Brief Clinical History

A 52-year-old male presented with weight loss, fever off and on with extensive *Lymphadenopathy* and massive Hepatosplenomegaly. Lymph node biopsy showed CLL/SLL.

CBC: Counts showed severe anemia, leukocytosis, and thrombocytopenia.

Hb: 3.9 g/dl; TLC: 5,45,870/cu mm and Platelets: 80,000/cu mm.

DLC: Neutrophils: 0.7%; Lymphocytes:87.5%; Eosinophils: 0.0%; Monocytes: 11.8%; Basophil: 0.0%.

Peripheral Blood Smear and Bone Marrow

Hyperleukocytosis with lymphocytosis, absolute lymphocyte count was 4,77,430/cu mm Majority were small mature lymphocytes with 3% prolymphocytes. Occasional smudge cells were seen. The picture was consistent with chronic lymphoproliferative disease.

Flow Cytometric-CLL/Lymphoma Panel: Immunophenotyping

B lymphoid cells (gated on CD19-SSC) (~92.8% of the acquired events).

CD19+, CD5+, CD23+, CD22+, CD200+, CD20+ (bright), CD11c+, Kappa+, CD38+, Lambda negative.

The immunophenotype was of a B cell, Chronic Lymphoproliferative Disorder, compatible with B-Chronic Lymphocytic Leukemia (CLL). Kappa light chain restricted.

Expression of CD200 (MRC OX-2) helps in differentiating CLL from non-CLL CLPDs. Expression of CD38 is a prognostic marker.

FISH-CLL Panel on Peripheral Blood

Fish results showed the absence of the TP53 gene, ATM gene, and RB1gene deletions. Interphase cells showed positivity for IgH gene rearrangement in 64% cells and Trisomy 12 in 80% cells.

Case Study 10.5 (1) Interphase nuclei showing 1F1G1R signal pattern using *IGH* breakapart probe indicating *IGH* rearrangement

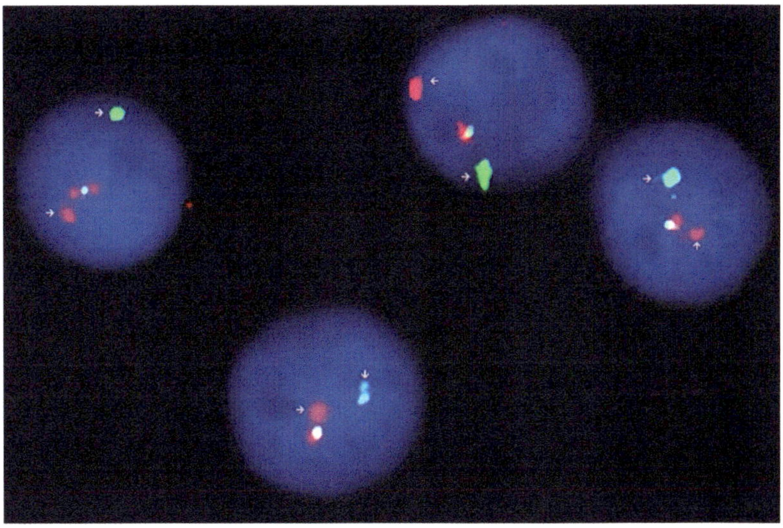

Case Study 10.5 (2) Interphase nuclei showing 2R3G signal pattern using DLEU/ LAMP1/CEN 12 probe indicating trisomy 12

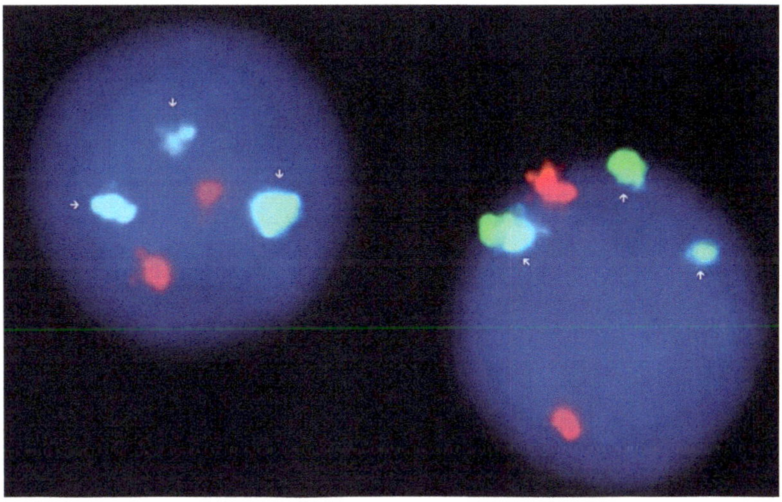

Discussion

This patient morphologically and immuophenotypically CLL on Peripheral blood smear presented with advanced disease, severe anemia with thrombocytopenia and hyperleukocytosis together with extensive lymphadenopathy and massive hepatosplenomegaly. At present, FISH remains the most important outcome predictor targeting chromosomal abnormalities such as deletion 13q14, del 11q22 (ATM), del 17p13 (TP53), and trisomy 12. In CLL, translocations involving the immunoglobulin gene are regularly identified. In a study of 369 patients, Brackeleer et al. [11] identified IG gene rearrangement in 18 patients. The IGH was involved in 77.1%. Associated Chromosome band partners were 18q21, 11q, 19q13, 8q24, and 7q21. IGHV mutational status various between translocations, those with unmutated IGHV involving cells at an earlier stage of B cell lineage. All partner genes identified so far control cell proliferation and/or apoptosis. In one of the studies (cases = 392) Kadam Amare et al. [12], IgH (chromosome band 14q32) was involved in 77.1% of cases and trisomy 12 in 48%. Deletions of 11q22 and 17p13 were found in 4% and 12% of patients, respectively. In a meta-analysis, IgH rearrangements were found in 8.3% [11]. IgH translocation in B-CLL has shown an association with unmutated IgVH, ZAP70 expression, CD38 positivity, atypical nuclear morphology, and atypical immunophenotype. Trisomy 12 is seen in ~15% of all CLLs and is associated with an intermediate prognosis.

10.6 Case Study 10.6: Chronic Lymphocytic Leukemia with Isolated RB 1 Gene Deletion (13q14)

Brief Clinical History
A 72-year-old male presented with a long history of malignancies, oral cavity cancer in 2004, prostatic cancer in 2009 with widespread metastatic disease. Investigated further for lymphocytosis. CECT did not show any significant lymphadenopathy.

CBC: *Hb*: 11.4 g/dl; TLC 15,350/cu mm and Platelets: 3,53,000/cu mm.

DLC: Neutrophils: 41.4%; Lymphocytes: 52.2%; Eosinophils: 2.1%; Monocytes: 4.0%; Basophil: 0.3%.

Peripheral Blood Smear and Bone Marrow
Peripheral smear showed Absolute lymphocytosis (ALC ~ 8010 cells/cu mm). Lymphocytes were small and mature. Few smudge cells were seen.

Bone marrow not done.

Flow Cytometric-CLL/Lymphoma Panel: Immunophenotyping
B lymphoid cells (gated on CD19-SSC) (~22.0% of the acquired events).

CD19+, CD5+, CD23+, CD22+, CD200+, CD20+, Kappa+ (moderate), Lambda−.

The immunophenotype was that of a B cell, Chronic Lymphoproliferative Disorder, compatible with Chronic Lymphocytic Leukemia. Kappa light chain restricted. There was an expression of CD200 (MRC OX-2), which helps to differentiate CLL from non-CLL CLPDs.

FISH-CLL Lymphoma Panel on Peripheral Blood
FISH showed RB1 gene deletion in 60% of cells. There was no evidence of TP53 gene deletion, ATM gene deletion, IgH gene rearrangement, or Trisomy 12.

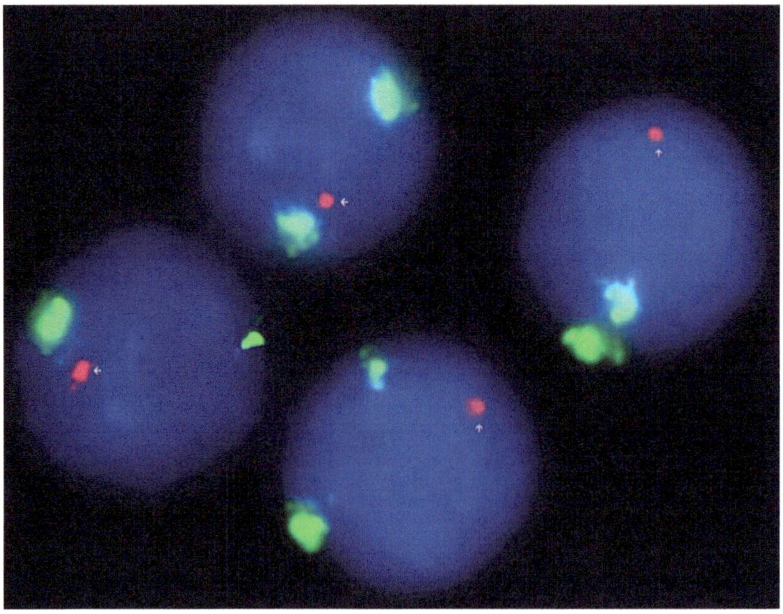

Case Study 10.6 Interphase nuclei showing 1R2G signal pattern using DLEU/LAMP1/CEN 12 probe indicating *RB1* gene deletion

Discussion

This patient of classic CLL morphologically and immunophenotypically showed an isolated RB1 gene deletion (13q14) on FISH. del (13q14) is found in more than 50% of CLL cases and is associated with a good prognosis [13]. It has been demonstrated that large 13q losses involving the RB1 gene are related to shorter time to first treatment (TTFT) and overall survival than those involving small deletions encompassing only miR-15a and miR 16-1. The pathogenic role of del(13q) in CLL has been related to lack of B cell proliferation control determined by deletion of DLEU2/MIR15A/MIR16-1 locus [9]. The RB1 gene is also involved in the regulation of cell cycle progression and genomic stability. Although CLL with sole 13q deletion abnormality usually has a good prognosis, a more aggressive clinical course has been documented in those carrying a higher percentage of 13g deleted nuclei [14].

10.7 Case Study 10.7: Chronic Lymphocytic Leukemia with Partial deletion of IgH (14q32) Gene Rearrangement

Brief Clinical History

A 78-year-old male presented with past history of axillary lump, treated on steroids. PET showed multiple enlarged lymph nodes diagnosed as DLBCLL outside.

CBC: *Hb*: 10.7 g/dl; TLC 26,000/cu mm and Platelets: 2,28,000/cu mm.

DLC: Neutrophils: 16.0%; Lymphocytes: 83.0; Eosinophils: 1.0%; Monocytes: 0.0%; Basophil: 0.2%.

Peripheral Blood Smear and Bone Marrow

Peripheral smear showed Leukocytosis with absolute lymphocytosis. Lymphoid cells were mature having clumped chromatin. Occasional smudge cells were seen. The findings were suggestive of a chronic lymphoproliferative disorder (CLPD), CLL.

Bone marrow not done.

Flow Cytometric-CLL/Lymphoma Panel: Immunophenotyping

B cells (gated on CD19-SSC) (~64% of the acquired events).

CD5+, CD23+, CD22+, CD200+, CD20+, CD11c+ (heterogenous), kappa+, lambda−, CD38−.

The immunophenotype was that of a B cell, Chronic Lymphoproliferative Disorder, compatible with Chronic Lymphocytic Leukemia. Kappa light chain restricted. CD200 positivity helps in differentiating CLL from other CLPDs.

FISH-CLL Lymphoma Panel on Peripheral Blood

Fish results showed the absence of RB1 gene, TP53 gene, ATM gene deletions, and Trisomy 12. However, IGH partial deletion (14q 32) gene rearrangement was seen in 87% of interphase cells analyzed.

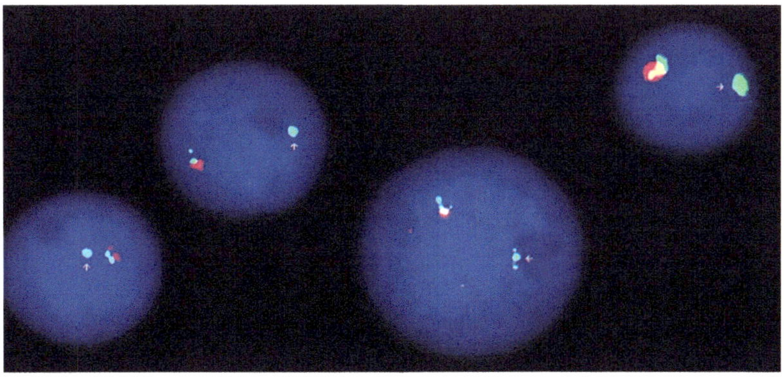

Case Study 10.7 Interphase nuclei showing 1F1G signal pattern using *IGH* probe indicating *IGH* partial deletion

Discussion

IgH gene rearrangement is seen in lymphoid malignancies of B cell origin such as B-ALL, CLL, lymphomas, and multiple myeloma. This patient of CLL showed a partial deletion of IgH at the 5′ region. Deletion of 5′ IgH corresponding to the variable IgH segment (VJ) was found to be the most recurrent aberration (82% cases) in a study by Quintero-Rivera et al. [15] amongst 291 patients and was associated with a good prognosis. IgH deletion may be hetero- or homozygous. A 14q32 involvement is considered a good prognostic feature (median > 15 years) in CLL. Huh et al. [16] studied 78 ALL patients by FISH to look for submicroscopic IgH deletions. Among them, two showed deletion of the 5′ variable region of IgH. Two kinds of abnormal cells were seen one had IgH gene rearrangement without submicroscopic deletions and the other with submicroscopic deletion, seen in FISH as one normal fusion signal and one 3′ IgH signal was deleted. The 5′ deletion may have occurred in either the native or rearranged chromosome 14.

10.8 Case Study 10.8: Chronic Lymphocytic Leukemia with Biallelic and Monoallelic RB 1 Gene Deletion (13q14)

Brief Clinical History

A 53-year-old female investigated for lymphocytosis and weight loss. No hepatosplenomegaly.

CBC: *Hb*: 14.5 g/dl; TLC: 40,460/cu mm and Platelets: 3,47,000/cu mm.

DLC: Neutrophils: 27.8%; Lymphocytes: 69.5; Eosinophils: 0.3%; Monocytes: 2.1%; Basophil: 0.3%.

Peripheral Blood Smear and Bone Marrow

Smear showed absolute lymphocytosis, comprised mainly of small lymphocytes. The picture is consistent with chronic lymphoproliferative disorder (CLL).

Bone marrow not done.

Flow cytometric immunophenotyping not done.

FISH-CLL/Lymphoma Panel on Peripheral Blood

FISH results showed the absence of TP53 gene deletion, ATM gene deletion, Trisomy 12, and IgH gene rearrangement. The interphase cells showed positivity for biallelic RB1(13q14) gene deletion in 78% cells and monoallelic RB1 (13q14) gene deletion in 15% cells analyzed.

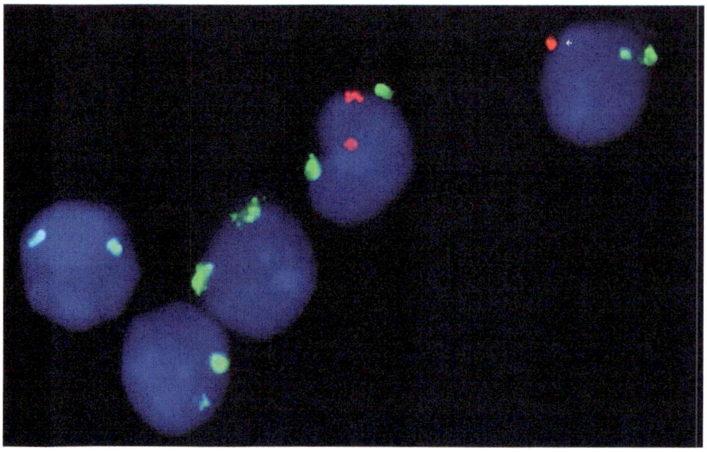

Case Study 10.8 Interphase nuclei showing 0-1R2G signal pattern using DLEU/LAMP1/CEN 12 probe indicating Biallelic and Monoallelic deletion of *RB1* gene. One nucleus is showing a 2G2R signal pattern indicating normal cell

Discussion

This patient showed mainly a biallelic RB1 (13q) gene deletion (78%) and monoallelic gene deletion in 15% cells. del (13q) involving band q14 is seen in more than 50% of patients of CLL and is generally associated with a good prognosis. In a study of 176 patients [17], 143 patients had monoallelic and 33 had a biallelic 13q deletion. No difference was observed in the diagnostic significance between the two groups including baseline characteristics or time to treatment. In another study Grygalewicz [18], it was shown that bigger deletion including RB1 or presence of biallelic 13q deletion is not sufficient to be considered as an adverse prognostic factor in CLL/SLL.

10.9 Case Study 10.9: Chronic Lymphocytic Leukemia with Sole Abnormality of ATM Gene Deletion (11q22)

Brief Clinical History

A 54-year-old male was diagnosed to have lymphoblastic lymphoma in January 2017. Received chemotherapy 4 cycles. Has now presented with generalized lymphadenopathy.

CBC: *Hb*: 9.7 g/dl; TLC: 17.540/cu mm and Platelets: 1,89,000/cu mm.

DLC: Neutrophils: 26.3%; Lymphocytes: 69.6%; Eosinophils: 0.2%; Monocytes: 3.8%; Basophil: 0.1%.

Peripheral Blood Smear and Bone Marrow

Smear showed leukocytosis with absolute lymphocytosis.

Bone marrow not done.

Flow Cytometric Immunophenotyping CLL/Lymphoma Panel on Peripheral Blood

B cells (gated on CD19-SSC) (~43.0% of the acquired events).

CD19+, CD5+, CD23+, CD20+, CD22+, CD200+, Kappa+, Lambda−.

The immunophenotype showed chronic lymphocytic leukemia, kappa light chain restricted.

FISH-CLL/Lymphoma Panel on Peripheral Blood

ATM gene deletion was seen in 90% interphase cells. There were no TP53 gene deletion, RB1 gene deletion, or trisomy 12 or IgH gene rearrangement. There was *ATM Gene Deletion (11q22)* generally associated with a poor prognosis and relatively aggressive disease.

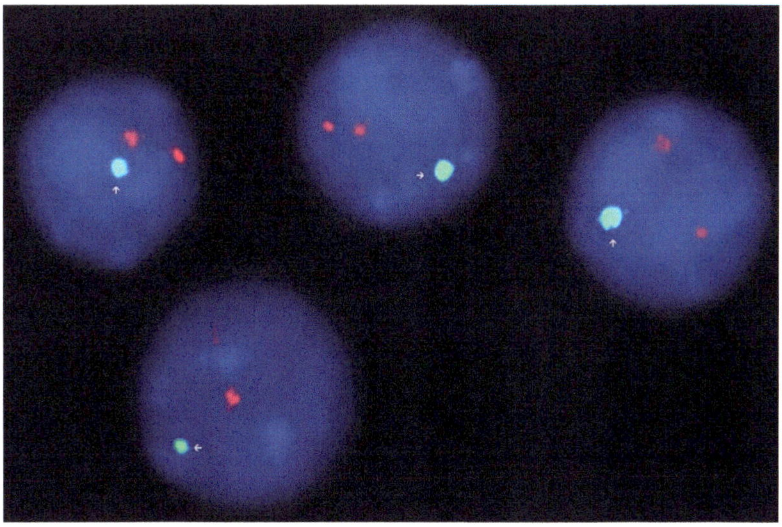

Case Study 10.9 Interphase nuclei showing 2R1G signal pattern using *TP53/ATM* dual-color probe indicating *ATM* gene deletion

Discussion

Deletions of 11q23-q23 almost invariably include the ATM gene. Guarini et al. [19] studied CLL patients with and without deletion of 11q22–23 and found that combining deletions and mutations, genetic lesions of ATM occur in 25% of CLL patients. For TP53 and ATM, both deletions and mutations affect prognosis. While deletions can be picked by FISH, mutations require sequencing. Like TP53 alterations, ATM alterations too are poor prognostic markers in CLL [20].

This patient with morphologic and immunophenotypic evidence of CLL showed an isolated ATM gene (11q22) gene deletion. CLL is characterized by a clonal proliferation and accumulation of mature, typically CD5 positive B cells in the blood, BM, lymph nodes, and spleen. Often the disease may be initiated by loss or addition of chromosome material such as deletion 13q, deletion 11q, trisomy 12 followed by additional mutation such as TP53 mutation making the disease more aggressive [21]. Deletions of the long arm of chromosome 11, del (11q) are seen in about 25% of advanced disease stages and 10% with early-stage disease. They typically show bulky lymphadenopathy, rapid progression, and reduced survival. Some studies show that ATM gene behaves like the TP53 gene. The majority of patients with ATM gene deletion have *unmutated* [19] IGHV, ZAP70, and CD38 expression.

10.10 Case Study 10.10: Leukemic Mantle Cell Lymphoma with Rearranged IgH gene, t(11;14) and Trisomy 12

Brief Clinical History

A 77-year-old male investigated for leukocytosis and weight loss for 6 months.

CBC: *Hb*: 10.1 g/dl, TLC: 95,680/cu mm and Platelets: 140,000/cu mm.

DLC: Neutrophils: 7.0%, Lymphocytes: 88.0%, Monocytes: 5.0%, Eosinophils: 0.0%, Basophils: 0.0%.

Peripheral Blood Smear and Bone Marrow

Smear showed Absolute lymphocytosis (ALC ~ 84,198 cells/cu mm). Small lymphoid cells with scant cytoplasm, round nucleus, and condensed chromatin with many showing deep nuclear clefts. Few larger blastoid cells were also seen constituting ~7% of all lymphoid cells. Similar cells infiltrated the bone marrow (85%). Bone marrow trephine biopsy showed infiltration by a lymphoma.

Peripheral blood and bone marrow aspirate findings were consistent with bone marrow infiltration by a lymphoma with a spill, likely leukemic mantle cell lymphoma.

Cervical lymph node showed mantle cell lymphoma with IHC CD20+, CD23+, BCL-2+, cyclin D1−, ki67 12–50%.

Flow Cytometric Immunophenotyping

B cells gated on CD19-SSC (~65% of the acquired events).

CD19+, CD5+, CD23-, CD22+, CD20 + (bright), kappa+, lambda−, CD200−, CD23−. The immunophenotype was that of CD5 positive, CD10-negative B cell, Chronic Lymphoproliferative Disorder CD200 negativity with bright CD20 positivity together with peripheral smear and bone marrow picture was suggestive of leukemic mantel cell lymphoma.

FISH-CLL/Lymphoma Panel on Peripheral Blood

FISH results on interphase cells showed no TP53 gene deletion, ATM gene deletion, or RB1 gene deletion. Positivity was seen for IgH gene rearrangement and t(11;14). In addition, 75% of interphase cells showed Trisomy 12.

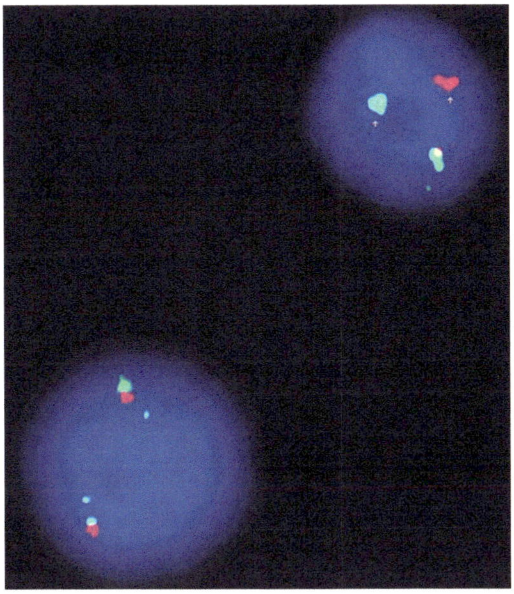

Case Study 10.10 (1) Interphase nuclei showing 1F1G1R signal pattern using *IGH* probe indicating *IGH* rearrangement. One nucleus showing a 2F (yellow) signal indicating a normal nucleus

Case Study 10.10 (2) Interphase nuclei showing 1F1G1R aberrant signal pattern using *IGH/CCND1* probe indicating *IGH/CCND1* fusion, t(11;14)

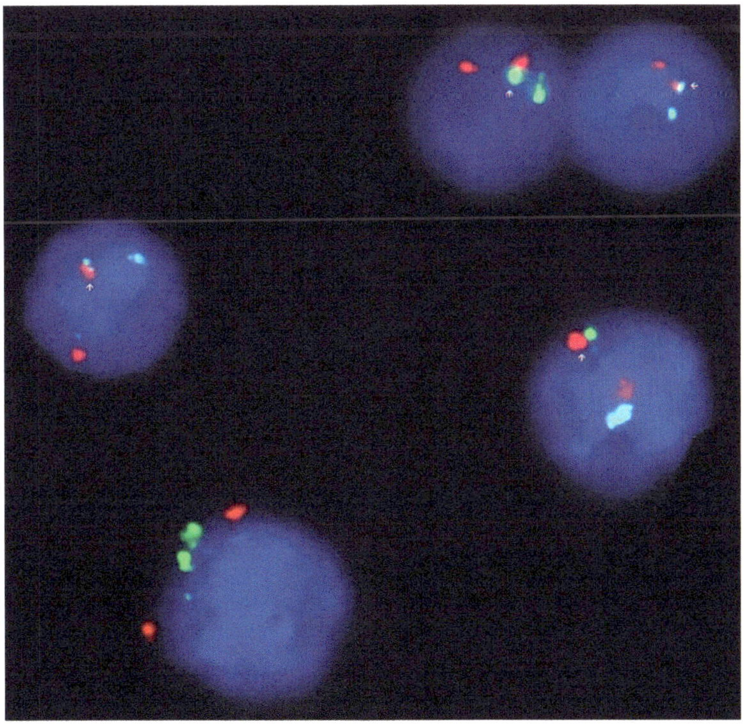

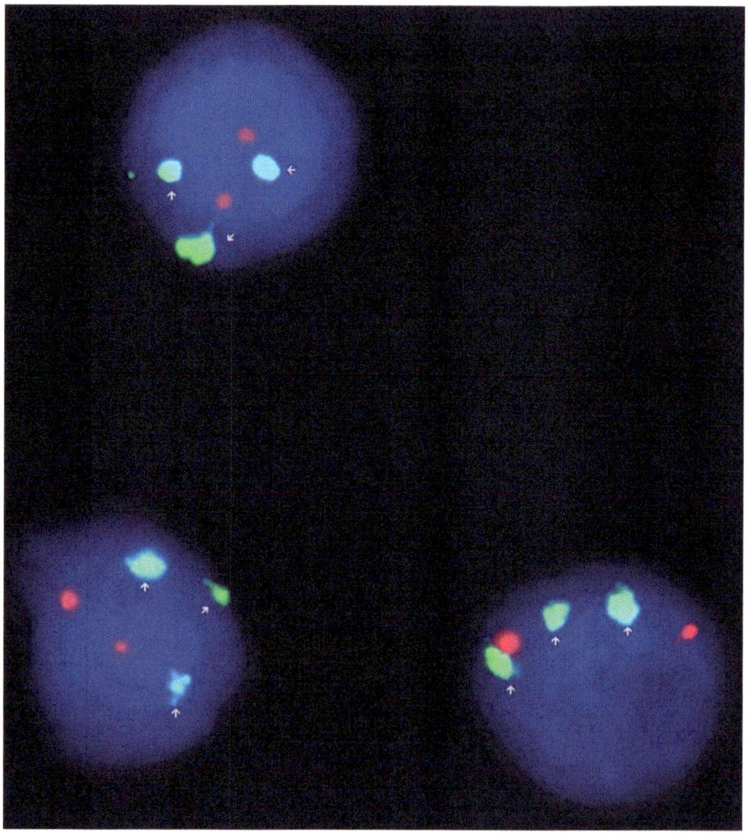

Case Study 10.10 (3) Interphase nuclei showing 2R3G signal pattern using DLEU/LAMP1/CEN 12 probe indicating trisomy 12

Discussion

This patient morphologically and by flow cytometry immunophenotype was commensurate with diagnosis of a mantle cell lymphoma in the leukemic phase. High leukocyte count in MCL is uncommon and needs differentiation from other CLPDs. FISH results t(11;14) confirmed the diagnosis of MCL. On IHC Cyclin D1 was negative. In addition, there was trisomy 12 and IgH gene rearrangement showing overlapping features with CLL. Other affected genes can be ATM (40%–75%) and CCND 1 (35%). The latter is often present in cyclin D1 negative cases [22]. Two MCL negative subtypes are recognized which have different clinicopathological and molecular pathogenetic pathways, one with unmutated IgHV and SOX11+ and the other with mutated IgHV and negative SOX11 (indolent leukemic nonnodal MCL with PB, BM, and ±splenic involvement) [23]. This may become more aggressive. This was the case under study here. Mutations of TP53, NOTCH 1/2 are seen in a small number of cases.

10.11 Case Study 10.11: Chronic Lymphocytic Leukemia with Trisomy 12 and RB1 (13q14) Gene Deletion

Brief Clinical History

A 66-year-old male presented with epididymo orchitis, anemia, leukocytosis, splenomegaly, and weight loss for 1 year. Investigated further for Chronic lymphoproliferative disorder.

CBC: *Hb*: 7.8 g/dl; TLC: 61,400/cu mm and Platelets: 1,22,000/cu mm.

DLC: Neutrophils: 2.6%, Lymphoid cells: 96.7%, Eosinophils: 0.1%, Monocytes: 0.4%, Basophils: 0.2%.

Peripheral Blood and Bone Marrow

Smear showed Leukocytosis with absolute lymphocytosis. Lymphocytes were small and mature. Occasional prolymphocyte were seen.

Bone marrow smear showed a virtual replacement by small mature lymphocytes (~85%).

The bone marrow and peripheral blood findings were consistent with CLPD (CLL).

Flow Cytometric-CLL/Lymphoma Panel: Immunophenotyping

B cells (gated on CD19-SSC) (~83% of the acquired events).

CD5+, CD23+, CD22+, CD200+, CD20+, CD11c+, CD38+, kappa+, lambda−.

The immunophenotype was compatible with Chronic Lymphocytic Leukemia. CD38 expression is a prognostic marker. Kappa light chain restricted. Expression of CD200 (MRC OX-2) helps to differentiate CLL from non-CLL CLPDs.

FISH-CLL Panel on Peripheral Blood

FISH results showed the absence of TP53 gene deletion, ATM gene deletion, and IgH gene rearrangement. Interphase cells showed positivity for trisomy 12 in 80% and RB1 (13q14) gene deletion in 10% of cells.

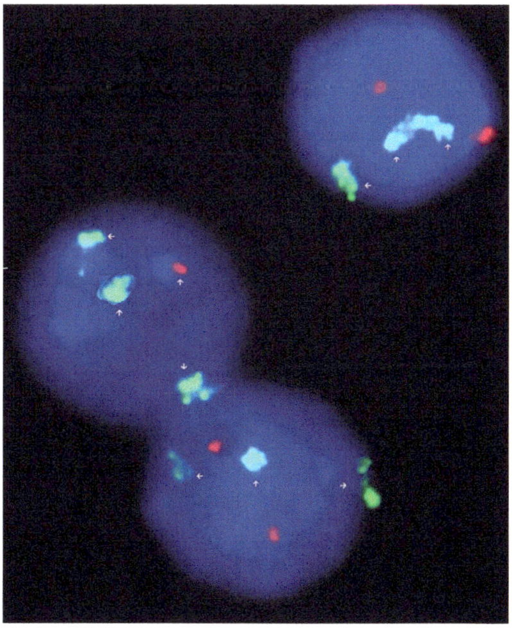

Case Study 10.11 Interphase nuclei showing 1-2R3G signal pattern using DLEU/LAMP1/CEN 12 probe indicating *RB1* gene deletion and Trisomy 12

Discussion

The coexistence of trisomy 12 and deletions of chromosome 13 (13q12-q32) have rarely been observed in chronic lymphocytic leukemia (CLL) and is a secondary event rather than a primary event in CLL leukemogenesis Navarro et al. [24]. The coexistence of the two on prognosis is not very clear. Isolated del(13q) involving the band q14, found in more than 50% of CLL patients, is the most common cytogenetic abnormality detected by FISH and has been associated with a good prognosis [25]. Trisomy 12 by FISH is found in around 15% of all Chronic Lymphocytic Leukemia (CLL) cases and is generally associated with good to intermediate median survival [26].

10.12 Case Study 10.12: Non-Hodgkin's lymphoma (Follicular Lymphoma)—Double-Hit Extra Copy Lymphoma (Atypical Double-Hit Lymphoma)

Brief Clinical History

A 68-year-old male diagnosed outside as a case of Follicular lymphoma (2017). Now presented with thrombocytopenia. Bone marrow was done for evaluation of the disease.

CBC: *Hb*: 12.2 g/dl, TLC: 8770/cu mm and Platelets: 24,000/cu mm.

DLC: Neutrophils: 43.0%, Lymphoid cells (including abnormal forms): 54.0%, Eosinophils: 0.0%, Monocytes: 3.0%, Basophils: 0.0%.

Peripheral Blood Smear and Bone Marrow

Smear showed 54% lymphoid cells including abnormal forms. Bone marrow aspirate showed abnormal lymphoid cells (60.0%) infiltrating the BM. Bone marrow trephine biopsy together with IHC (CD10+, CD20+, PAX5+) showed infiltration by a lymphoma in a known case of follicular lymphoma. Peripheral blood and bone marrow were consistent with infiltration of the bone marrow by a non-Hodgkin's lymphoma (Follicular Lymphoma) with peripheral blood spill.

Cytogenetics

Conventional Karyotype

45,X,-Y[15]/46,XY[5]

Cytogenetic analysis of 20 metaphases showed loss of chromosome Y in 15 metaphases analyzed. The loss of the Y chromosome was confined to the malignant cells.

FISH (Lymphoma Panel) on Bone Marrow Aspirate

Positive for BCL-6 gene rearrangement in 38% cells. Positive for 3–4 copies of MYC in 26% cells.

Findings consistent with double-hit extra copy lymphoma (BCL-6 rearrangement and MYC extra copies).

BCL-2 was negative.

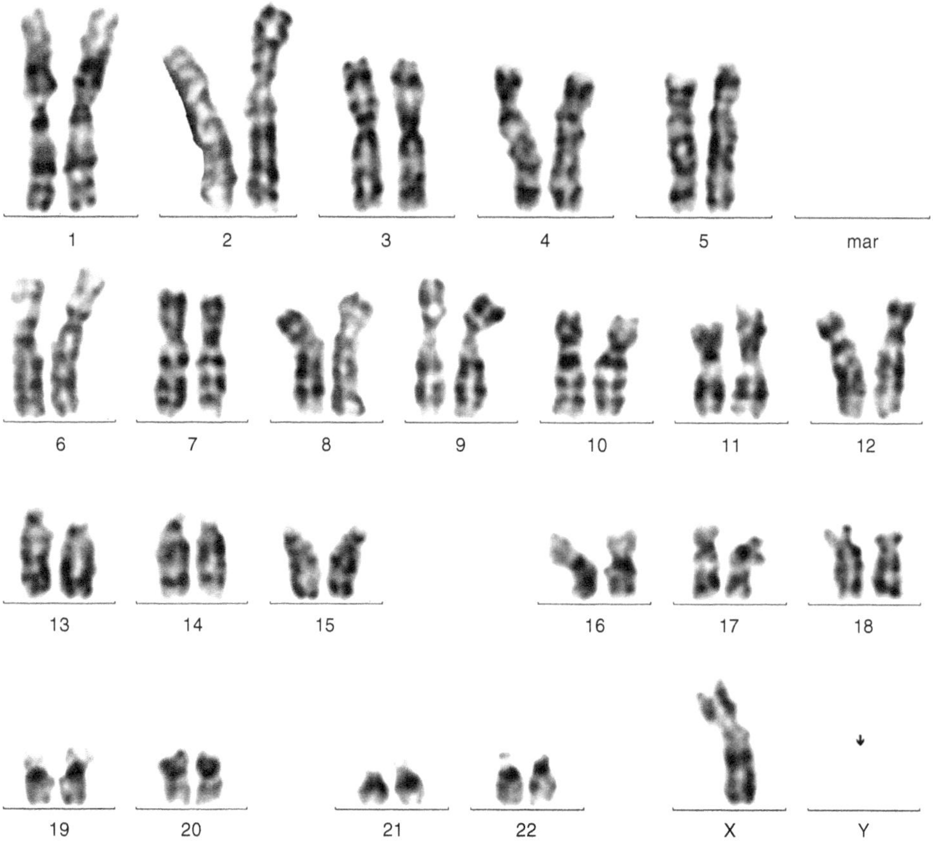

Case Study 10.12 (1) Karyotype showing loss of chromosome Y, i.e., 45,X,-Y

Case Study 10.12 (2) Interphase nuclei showing four copies of the *C-MYC* allele

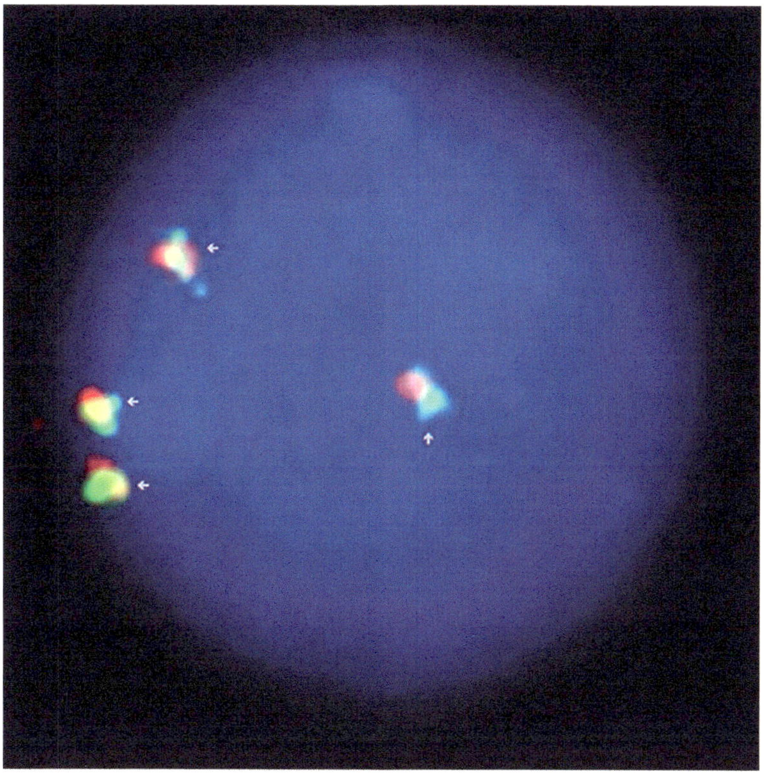

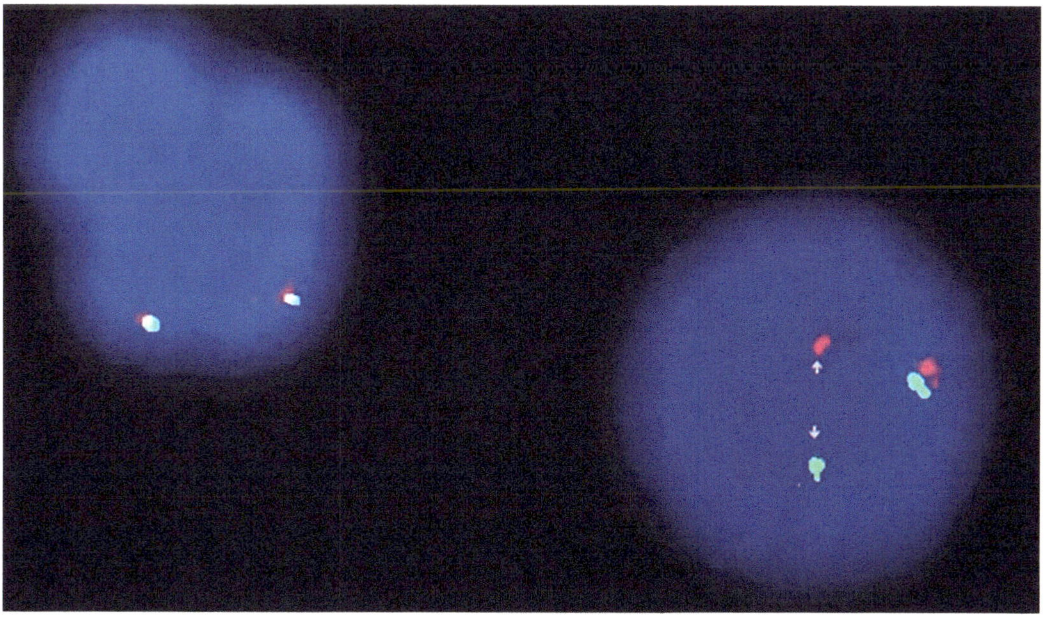

Case Study 10.12 (3) Interphase nuclei showing 1F1G1R signal pattern using *BCL-6* probe indicating *BCL-6* rearrangement. One nucleus showing 2G2R signal pattern indicating a normal nucleus

Discussion

This patient, an already diagnosed case of follicular lymphoma, showed involvement of the peripheral blood and bone marrow with IHC showing CD10+, CD20+, and PAX5 positive. FISH studies on BM aspirate showed positivity for BCL-6 rearrangement in 38% cells and 3–4 extra copies of MYC. BCL-2 was negative. It was labeled as double-hit extra copy lymphoma. Recently, studies from The University of Texas MD Anderson Cancer Center described 40 patients with other MYC/BCL-2 structural abnormalities, a subset labeled as "atypical double-hit lymphomas" [27]. Most (n22) had BCL-2 translocation with extra copies of MYC; others (n15) had extra copies of MYC and BCL-2 without translocations and a few (n3) had MYC translocations with extra copies of BCL-2 [28, 29]. Compared to double-hit lymphomas, atypical cases like these are more likely to have a DLBCL morphology and normal LDH as was the case in the present patient. In about 20% of patients, double-hit lymphomas are seen as transformation from follicular lymphoma [30].

10.13 Case Study 10.13: Non-Hodgkin's Lymphoma—Double-Hit Diffuse Large B Cell Lymphoma

Brief Clinical History

A 62-year-old male presented with fever, night sweats, weight loss with generalized lymphadenopathy with hepatosplenomegaly. Suspected lymphoma for further evaluation.

CBC: *Hb*: 9.1 g/dl; TLC: 20,240/cu mm and Platelets: 69,000/cu mm.

DLC: Neutrophils: 2.0%, Lymphocytes: 15.0%, Eosinophils: 0.0%, Monocytes: 0.0%, Basophils: 0.0%, Myelocytes: 2.0%, Abnormal lymphoid cells/Blasts: 81.0%.

Peripheral Blood Smear and Bone Marrow

Smear showed Leukocytosis with 81% abnormal lymphoid cells/blasts. Bone Marrow showed near-total replacement by abnormal lymphoid cells/blasts (~90%). These cells were large with a high N/C ratio with scant light blue cytoplasm and 1–2 prominent nucleoli. Peripheral blood and bone marrow features were suggestive of high-grade lymphoma/acute leukemia (morphologically lymphoid). Bone marrow biopsy from a left supraclavicular lymph node with IHC ($CD20^+$, $CD10^+$, $BCL\text{-}2^+$, k167 ± 70%) was consistent with diffuse large B cell non-Hodgkin's lymphoma (Germinal center type).

Cytogenetics

Conventional Karyotype

46,XY,t(3;8)(q21;q24),t(14;18)(q32;q21)[18]/46,XY[2]

Cytogenetic analysis of 20 metaphases showed abnormal clone characterized by a balanced translocation between the long arm of chromosome 3 and the long arm of chromosome 8 as well as a balanced translocation between the long arm of chromosome 14 and the long arm of chromosome 18 in 18 metaphases analyzed.

These chromosomal alterations showed there is t(14;18), i.e., (IGH/BCL-2) and t(3;8) which shows involvement of C-MYC 8q 24.21 region. These findings were confirmed by FISH analysis showing IGH gene rearrangement 90% positive.

The above picture was consistent with a double-hit high-grade B cell lymphoma with MYC and BCL-2 gene arrangement. The prognosis in such cases is poor.

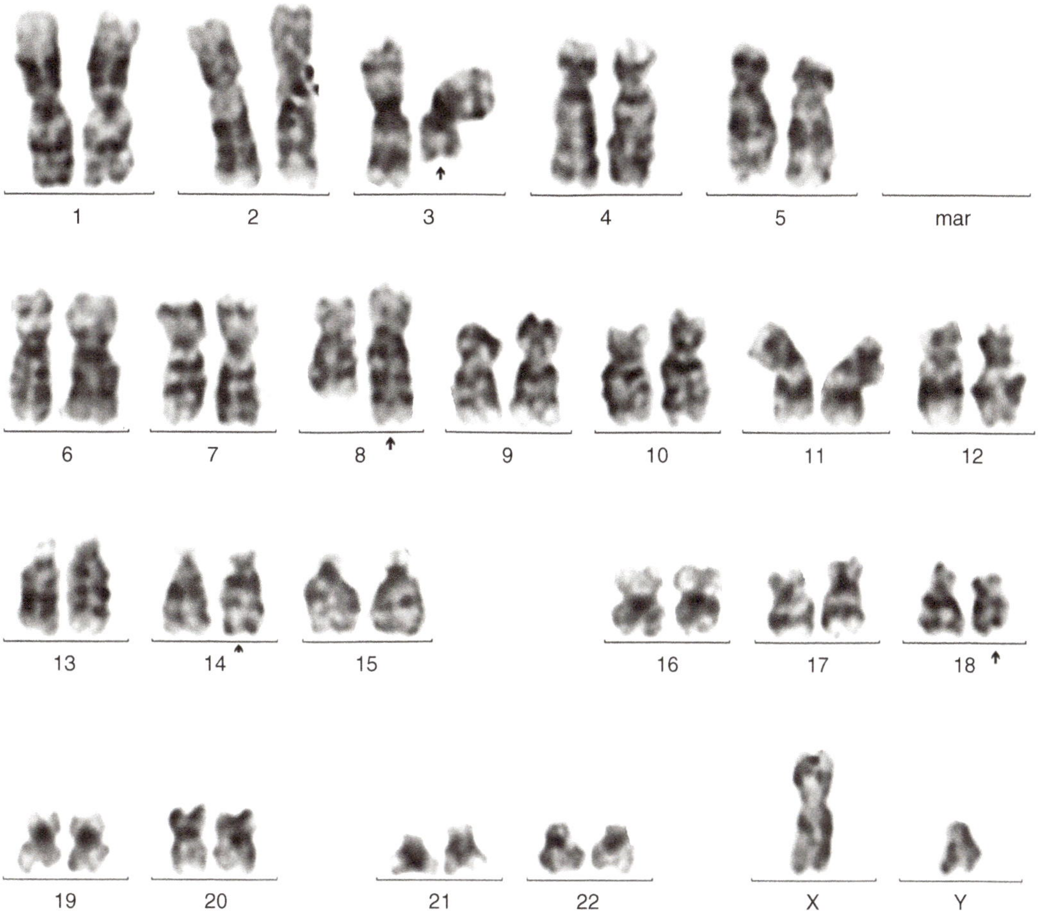

Case Study 10.13 (1) Karyotype showing translocation (3;8) and t(14;18). 46,XY,t(3;8)(q21;q24),t(14;18)(q32;q21)

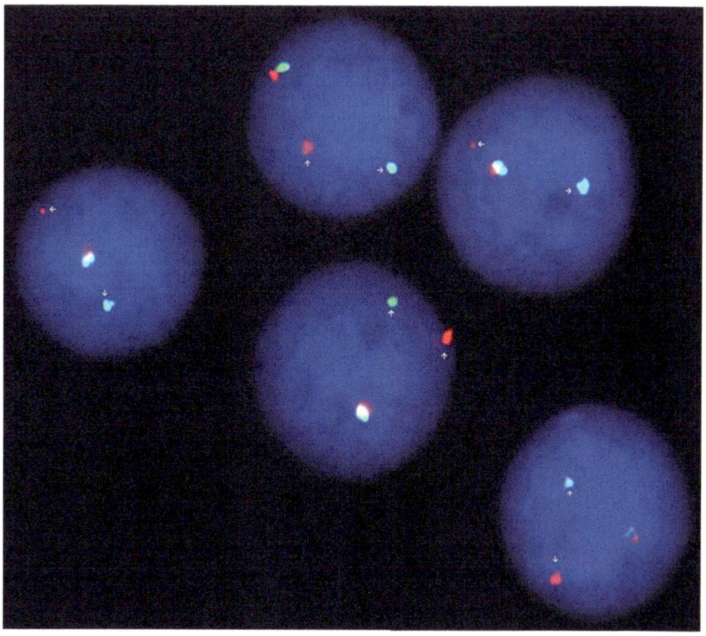

Case Study 10.13 (2) Interphase nuclei showing 1F1G1R signal pattern using *BCL-2* probe indicating *BCL-2* rearrangement

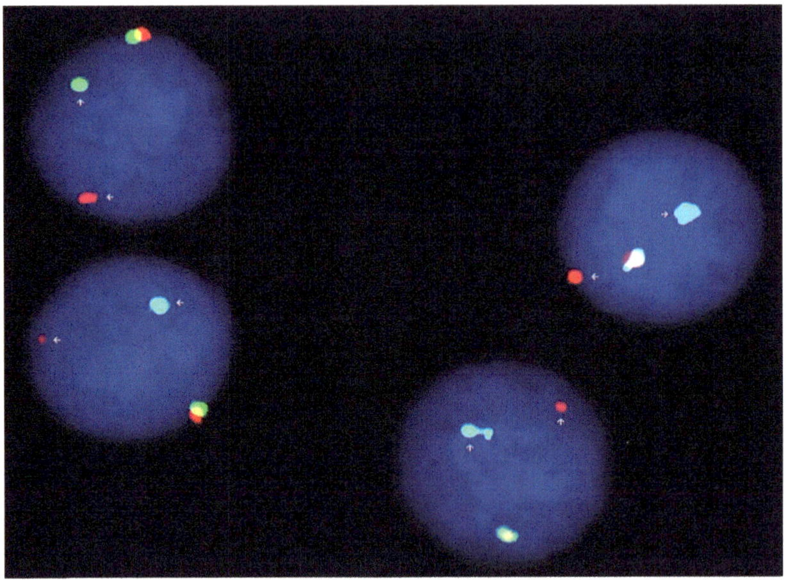

Case Study 10.13 (3) Interphase nuclei showing 1F1G1R signal pattern using *C-MYC* probe indicating *C-MYC* rearrangement

Discussion

This patient showing generalized lymphedenopathy with involvement of peripheral blood and bone marrow by a lymphoma showed DLBCL on cervical lymph node biopsy with IHC BCL-2+ (GCB). FISH studies on bone marrow aspirate showed BCL-2 positive in 96% cells, MYC in 92% with BCL-6 negative confirming the diagnosis of non-Hodgkin's lymphoma (double-hit large B cell lymphoma).

The WHO 2017 classification of tumors of hematopoietic and lymphoid tissues introduced a new entity of high-grade B cell lymphoma with MYC and BCL-2 and/or BCL-6 rearrangements (HGBL-DH/TH) with important therapeutic implications. The proportion of these among DLBCL morphology is about 1–12% [31].

It has been established that morphological features alone and proliferation rate cannot reliably identify HGBL (DH-TH) lymphomas as was the case in the present patient. A large registry-based study compared results of FISH, COO (cell of origin), and IHC for MYC and BCL-2 and it found that the only method that detects cases of HGBL (DH-TH) among tumors of DLBCL morphology would be FISH to detect first the MYC rearrangement in all cases followed by BCL-2 and BCL-6 rearrangements. This would allow for no misclassification if all three were not done and properly define treatment strategies [32].

10.14 Case Study 10.14: Non-Hodgkin's Lymphoma—Splenic Marginal Zone Lymphoma with Extra Copies of MYC gene (NOS)

Brief Clinical History

A 38-year-old male patient presented with fever, on and off for the past 9 months. On examination showed massive splenomegaly. PET-CT revealed splenic and skeletal hypermetabolism with abdominal lymph nodes. Bone marrow was done for further evaluation.

CBC: *Hb*: 9.4 g/dl; TLC: 4410/cu mm and Platelets: 1,20,000/cu mm.

DLC: Neutrophils: 51.1%, Lymphocytes: 22.2%, Eosinophils: 5.2%, Monocytes: 24.5%, Basophils: 0.0%.

Peripheral Blood Smear and Bone Marrow

Peripheral blood smear and bone marrow showed no specific abnormalities except for a lymphocyte count of 32% in BM. However, bone marrow trephine biopsy showed features consistent with marrow involvement by a low-grade mature CD5−, CD10− B-Lymphoproliferative disorder. The possibility of a splenic marginal zone lymphoma was considered in view of massive splenomegaly. IHC showed BCL-2 and CD23 positive.

FISH (Lymphoma Panel)

Lymphoma Panel performed on Bone marrow aspirate slides showed three copies of MYC in 20% cells consistent with Lymphoma with MYC extra copies (NOS).

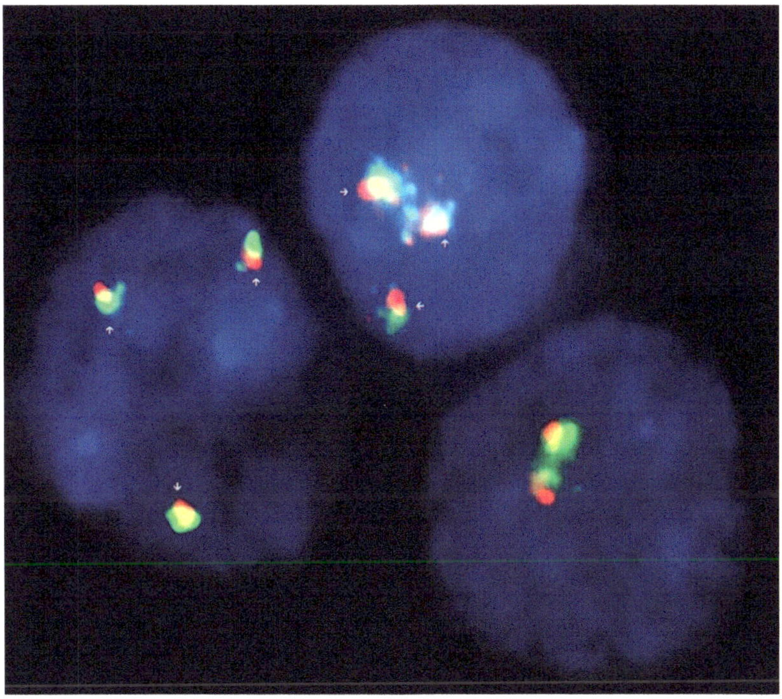

Case Study 10.14 Interphase nuclei showing three copies of *C-MYC* allele

Discussion

This patient with massive splenomegaly showed BM involvement by a low-grade CD5−, CD10− lymphoma. IHC showed CD23+ and BCL-2+. SMZL (Splenic marginal zone lymphoma) was suggested as a diagnosis of exclusion. Patients with double-hit or triple-hit lymphoma have a poorer prognosis compared to DLBCLs without MYC rearrangement. However, the importance of extra copies of MYC is not well understood. Quesada et al. [27] explored the importance of extra copies in 663 patients of DLBCL by FISH. Extra copy lymphoma was defined by the presence of extra copies of MYC or BCL-2 or BCL-6 rearrangement. A total of 76 patients of DLBCL had MYC extra copies including 43 cases of double-hit or triple-hit lymphoma. Patients with MYC extra copies, similar to MYC rearrangement cases had worse overall survival compared with MYC normal patients. Therefore, MYC extra copies in DLBCL are an independent poor prognostic factor similar to MYC rearrangement.

10.15 Case Study 10.15: Non-Hodgkin's Lymphoma—Anaplastic Large Cells Lymphoma

Brief Clinical History

A 14-year-old boy presented with chief complaints of abdominal pain, poor appetite, and vomiting for a month. On investigations, CT showed bilateral pleural effusion and CT abdomen showed generalized lymphadenopathy with hepatosplenomegaly.

CBC: *Hb*: 16.0 g/dl; TLC: 8250/cu mm and Platelets: 150,000/cu mm.

DLC: Neutrophils: 82.0%, Lymphocytes: 10.0%, Eosinophils: 1.0%, Monocytes: 7.0%, Basophils: 0.0%.

Peripheral Blood Smear and Bone Marrow

Smear showed total and absolute counts within normal limits. No abnormal cells were seen. Bone marrow showed infiltration by large abnormal lymphoid cells. Similar cells were seen infiltrating the BM trephine biopsy IHC showed CD45+, CD5+, CD30+, CD3−, CD20−, CD34−, PAX5−, TDT-CD68−, CD33−, CD56−. The bone marrow aspirate and biopsy together with the immunohistochemistry showed marrow infiltration by T lineage, Non-Hodgkin's Lymphoma.

A left supraclavicular lymph node biopsy on IHC was: CA+, CD30+, DCL2+, CD5+, CD4+, and a diagnosis of ALCL was suggested. IHC for ALK was negative.

FISH (Lymphoma Panel)

Lymphoma panel performed on FFPE tissue showed 3–4 copies of ALK in 40% cells confirming the diagnosis of ALCL.

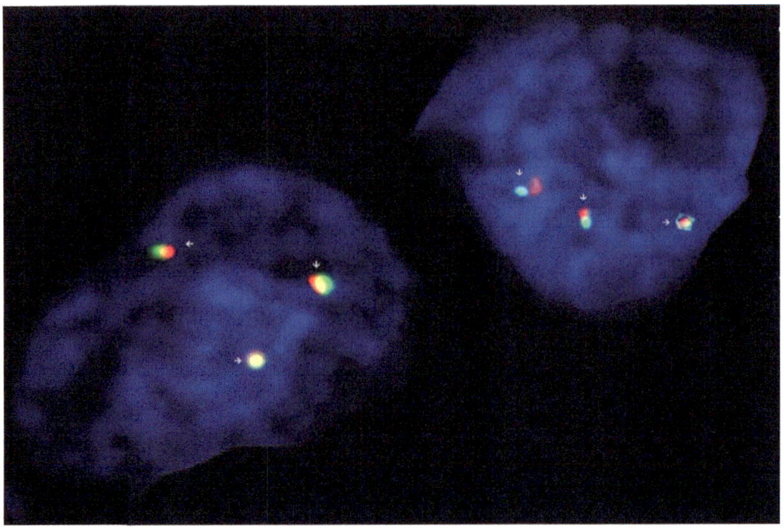

Case Study 10.15 (1) Interphase nuclei showing three copies of *ALK* allele

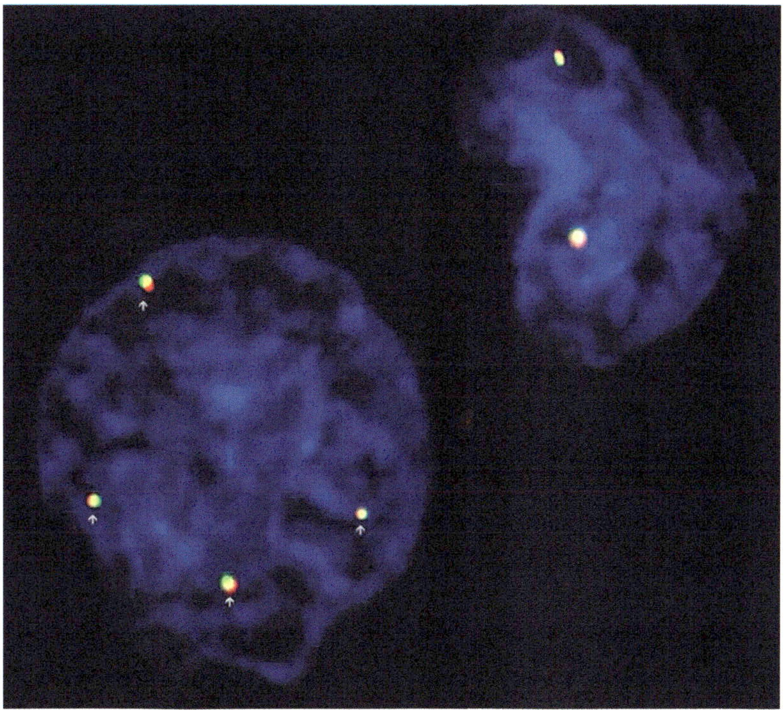

Case Study 10.15 (2) Interphase nuclei showing four copies of *ALK* allele

Discussion

This 14-year-old patient with a characteristic clinical picture of a lymphoma showed no histopathology of a lymph node—non-Hodgkin's lymphoma with IHC suggestive of a T cell lymphoma—ALCL. FISH studies showed ALK gene arrangement in 40% of cells. The ALK gene is fused to the NPM gene as a result of the t(2;5)(P23;q35) and is seen in almost 50–60% of the cases and has a good prognosis. ALCL comprises 10–15% of childhood NHL. In a comparative study of IHC and FISH, Perkins et al. [33] found ALK positivity by FISH in 83% cases and by IHC in 97% cases. However, ALK-negative cases on IHC were confirmed for translocation by FISH. Cataldo et al. [34] studied 27 patients of ALCL by IHC, FISH, and RT-PCR. IHC detected AIK in 14 of 27 patients. In 13 of the 14 cases, ALK was detected by P80 NDM/ALK antibody. In these cases, FISH was positive in all (100%). A translocation involving the ALK gene was detected in all 14 cases by RT-PCR.

10.16 Case Study 10.16: Non-Hodgkin's Lymphoma—Mantle Cell Lymphoma with t(11;14)

Brief Clinical History

A 60-year-old male incidentally detected to have leukocytosis with absolute lymphocytosis. Investigated for further evaluation of the disease and found to have cervical lymphadenopathy and multiple mediastinal lymph nodes on CECT. There was no organomegaly. PET showed generalized lymphadenopathy.

CBC: *Hb*: 11.6 g/dl; TLC: 46,460/cu mm and Platelets: 189,000/cu mm.

Bone Marrow not done.

Flow Cytometry Immunophenotyping Done Outside Had Shown CLL

Histopathology report on a tissue block received from outside of a lymph node showed CLL/SLL. IHC markers were positive for LCA, CD20, CD19, CD23, and CD5 and negative for CD10, cyclin D1, and MUM-1, ki67 was 5–10%.

FISH Lymphoma Panel

Studies on peripheral blood. Suggestive of a mantle cell lymphoma. Positive for t(11,14) in 60% cells and RB1 gene deletion in 40% cells.

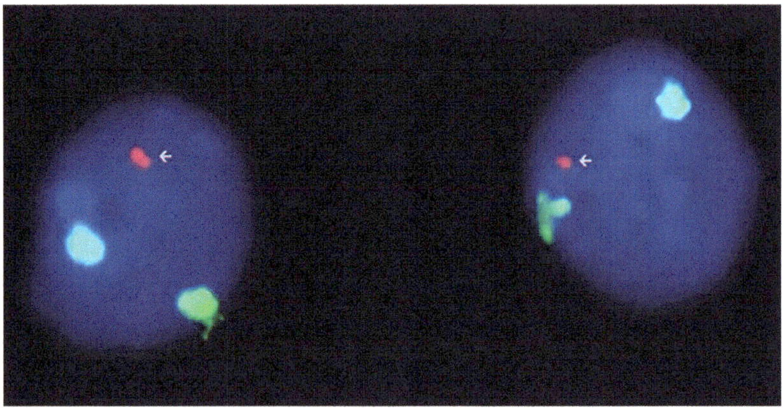

Case Study 10.16 (1) Interphase nuclei showing 1R2G signal pattern using DLEU/LAMP1/CEN 12 probe indicating *RB1* gene deletion

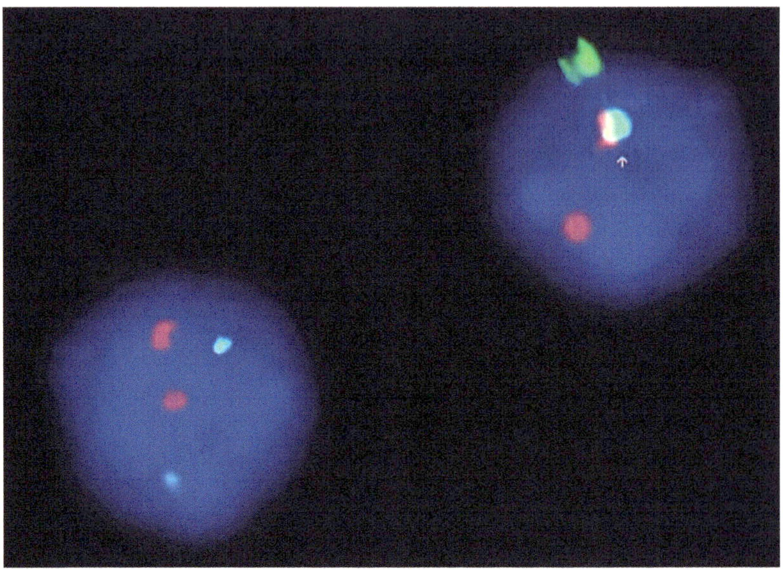

Case Study 10.16 (2) Interphase nuclei showing 1F1G1R aberrant signal pattern using *IGH/CCND1* probe indicating *IGH/CCND1* t(11;14)

Discussion

This patient of leukocytosis with absolute lymphocytosis exhibited a monoclonal B cell lymphocytosis (MBL) both on flow cytometry immunophenotype and IHC in a case of CLL/SLL. MBL precedes virtually all cases of CLL/SLL [35].

In addition, proliferation centers (PCs) in overt CLL/SLL can have cyclin D1 expression in about 30% of cases making it difficult to differentiate it from mantle cell lymphoma, another small B cell lymphoma. In this case, FISH showed t(11;14), a marker for MCL in 95% cases [36]. However, cyclin D-1 on IHC was negative. Also, some MCLs are known to be cyclin D1 negative. Small to intermediate cell lymphomas, including SLL, MCL, follicular lymphoma, and marginal zone lymphomas can show overlapping features with one another as well as reactive lymphoid proliferations. Distinction of MCL from other lymphomas is essential because they are aggressive and response to conventional therapy is poor. In this case, therefore, the presence of t(11;14) by FISH confirmed the diagnosis of MCL in the face of a morphologic diagnosis of CLL/SLL and cyclin D1 negativity.

Two other features of importance in this case were that FISH could be done on peripheral blood (PB) indicating some spill of MCL cells in the PB. There was an additional abnormality of RB1 gene deletion which is of ancillary help toward the diagnosis of MCL in 43–51% cases.

10.17 Case Study 10.17: Non-Hodgkin's Lymphoma—High-Grade B Cell Lymphoma (NOS) with BCL-2 Extra Copies

Brief Clinical History

A 70-year-old male presented with an adrenal mass, diagnosed as DLBCL (GCB) Stage III E with generalized lymphadenopathy. Bone marrow was done for evaluation of the disease. Bone marrow aspirate and trephine biopsy showed no evidence of infiltration by a lymphoma.

CBC: *Hb*: 8.5 g/dl; TLC: 7480/cu mm and Platelets: 89,000/cu mm.

DLC: Neutrophils: 63.3%, Lymphocytes: 16.3%, Eosinophils: 0.7%, Monocytes: 19.7%, Basophils: 0.0.

Peripheral Blood Smear and Bone Marrow

Smear showed mild monocytosis (1470/uL). No abnormal cells were seen. Bone marrow showed no infiltration by a lymphoma. Slides and blocks for review (labeled left adrenal mass): High-grade B cell Non-Hodgkin's Lymphoma, LCA+, CD20+, MUM 1+, ki67 70–80%, CD10−.

FISH Results

On FFPE, tissue showed MYC-negative, BCL-6 negative, and BCL-2—3 copies in 30% of the cells.

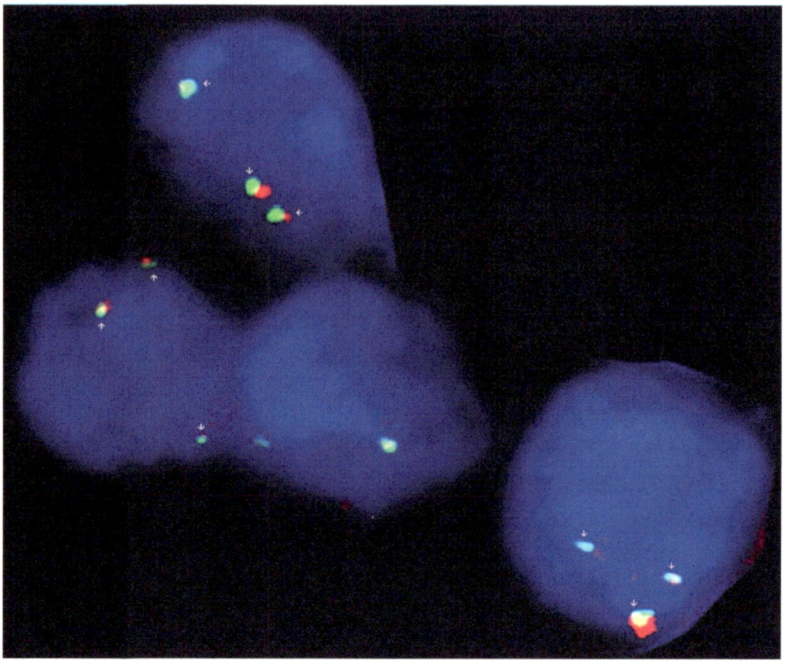

Case Study 10.17 Interphase nuclei showing three copies of *BCL-2* allele

Discussion

On this patient non-Hodgkin's lymphoma, DLBCL was diagnosed on an adrenal mass without peripheral blood or BM involvement (high-grade B cell non-Hodgkin's lymphoma). The tumor was LCA+, CD20+, MUM 1+, CD10−, ki 75–80%. FISH studies on FFPE tissue showed BCL-2, 3 copies in 30% of the cells. BCL-6 and MYC were negative. It was labeled as DLBCL, BCL-2 extra copies (NOS). In a broadening of definition, studies have suggested the inclusion of extra copies of the rearranged genes with MYC or BCL-2 rearrangements [37]. In this case, no rearrangements were detected (NOS cases lacking BCL-2, BCL-6, MYC rearrangements). However, extra copies of BCL-2 were seen. The poor outcome of Double-hit and Triple-hit lymphomas is well documented. However, the clinical significance of extra copies of MYC, BCL-2, or BCL-6 is not well understood. In a study of 130 cases of DLBCL by FISH [38], the prevalence rate of extra copies of MYC, BCL-2, and BCL-6 were 10.8, 20.0, and 16.9, respectively. It was suggested that patients with MYC and BCL-2 and/or BCL-6 gene extra copies might show a trend toward poor prognosis [39].

Annexure

- Case Study 10.1: Chronic Lymphocytic Leukemia with ATM Gene (11q22) Gene Deletion and Trisomy 12
- Case Study 10.2: Chronic Lymphocytic Leukemia with Isolated Trisomy 12
- Case Study 10.3: Chronic Lymphocytic Leukemia with TP53 (17p13) and RB 1 (13q14) Gene Deletions
- Case Study 10.4: Chronic Lymphocytic Leukemia with ATM (11q22) and RB 1 (13q14) Gene Deletions
- Case Study 10.5: Chronic Lymphocytic Leukemia with IgH and Gene Rearrangement and Trisomy 12
- Case Study 10.6: Chronic Lymphocytic Leukemia with isolated RB1 Gene Deletion (13q 14)
- Case Study 10.7: Chronic Lymphocytic Leukemia with Partial deletion of IgH (14q32) Gene Rearrangement
- Case Study 10.8: Chronic Lymphocytic Leukemia with Bialletic and Monoalletic RB1 Gene Deletion (13q14)

- Case Study 10.9: Chronic Lymphocytic Leukemia with Sole Abnormality of ATM Gene Deletion (11q22)
- Case Study 10.10: Leukemic Mantle Cell Lymphoma with Rearranged IgH gene, t(11;14) and Trisomy 12
- Case Study 10.11: Chronic Lymphocytic Leukemia with Trisomy 12 and RB1 (13q14) Gene Deletion
- Case Study 10.12: Non Hodgkin's lymphoma (Follicular Lymphoma) Double Hit Extra Copy Lymphoma (Atypical double hit lymphoma)
- Case Study 10.13: Non Hodgkin's Lymphoma—Double Hit Diffuse Large B Cell Lymphoma
- Case Study 10.14: Non Hodgkin's Lymphoma—Splenic Marginal Zone Lymphoma with Extra Copies of MYC gene (NOS)
- Case Study 10.15: Non Hodgkin's Lymphoma—Anaplastic Large Cells Lymphoma
- Case Study 10.16: Non Hodgkin's Lymphoma: Mantle Cell Lymphoma with t(11;14)
- Case Study 10.17: Non Hodgkin's Lymphoma: High Grade B- Cell Lymphoma(NOS) with BCL -2 Extra Copies

References

1. Puiggros A, et al. Genetic abnormalities in chronic lymphocytic leukemia: where we are and where we go. Biomed Res Int. 2014;435983. https://doi.org/10.1155/2014/435983
2. Guarini A, Marinelli M, Foe R, et al. ATM gene alterations in chronic lymphocytic leukemia patients induce a distinct gene expression profile and predict disease progression. Haematologica. 2012;97(1):47–55.
3. Kadam Amare, Gadage V, Jain H, et al. Clinicopathological impact of cytogenetic subgroups in B-cell chronic lymphocytic leukemia: experience from India. Indian J Cancer. 2013;50(3):261–7.
4. Autore F, Strati P, Laurenti L, et al. Morphologic immunophenotypic and genetic features of chronic lymphocyic leukemia with trisomy 12: a comprehensive review. Haematologica. 2018;103(6):931–8.
5. Lopez C, Delgado J, Costa D, et al. Different distribution of NOTCH1 mutations in chronic lymphocytic leukemia with isolated trisomy 12 or associated with other chromosomal aberrations. Genes Chromosomes Cancer. 2012;51(9):881–9.
6. Dohner H, Stilgenbauer S, Dohner K, et al. Chromosome aberrations in B-cell chronic lymphocytic leukemia: reassessment based on molecular genetic analysis. J Mol Med. 1991;77:266–81.
7. Fabbri G, Khiabanian H, Holmes AB, et al. Genetic lesions associated with chronic lymphocytic leukemia transformation to Richter syndrome. J Exp Med. 2013;210(11):2273–88.
8. Streford J. The genomic landscape of chronic lymphocytic leukemia: biological and clinical implications. Br J Haemat. 2015;169(1):14–31.
9. Puiggros A, Blanco G, Espinet B. Genetic abnormalities in chronic lymphocytic leukemia: where we are and where we go. Biomed Res Int. 2014;435983. https://doi.org/10.1155/2014/435983
10. Stilgenbauer S, Bullinger L, Lichter P. The German CLL study genetics of chronic lymphocyric leukemia: genomic aberrations and VH gene mutation status in pathogenesis and clinical course. Leukemia. 2002;16:993–1007.
11. De Braekeleer M, Le Bris MJ, Basinko M, et al. Incidence of chromosomal anomalies detected by interphase fluorescent in situ hybridisation in chronic lymphoid leukemia. Int J Hematol Oncol. 2015;4:133–41.
12. Kadam Amare PS, Kakade S, Chopra K, et al. Comprehensive genetic analysis by integration of conventional karyotyping and interphase FISH helps refinement of biological subclasses with clinical impact in chronic lymphocytic leukemia. J Cancer Therapy. 2016;7:427–38.
13. Salcedo LMJ, Dalia S. del (13q) chronic lymphocytic leukemia. Atlas Genet Cytogenet Oncol Haematol. 2017;21(8):292–3.
14. Dohner H, Stilgenbauer S, Benner A, et al. Genomic aberrations and survival in chronic lymphocytic leukemia. N Engl J Med. 34:1910–6.
15. Quintero-Rivera F, Nooraie F, Rao PN, et al. Frequency of 5′ IGH deletions in B-cell chronic lymphocytic leukemia. Cancer Genet Cytogenet. 2009;190(1):33–9.
16. Huh J, Mun YC, Yoo ES, et al. Submicroscopic deletions of immunoglobulin heavy chain gene (IGH) in precursor B lymphoblastic leukemia with IGH rearrangements. Ann Lab Med. 2015;35(1):128–31113.
17. Garg R, Wierda W, Ferrajoli A. The prognostic difference of monoallelic versus biallelic deletion of 13q in chronic lymphocytic leukemia. Cancer. 2012;118(14):3531–7.
18. Grygalewicz B, Woroniecka R, Rygier J, et al. Monoallelic and biallelic deletion of 13q14 in a group of CLL/SLL patients investigated by CGH hematological cancer and SNP array (8x60k). Mol Cytogenet. 2016;9:1.
19. Guarini A, Marinelli M, Robin F, et al. ATM gene alterations in chronic lymphocytic leukemia patients induce a distinct gene expression profile and predict disease progression. Haematologica. 2012;97(1):47–55.

20. Rossi D, Gaidano G. ATM and chronic lymphocytic leukemia: mutations, and not only deletions, matter. Haematologica. 2012;97(1).
21. Hallek M. Chronic lymphocytic leukemia: 2020 update on diagnosis, risk stratification and treatment. Am J Hemat. 94(11):1266. https://doi.org/10.1002/ajh.25595.
22. Neilson JR, Cai M, Bienz N. Leukemic mantle cell lymphoma with t(11;14) and trisomy 12 showing clinical features of stage AoB cell chronic lymphocytic leukemia. J Clin Path: Mol Pathol. 1995;48:M163–6.
23. Jares P, Colomer D, Campo E. Molecular pathogenesis of mantle cell lymphoma. J Clin Invest. 2012;122(10):3416–23.
24. Navarro B, Gracia Marco JA, Jones D. Association and clonal distribution of trisomy 12 and RB1 (13q14) deletions in chronic lymphocytic leukemia. Br J Haemat. 1998;102(5):1330–4.
25. Fekete MD, Mrsic SD, Autic D, et al. Co-existence of trisomy 12 and del (13q) in two patients with chronic lymphocytic leukemia. Arch Biol Sci. 2009;61(4):607–11.
26. Amare K, Gadage V, Jain H, et al. Clinico-pathological impact of cytogenetic subgroups in B-cell chronic lymphocytic leukemia: experience from India. Indian J Cancer. 2013;50(3):261–7.
27. Quesada A. Increased MYC number is a independent prognostic factor in patient with diffuse large B-cell lymphoma. Mod Pathol. 2017;30(12):1688–97.
28. Li S, et al. Advances in pathological understanding of high-grade B cell lymphomas. Expert Rev Hematol. 2018;11(8):637–48.
29. Li S, Seegmiller AC, Lin P, et al. B-cell lymphomas with concurrent MYC and BCL 2 abnormalities other than translocations behave similarly to MYC/BCL 2 double hit lymphomas. Mod Pathol. 2015;28:208–17.
30. Guo Y, Karube K, Kawano R. BCL −2 negative follicular lymphomas frequently have BCL-6 translocation and / or. Pathol Int. 2007;57(3):148–52.
31. Swerdlow SH, Campo E, Harris NL, Jaffe ES, Pileri SA, Stein H, Thiele J (Eds). WHO classification of tumours of haematopoietic and lymphoid tissues (revised 4th ed). IARC: Lyon; 2017.
32. Scott DW, King RL, Staiger AM, et al. High grade B cell lymphoma with MYC and BCL2 and / or BCL6 rearrangements with diffuse large B cell lymphoma morphology. Blood. 2018;131(18):2060–4.
33. Perkins SL, Pickering D, Lowe EJ, et al. Childhood anaplastic large cell lymphoma has a high incidence of ALK gene rearrangement as determined by immunohistochemical staining and fluorescent in situ hybridisation. Br J Haemat. 2005;131:624–7.
34. Cataldo KA, Jalal SM, Low ME, et al. Detection of t(2;5) in anaplastic large cells lymphoma: comparison of immunohistochemical studies, FISH and RT-PCR in paraffin-embedded tissue. AM-J Surg Path. 1999;23(11):1386–92.
35. Swerdlow SH, Campo E, Harris NL, Jaffe ES, Pileri SA, Stein H, Thiele J, (Eds). WHO classification of tumours of haematopoietic and lymphoid tissues (revised 4th ed). IARC: Lyon; 2017.
36. Bentz JS, Rowe LR, Anderson SR. Rapid detection of the t(11;14) translocation in mantle cell lymphoma by interphase fluorescence in situ hybridization on archival cytopathologic material. Cancer. 2004;102(2):124–31.
37. Swerdlow SH, Campo E, Pileri S, et al. The 2016 revision of the World Health Organization classification of lymphoid neoplasms. BLOOD First Edition Paper; 2016.
38. Huang S, Nong L, Wang W, et al. Prognostic impact of diffuse large B cell lymphoma with extra copes of MYC / BCL and or BCL 6: comparison with double/triple hit lymphoma and double expressor lymphoma. Diagn Pathol. 2019;14(1):81.
39. Swerdlow SH. Diagnosis of 'Double Hit' diffuse large B-cell lymphoma and B-cell lymphoma, unclassifiable, with features intermediate between DLBCL and Burkitt lymphoma: when and how, FISH versus IHC. Hematology Am Soc Hematol Educ Program. 2014;2014(1):90–9.

Part III

Case Studies in Multiple Myeloma

11 Multiple Myeloma

11.1 Case Study 11.1: Multiple Myeloma with IgH Gene Rearrangement, t(11;14)

Brief Clinical History
A 77-year-old male patient presented with complaints of backache for a few weeks. Bone marrow was done for evaluation of suspected myeloma.

CBC: *Hb*: 11.8 g/dl, TLC: 4770 /cu mm and Platelets: 2,50,000 /cu mm.

DLC: Neutrophils: 58%, Lymphocytes: 31%, Plasma cells: 1%, Eosinophils: 2%, Monocytes: 8%.

Peripheral Blood Smear and Bone Marrow
Smear showed Absolute blood counts within normal limits. Bone Marrow aspirate showed an increase in Plasma cells which constituted ~37.8% of total nucleated cells with a few binucleate Plasma cells and immature plasma cells. Morphological features were compatible with a diagnosis of Plasma cell Myeloma. Bone biopsy showed a similar picture with IHC, 138+, CD56+, kappa light chain+, lambda light chain negative. Picture consistent with plasma cell myeloma lambda light chain restricted.

FISH Multiple Myeloma Panel on Bone Marrow Aspirate
FISH with IGH breakapart probe showed IGH rearrangement. Further characterization showed t(11;14) in 95% cells. There was no evidence of TP53 gene deletion and RB1 gene deletion.

M. Bhargava, *Hematologic Malignancies*, https://doi.org/10.1007/978-981-33-4799-1_11

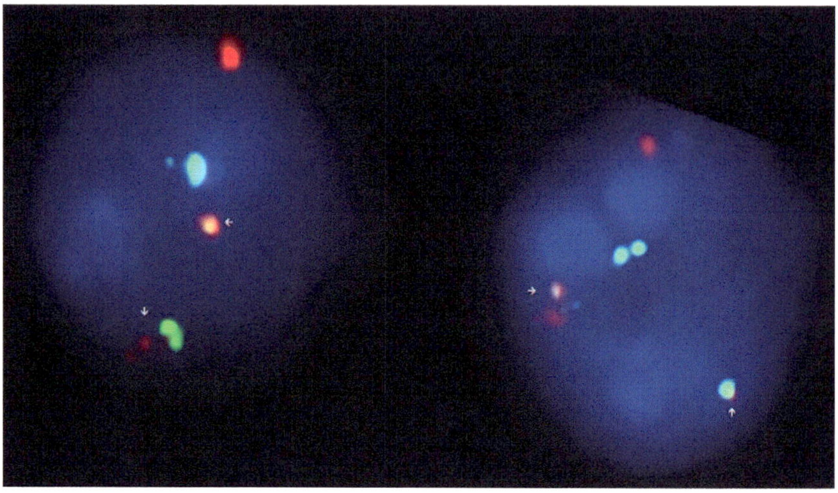

Case Study 11.1 Interphase nuclei showing 2F1G1R signal pattern using *IGH/CCND1* LSI probe indicating *IGH/CCND1*, t(11;14)

Discussion

Cytogenetic abnormalities in multiple myeloma are complex and occur at various time points in the disease course, from smoldering multiple myeloma (SMM) to MM. Specific abnormalities define clinical presentation, progression of the disease, and management strategies. Primary molecular cytogenetic abnormalities classify MM into Trisomic MM, IgH translocated MM including t(11;14), t(4;14), t(14;16), t(14;20) and other IgH translocations as well as combined IgH translocated/trisomic myelomas, isolated monosomy 14 or any other cytogenetic abnormalities. Secondary abnormalities of 13q, 17p gain of 1q, del 1p, and MYC translocations can also occur [1].

Immunoglobulin heavy chain (IgH) gene rearrangements are used as molecular markers in ~80% of lymphomas and multiple myelomas since they represent lineage-specific markers and the third complementarity determining region (CDR3) is unique to each clone [2]. Translocations involving the IGH locus are reported to be the most common genetic lesions in MM including t(11;14) in 15–21%, t(4;14) in 10–14%, t(16,14) in less than 5% and t(6;14) in 5% of all cases. The t(11;14)(q13;q32) as seen in this case leads to the overexpression of cyclin D1. Deregulation of MYC by (8:14) (8q24;q32) is more of a secondary event associated with tumor progression. IgH translocated MM with t(11;14)(q13;q32) shows good prognosis, standard risk MM median OS 7–10 years [1, 3].

11.2 Case Study 11.2: Multiple Myeloma with Partial Deletion of IgH Gene Rearrangement, Monosomy Chromosome 13, Trisomy Chromosome 11 and Gain of Chromosome 1q

Brief Clinical History

A 44-year-old male presented with weakness and acute kidney injury. Investigations done outside had shown an SPE band of 2.6 g/dL. Evaluated for multiple myeloma.

CBC: *Hb*: 5.8 g/dl; TLC: 6490 /cu mm and Platelets: 1,68,000 /cu mm.

DLC: Myelocytes: 3.0%, Metamyelocytes: 2.0%, Neutrophils: 69.0%, Lymphocytes: 22.0%, Eosinophils: 0.0%, Monocytes: 4.0%, Basophils: 0.0%.

Peripheral Blood Smear and Bone Marrow
Smear showed blood counts within normal limits. Neutrophils showed a left shift. Bone marrow aspirate and bone biopsy showed a striking increase in plasma cells (50%) IHC showed 138+, CD56+, lambda light chain+, kappa light chain−. Picture consistent with plasma cell myeloma lambda light chain restricted.

Serum-free Light Chain Assay
Kappa free light chain 24.70, Lambda free light chain 2860, kappa/lambda ratio 0.009, SPE Band 2.6 g/dL. The picture was consistent with lambda light chain restricted.

FISH Multiple Myeloma Panel on Bone Marrow
FISH with IGH breakapart probe showed no IGH translocation t(11;14), t(4;14), and t(14;16). However, partial deletion of IGH was seen in 70% interphase cells. Monosomy of chromosome 13 in 48% cells and trisomy of chromosome 11 in 95% interphase cells each was seen. In addition, there was 1q gain.

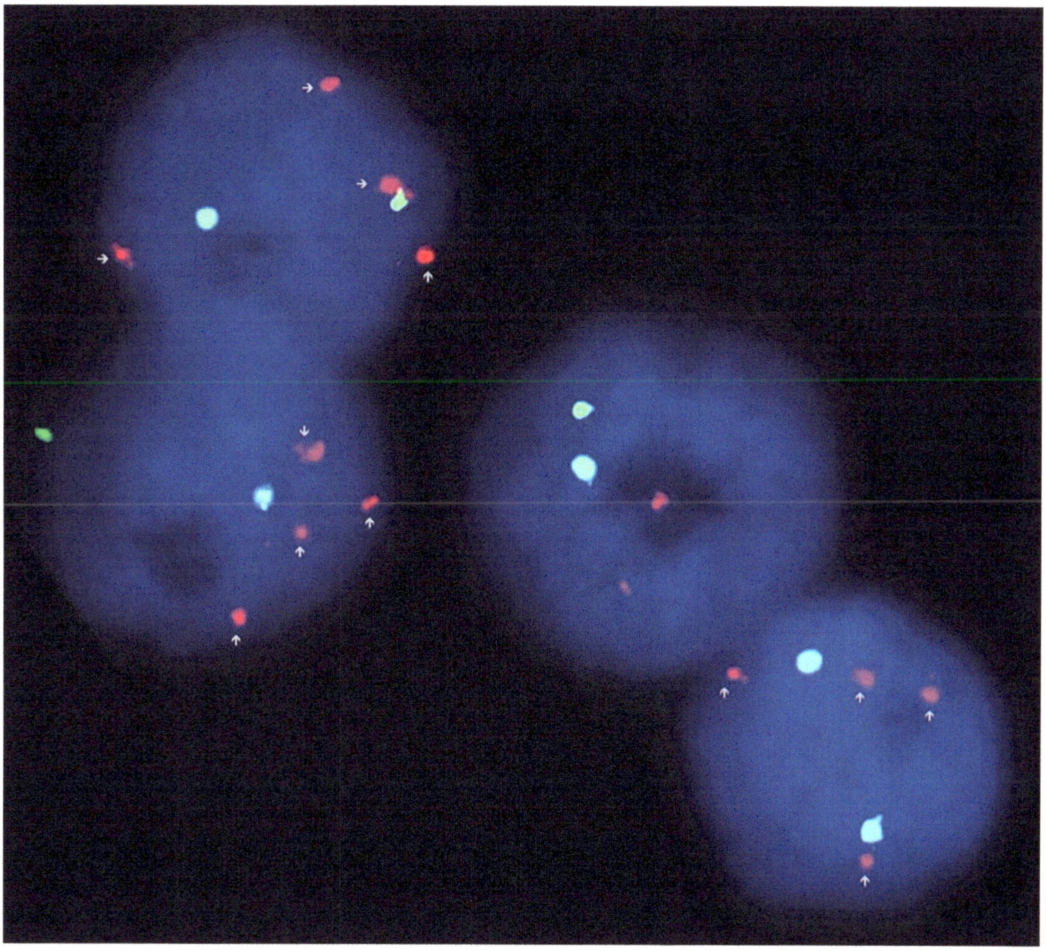

Case Study 11.2 Interphase nuclei showing 2G4R signal pattern using 1p/1q LSI probe indicating 1q gain. One nucleus is showing 2G2R signal pattern indication normal cells

Discussion

In this patient, FISH with IgH breakapart probe showed partial deletion of IgH at the 5′ end. In addition to IgH rearrangements, IgH submicroscopic deletions are seen in 14–21% of patients with multiple myeloma and in B-ALLs [4]. In another study Hwang et al. [5], IgH deletion was found in 14% of patients of MM. FISH can identify submicroscopic deletion adjacent to the breakpoints of rearrangements. The significance, however, in this patient was of monosomy 13, trisomy of chromosome 11, and 1q gain. Isolated monosomy 13 has a standard risk of progression and gain of 1q21 has a high risk of progression. In another study Shin et al. [6], to determine the prognostic effect of monosomies in MM, monosomies of chromosomes 13 and 16, deletion or monosomy of 13q 14, and loss of X were associated with reduced progression-free survival.

Trisomies of chromosomes 1q, 9q, and 11q are seen in 30–50% of patients. In a study in 484 patients, among patients with high-risk FISH (as in this patient with 1q gain), 49 patients who had at least 1 trisomy (as in this patient) a median overall survival was not reached [7]. It is suggested that trisomies in high-risk patients ameliorate the usual adverse impact [8].

11.3 Case Study 11.3: Multiple Myeloma with IgH Gene Rearrangement with t(14;16) and Monosomies of Chromosomes 13 and 16

Brief Clinical History

A 61-year-old man presented with lytic lesions and pain lower chest for 2 months, weakness, confusion, and renal impairment. Bone marrow was done for evaluation.

CBC: *Hb*: 8.8 g/dl; TLC: 4680 /cu mm and Platelets: 1,68,000 /cu mm.

DLC: Neutrophils: 50.0%, Lymphocytes: 32.7%, Eosinophils: 2.8%, Monocytes: 14.5%, Basophils: 0.0%.

Peripheral Blood Smear and Bone Marrow

Smear showed blood counts within normal limits. Bone marrow aspirate and bone biopsy showed almost a complete replacement by plasma cells and a few plasmablasts. IHC showed 138+, CD56+, kappa−, lambda+. The picture was consistent with the lambda light chain restricted plasma cell myeloma.

Serum-free Light Chain Assay

Kappa free light chain 21.90. Lambda free light chain 2760.00. Kappa/lambda ratio 0.008.

The picture was consistent with lambda chain restricted multiple myeloma.

FISH Multiple Myeloma Panel on Bone Marrow

FISH with IGH breakapart probe showed IGH rearrangement. Further characterization showed t(14;16) in 90% cells, Monosomy of chromosome 13 in 80%, of chromosome 16 in 90% of interphase cells each. Monosomy of chromosomes 13 and 16 reveal hypodiploidy.

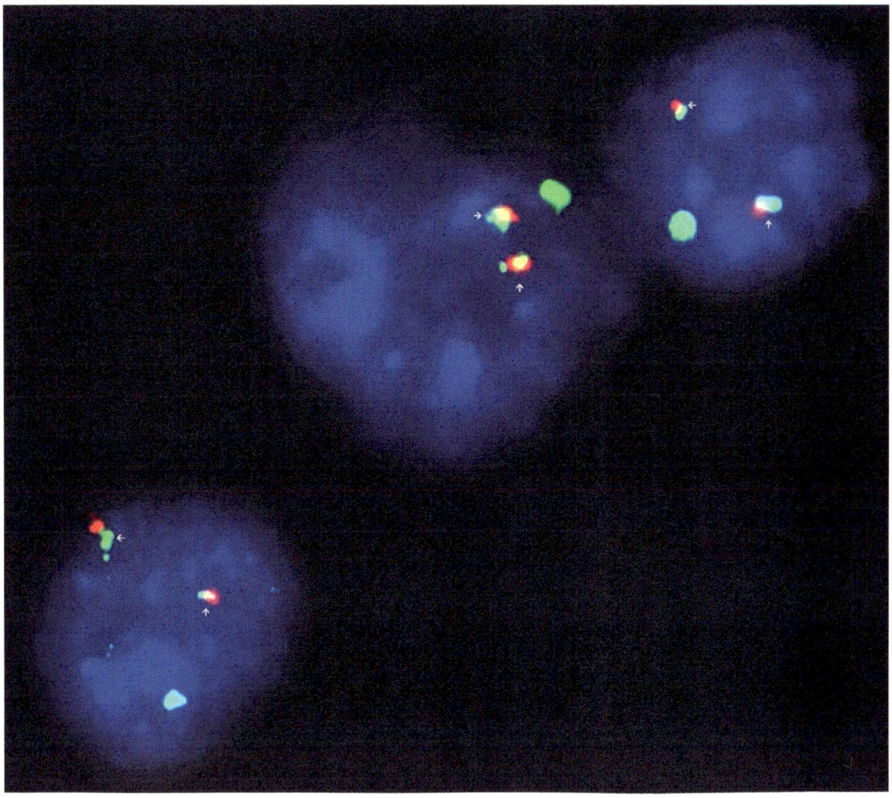

Case Study 11.3 Interphase nuclei showing 2F1G aberrant signal pattern using *IGH/MAF* LSI probe indicating *IGH/MAF* fusion t(14;16)

Discussion

FISH characterized this patient to have IgH translocated multiple myeloma t(14;16) with secondary changes of monosomies of chromosome 13 and chromosome 16. There are two broad types of cytogenetic abnormalities in MM: Primary and Secondary. Primary abnormalities are thought to occur at the MGUS level. Secondary abnormalities can occur in any of the primary subtypes and can affect disease outcome. The MM is first established by any one IgH translocations and/or trisomies (as clone, e.g., in this case). IgH translocated forms commonly are t(11;14), t(4;14), t(14;16), and t(14;20). In general, primary cytogenetic abnormalities are non-overlapping. Occasional patient may have isolated monosomy 14. Secondary abnormality of monosomy chromosome 13/del (13q), as in this case, has an adverse prognosis [1].

11.4 Case Study 11.4: Multiple Myeloma with Biallelic IgH Gene Rearrangement, t(4;14)

Brief Clinical History

A 57-year-old male presented with lower backache for the last 5–10 days. He was found to have generalized osteopenia with a PET-CT scan showing the collapse of D12 vertebrae. Serum-free light chain assay (SFLC) showed an increase in the kappa light chain. Bone marrow was done for suspected myeloma.

CBC: *Hb*: 8.4 g/dl, TLC: 9900 /cu mm and Platelets: 3,11,000 /cu mm.

DLC: Neutrophils: 64.9%, Lymphocytes: 23.9%, Eosinophils: 1.9%, Monocytes: 9.1%, Basophils: 0.2%.

Peripheral Blood Smear and Bone Marrow

Smear showed total and absolute counts within the normal reference range. Bone marrow aspirate and imprint smears showed an increase in mature and immature plasma cells, constituting ~40.0% of total nucleated cells. There were a few binucleate forms with an occasional plasmablast. The picture was consistent with a plasma cell myeloma. The bone marrow biopsy showed plasma cell myeloma IHC 138+, CD56+ kappa, chain restricted lambda negative.

Serum-free Light Chain Assays

Kappa free light chain >14,000, Lambda, free light chain 12.10. Kappa/lambda ratio cannot be calculated as the value of kappa was >14,000.

FISH-Multiple Myeloma Panel on Bone Marrow

FISH with IGH breakapart probe showed biallelic IGH rearrangement. Further characterization showed t(4;14) in 62% cells. However, there was no evidence of TP53 gene deletion and RB1 gene deletion.

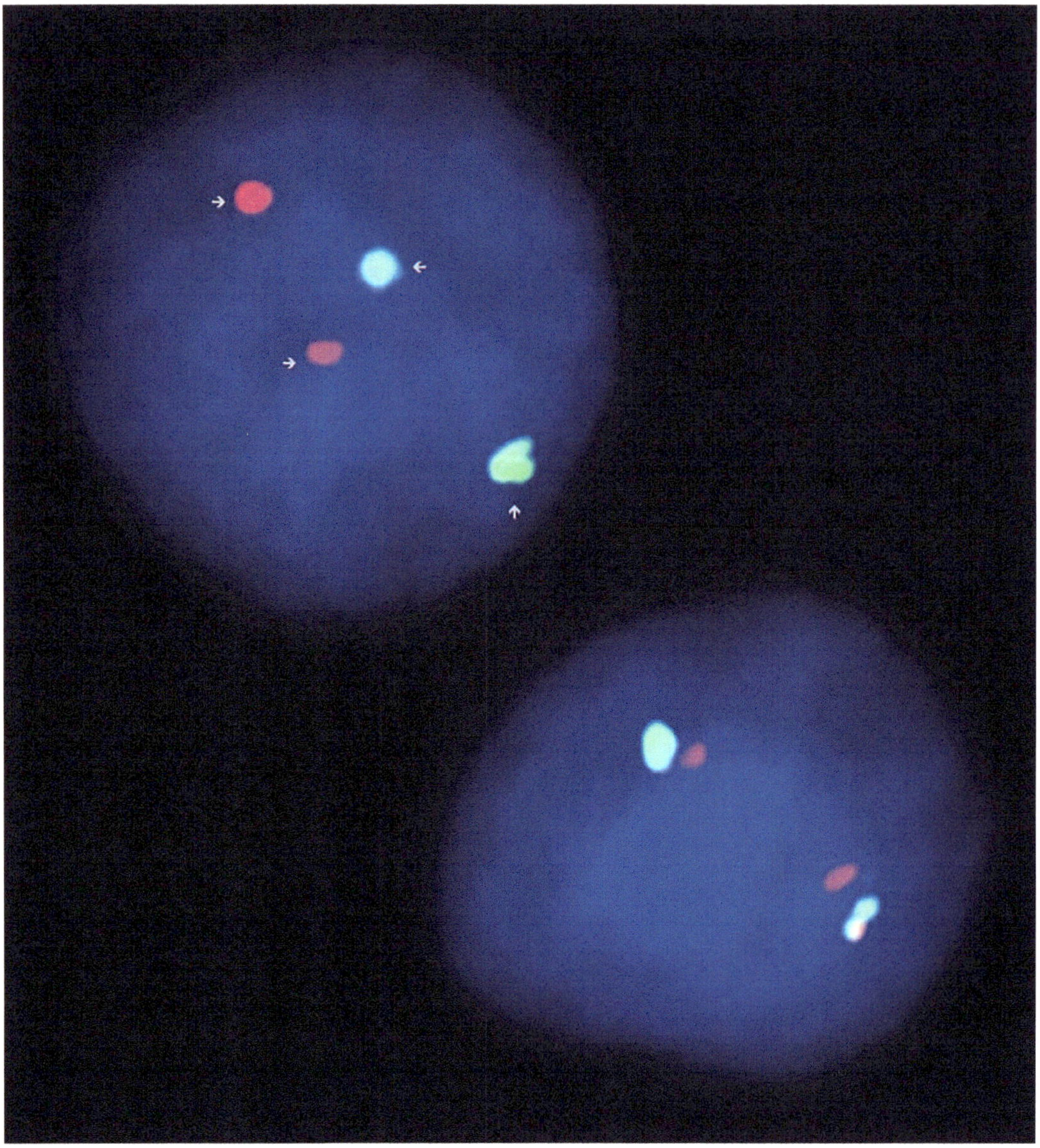

Case Study 11.4 (1) Interphase nuclei showing 2G2R signal pattern using *IGH* probe indicating biallelic *IGH* rearrangement

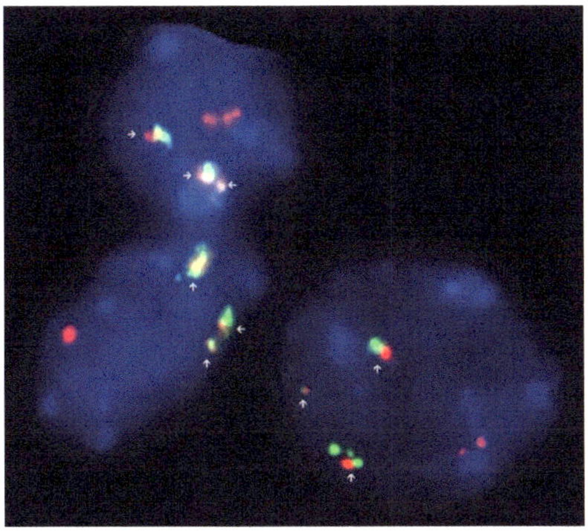

Case Study 11.4 (2) Interphase nuclei showing 3F1R aberrant signal pattern using *IGH/FGFR3* LSI probe indicating *IGH/FGFR3* fusion, t(4;14)

Discussion

This patient had an IgH translocated multiple myeloma, t(4;14) in whom there was biallelic IgH rearrangement. In mature B cell malignancies, up to 80% of patients show biallelic IgH rearrangements [9, 10]. t(4;14) is MM specific and is seen in 12–15% of patients and is one of the high-risk factors in risk stratification of multiple myeloma as per Mayo Clinic risk stratification for multiple myeloma (mSMART) as also by other studies [11]. The presence of del(17p), t(4;14),t(14;16), gain 1q or p53 mutation is considered high-risk multiple myeloma. Patients with t(4;14) translocation, del(17p), and gain (1q) are at a higher risk of progression from MGUS or SMM to multiple myeloma. The presence of any two high-risk factors is considered double-hit multiple myeloma, three or more high-risk factors as triple-hit myeloma [7].

11.5 Case Study 11.5: Multiple Myeloma with del(17) (p13.1) and Trisomy of Chromosome 11

Brief Clinical History

A 66-year-old female presented with back pain for one and half months, with multiple FDG avid osteolytic lesions involving bilateral ribs and an M spike of 5.8 g/dL. Bone marrow was done for evaluation.

CBC: *Hb*: 7.7 g/dl, TLC: 8050 /cu mm and Platelets: 1,89,000 /cu mm.

DLC: Neutrophils: 55.0%, Lymphocytes: 36.0%, Eosinophils: 0.0%, Monocytes: 7.0%, Basophils: 1.0%.

Peripheral Blood Smear and Bone Marrow

Smear showed absolute counts within normal limits and few plasmacytoid lymphocytes. Bone marrow aspirate smears showed replacement of the marrow by mature and immature plasma cells constituting ~86.7% of total nucleated cells. There were a few binucleate forms with occasional plasmablast. The picture was consistent with plasma cell myeloma. Bone biopsy showed a similar picture with IHC showing CD138+, CD56+, Kappa light chain+, lambda light chain negative.

Serum-free Light Chain Assay

Serum M Spike SPE 5.0. Kappa light chain 546. Lambda light chain 9.4. Kappa/lambda ratio 57.9. Beta 2 microglobulin 5533. The picture was consistent with the kappa light chain restricted multiple myeloma.

FISH-Multiple Myeloma on Bone Marrow

del(17)(p13.1) (TP53 deletion) was seen in 20% cells. Also, there were three copies of MYEOV, 11q13/trisomy of chromosome 11 in 90% interphase cells each. IGH breakapart probe showed no IGH translocation t(11;14), t(4;14), and t(14;16) and no RB1 gene deletion.

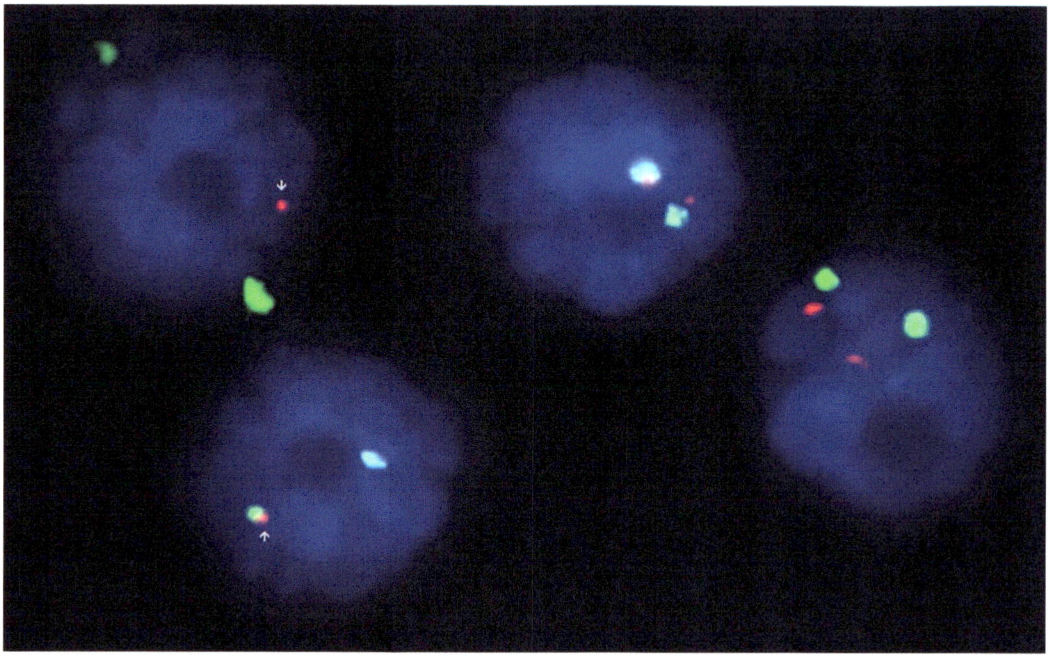

Case Study 11.5 Interphase nuclei showing 2G1R signal pattern using *TP53* LSI probe indicating deletion of *TP53* gene

Discussion

This patient with multiple myeloma showed TP53 gene deletion by FISH as well as trisomy of chromosome 11. Multiple myeloma is clinically and biologically heterogeneous, the outcome being determined by patient characteristics, tumor load, and genomic abnormalities in malignant plasma cells. Recurrent abnormalities on interphase FISH give key prognostic information [12]. Deletion of chromosome 17(17p 13.1) is an adverse cytogenetic marker at diagnosis. FISH detects del(17p) in 5–20% patients with newly diagnosed MM (NDMM). It is seen more frequently at relapse and more aggressive disease. It is a critical regulator of the cell cycle and apoptosis. Being a secondary cytogenetic abnormality, del(17p) coexists with other FISH abnormalities as in this case (trisomy 11) which may also determine outcome [11]. A study of 310 pts of MM. Lakshman et al. [13] described multiple other abnormalities such as t(4;14), t(6;14), t(11;14), t(14;16), t(14;20), IgH gene rearrangement, trisomy/tetrasomy, hyperdiploidy, del(13q), and/or monosomy 13, del(13q), monosomy 13, 1q22 gain, del(1p31). In this patient, therefore, presence of trisomy 11 in addition to TP53 gene deletion may also determine the outcome.

11.6 Case Study 11.6: Multiple Myeloma with Trisomy Chromosome 17 and 1q Amplification

Brief Clinical History

A 66-year-old male presented with bilateral pneumonia and multiple lytic lesions involving almost all the bones on HRCT. Diagnosed outside as Multiple Myeloma.

CBC: *Hb*: 8.1 g/dl; TLC: 7910 /cu mm and Platelets: 2,06,000 /cu mm; TLC: 7910/uL.

DLC: Neutrophils: 67.4%, Lymphocytes: 24.5%, Eosinophils: 1.8%, Monocytes: 6.2%.

Peripheral Blood Smear and Bone Marrow
Smear showed blood counts within normal limits. Bone marrow aspirate and bone biopsy showed sheets and aggregates of plasma cells (40.0%). IHC showed CD138−, CD56+, kappa light chain+, lambda light chain−. The picture was consistent with kappa light chain restricted plasma cell myeloma.

Serum-free Light Chain Assay
Kappa free light chain 513.0, Lambda free light chain 19.3. kappa/lambda ratio 26.58. The picture was consistent with kappa light chain restricted multiple myeloma.

FISH Myeloma Panel on Bone Marrow
FISH with IGH breakapart probe showed no IgG translocation t(4;14), t(11;14), t(14;16). There was no TP53 gene deletion but there was a trisomy of chromosome 17. No RB1 gene deletion was seen. 1q amplification was seen. Using CKS1B/CDKN2C Dual-Color Probe in 98% cells.

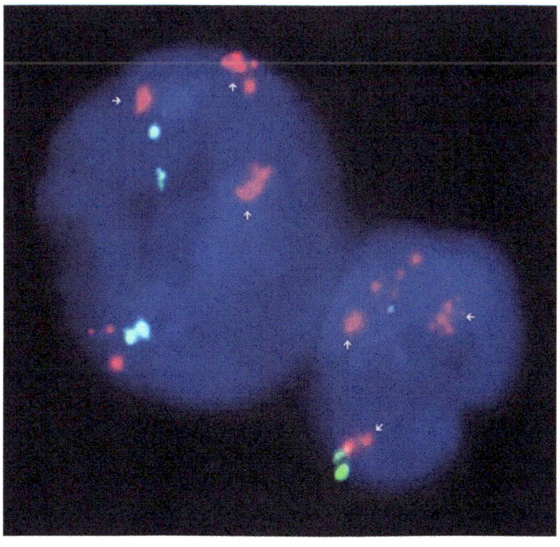

Case Study 11.6 Interphase nuclei showing 1q amplification, i.e., cluster of orange signal pattern using 1p/1q LSI probe

Discussion
In a study of trisomies in multiple myeloma in 484 patients, Rajkumar et al. [7] reported 57% patients to have trisomy of at least one odd-numbered chromosome 3, 7, 9, 11, 13, 15, 17 (chromosome 17 in this case). Among patients with high-risk FISH (1q amplification in this case), 49 patients who had at least 1 trisomy, median overall survival was not reached, compared with 3 years for high-risk patients without a trisomy indicating that trisomies may ameliorate the adverse prognostic effect of a high-risk marker. Forty-eight percent had trisomy of at least two odd number chromosomes called hyperdiploidy.

Gains in 1q are identified in 20–50% of patients of MM. The frequency increases from smoldering to relapsed disease. Studies have shown an association between +1q and other FISH abnormalities such as del17p and del13q and other markers of disease like high Beta 2, microglobulin, LDH, anemia, bone marrow plasmacytosis, and stage 3 disease. Patients with gains in chromosome 1 have a worse PFS and OS even after undergoing HCT [14].

11.7 Case Study 11.7: Multiple Myeloma (Plasmablastic) with TP53 Gene Deletion and Tetrasomies of Chromosomes 4, 11, 13, 14, and 1

Brief Clinical History
A 53-year-old female presented with pain in the ribs since the past 2–3 weeks. PET-CT revealed multiple punched out lesions. M spike was 1.6 g/dL. Bone marrow was done for further evaluation.

CBC: *Hb*: 8.9 g/dl, TLC: 11,850 /cu mm and Platelets: 2,48,000 /cu mm.

DLC: Neutrophils: 72.0%, Lymphocytes: 13.0%, Eosinophils: 1.0%, Monocytes: 9.0%, Basophils: 0.0%; Blasts: 1.0%; Myelocytes: 4.0%.

Peripheral Blood Smear and Bone Marrow

Smear showed Neutrophilic leukocytosis with mild left shift and an occasional plasma cell. Bone marrow aspirate smears showed replacement of the marrow by abnormal plasma cells constituting ~48% of the total nucleated cells, focal clusters of predominantly large abnormal plasma cells with high N:C ratio (plasmablastic morphology) with basophilic vacuolated cytoplasm and large immature nucleus, few showing nucleoli. A few mature forms and a few binucleate to multinucleate forms were also seen. The picture was consistent with Plasmoblastic multiple myeloma. The bone biopsy showed a similar picture with IHC, CD138+, CD56+, with lambda light chain restriction.

FISH-Multiple Myeloma Panel on Bone Marrow

FISH with IGH breakapart probe showed no IGH translocation t(11;14), t(4;14), and t(14;16). However, there were four copies of IGH, 14q32/ Tetrasomies of chromosome 14, chromosome 13, chromosome 17, chromosome 11, and of chromosome 4 in 70% interphase cells each analyzed. Also, del(17)(p13.1)(TP53 deletion) was seen in 80% of interphase cells analyzed. The above observations indicated a highly proliferative disease based on tetraploidy and del(17) (p13.1) (TP53 deletion) which has been shown to have a poor prognosis.

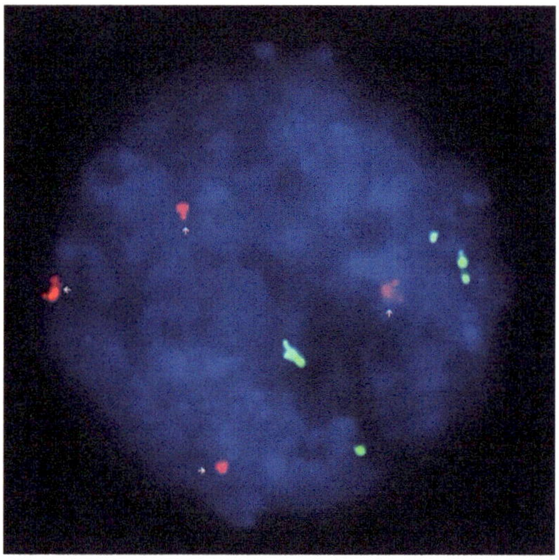

Case Study 11.7 (1) Interphase nuclei showing 4G4R signal pattern using IGH/FGFR3 DC DF probe indicating tetrasomy 4

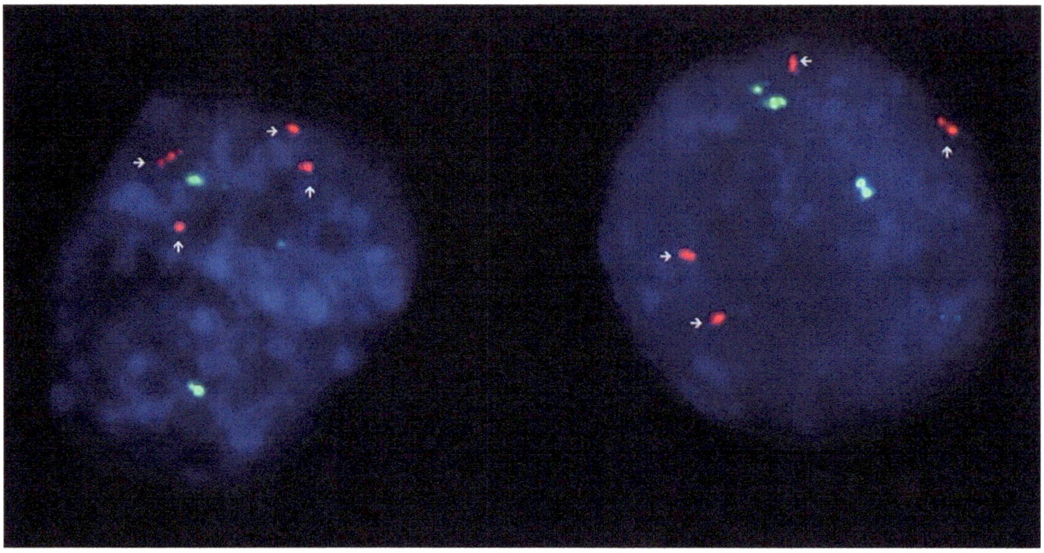

Case Study 11.7 (2) Interphase nuclei showing tetrasomy 11 using *IGH/CCND1* DC DF probe

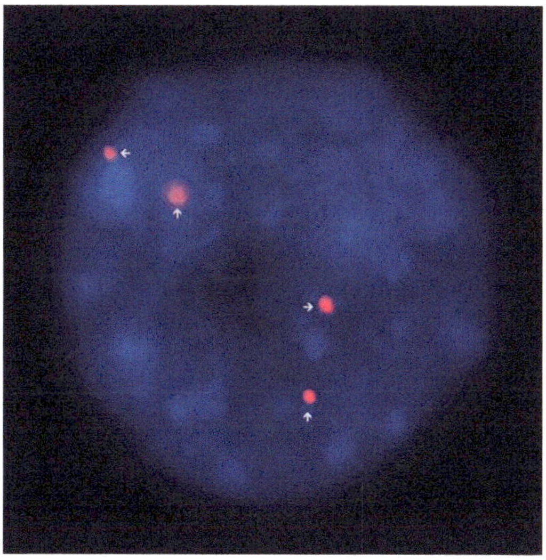

Case Study 11.7 (3) Interphase nuclei showing tetrasomy 13 using deletion 13q probe

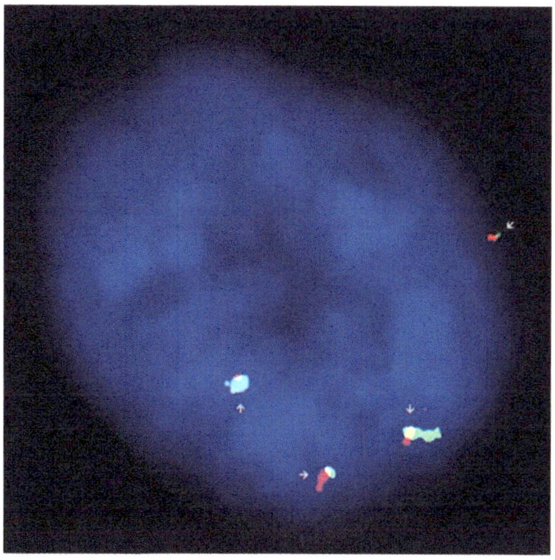

Case Study 11.7 (4) Interphase nuclei showing 4F signal pattern using *IGH* probe indicating tetrasomy 14

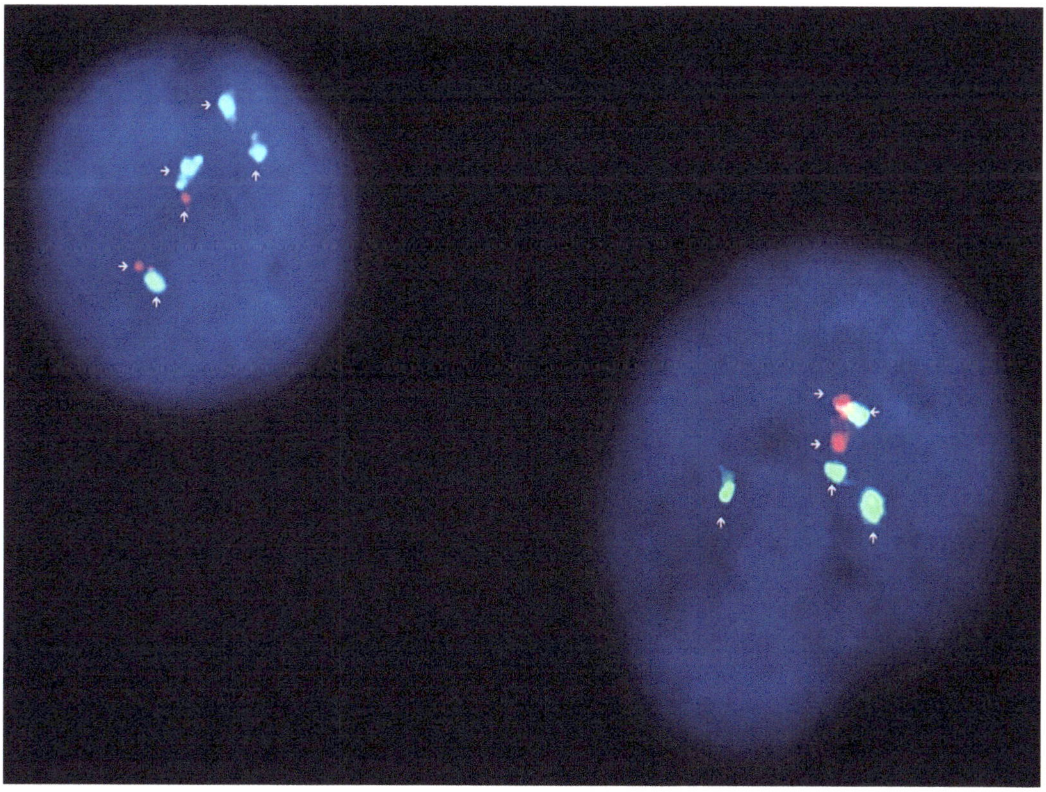

Case Study 11.7 (5) Interphase nuclei showing 4G2R signal pattern using *TP53* LSI probe indicating deletion of *TP53* gene along with tetrasomy 17

Discussion

This patient of a Plasmablastic multiple myeloma showed cytogenetically too, two high-risk factors of TP53 deletion and tetraploidy. *Aneuploidy* is a common prognostic factor in MM. Nonhyperdiploidy including near-tetraploidy (NT) is a poor prognostic indicator compared to hyperdiploidy [15], NT results from endoduplication of hypodiploidy. In a case report of a patient with advanced disease (as in this case of plasmablastic MM) with hyperdiploidy and t(4:14) (q24;q32), the patient responded well to *Lenalidomide and dexamethasone* but after progression over 1 year, she became unresponsive to treatment and was found to have NT and del13q indicating evolution from hyperdiploid to NT clone [16]. In our patient presence of TP53 may be a secondary abnormality to NT as the disease advanced. Del TP53 by itself is also a poor prognostic factor as is tetraploidy.

11.8 Case Study 11.8: Double-Hit Multiple Myeloma with TP53 Gene Deletion, Monosomy 13 and Gain of 1 q

Brief Clinical History

A 62-year-old woman diagnosed with multiple myeloma, treated for it has now presented with sensory neuropathy, backache, and anemia. Investigated for a suspected relapse.

CBC: *Hb*: 6.0 g/dl; TLC: 5230 /cu mm and Platelets: 1,00,000 /cu mm.

DLC: Plasma cells: 5.0%, Myelocytes: 5.0%, Neutrophils: 36.0%, Lymphocytes: 44.0%, Eosinophils: 1.0%, Monocytes: 9.0%, Basophils: 0.0%.

Peripheral Blood Smear and Bone Marrow

Smear showed counts within normal limits. Neutrophils showed left shift, 5% circulating plasma cells were seen. Bone marrow showed a striking increase in mature and immature plasma cells in sheets constituting ~87.0% of the total nucleated cells. Bone biopsy showed a similar picture with IHC showing CD138+, kappa+, lambda−.

Serum SPE, IFE, and Light Chain Assays

SPE Biclonal M band 2.7 and 1.4 g/dL. IFE IgA kappa, kappa light chain 1410, lambda light chain 1.63. kappa/lambda ratio 865. The picture was consistent with kappa light chain restricted multiple myeloma.

FISH-Multiple Myeloma Panel on Bone Marrow

FISH with IGH breakapart probe showed no IGH translocation t(4;14), t(11;14), t(14;16). However, there was TP53 gene deletion in 98% cells and Monosomy of chromosome 13 in 94% cells analyzed.

There were four copies of 1q showing 1q gain in 98% of the cells.

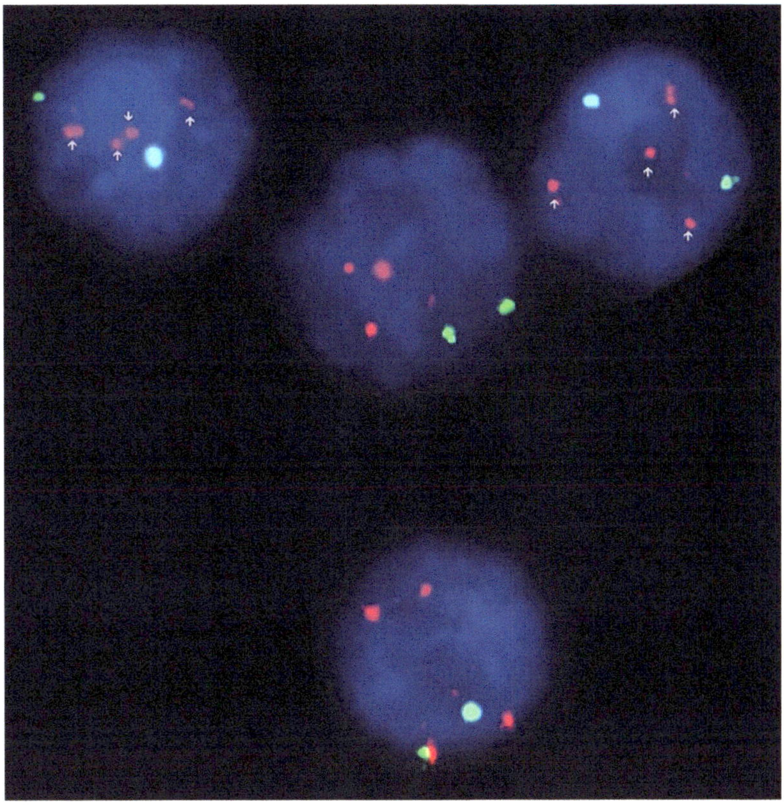

Case Study 11.8 (1) Interphase nuclei showing 2G4R signal pattern using 1p/1q LSI probe indicating 1q amplification

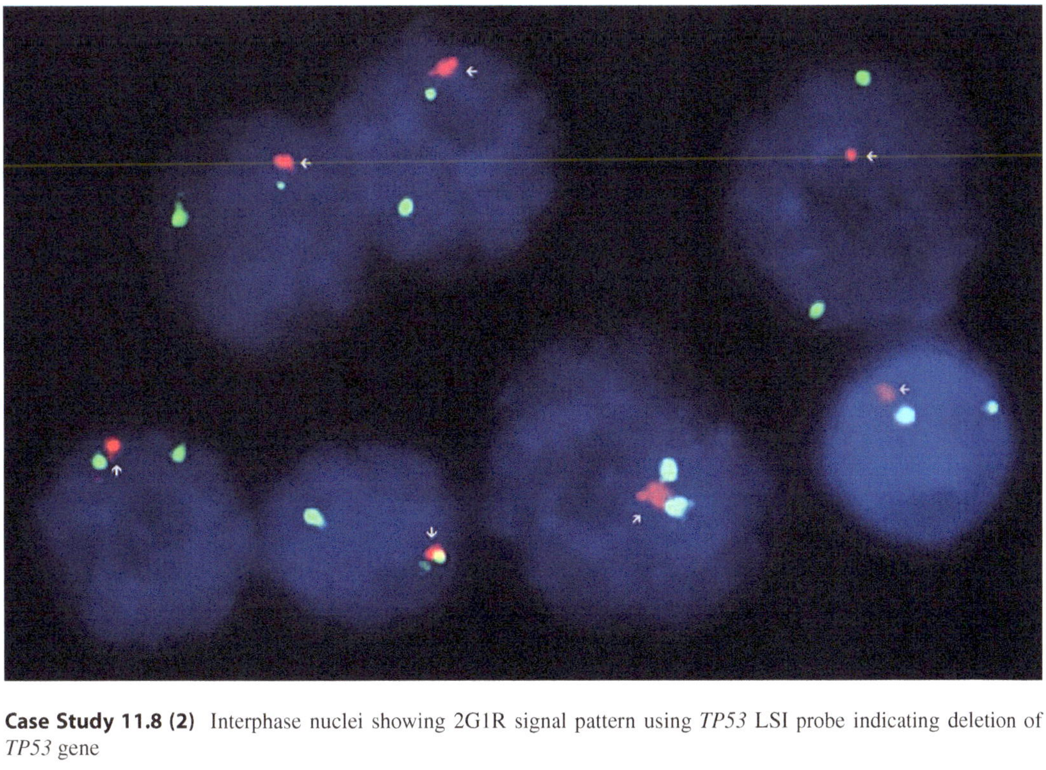

Case Study 11.8 (2) Interphase nuclei showing 2G1R signal pattern using *TP53* LSI probe indicating deletion of *TP53* gene

Discussion

The concept of double-hit and triple-hit myelomas, like lymphomas is of fairly recent origin based on the identification of high-risk factors such as multiple clinical cytogenetic or gene expression features. This patient had TP53 deletion and four copies of 1q signifying 1q gain, both high-risk findings on FISH analysis. In the Myeloma Genome Project by Walker et al. [17] in 1273 newly diagnosed multiple myeloma (NDMM) patients, DNA drivers of aggressive disease were reported. A high-risk subgroup was defined as biallelic TP53 gene inactivation or amplification of more copies of 1q21(CKS1B) on the background of International staging system III. As per the 2020 update on diagnosis, risk stratification and management, Rajkumar et al. [7] reported the presence of del(17p), t(4,14), t(14;16), t(14;20), gain 1q p53 mutation on cytogenetics as high-risk factors. Presence of any two as double-hit and presence of any three as triple-hit myelomas. In this patient, both 1q gain and TP53gene deletion signified a high risk of progression, median TTP (Time to Progression) of 2 years.

In another report, monosomy of chromosome 13 (as seen in this patient as an additional abnormality) was associated with significantly shorter survival and progression-free survival. On the other hand, del(13q) had longer OS but not progression-free survival [18].

11.9 Case Study 11.9: Triple-Hit Multiple Myeloma with TP53 Gene Deletion, t(4;14) with Gain of 1q and Deletion 1p

Brief Clinical History

A 47-year-old male, diagnosed to have multiple myeloma in March 2018, received chemotherapy and radiotherapy for a chest wall lytic lesion, and ABMT. Presented with relapsed myeloma. Evaluated for disease progression.

CBC: *Hb*: 9.1 g/dl; TLC: 4360 /cu mm and Platelets: 41,000 /cu mm; TLC: 4360/cu mm.

DLC: Plasma cells 2.0%, Neutrophils: 19.5%, Lymphocytes: 68.2%, Eosinophils: 0.9%, Monocytes: 9.2%, Basophils: 0.2%.

Peripheral Blood Smear and Bone Marrow

Smear showed neutropenia, ANC: 850/uL with the presence of 2% plasma cells. Bone marrow imprint smears showed an increase in Plasma cells which constituted ~60% of total nucleated cells comprising immature and few mature plasma cells. Bone biopsy showed a similar picture with IHC CD138+, CD56+ with lambda light chain restriction. The picture was consistent with Plasma cell myeloma.

FISH: Multiple Myeloma Panel on Bone Marrow

FISH with IGH breakapart probe showed IGH rearrangement. Further characterization showed t(4;14) in 99% cells. Also, there was TP53 gene deletion in 96% cells and Monosomy 13 in 97% cells analyzed. In addition three copies of 1q were seen (1q gain) plus 1p deletion in 98% cells.

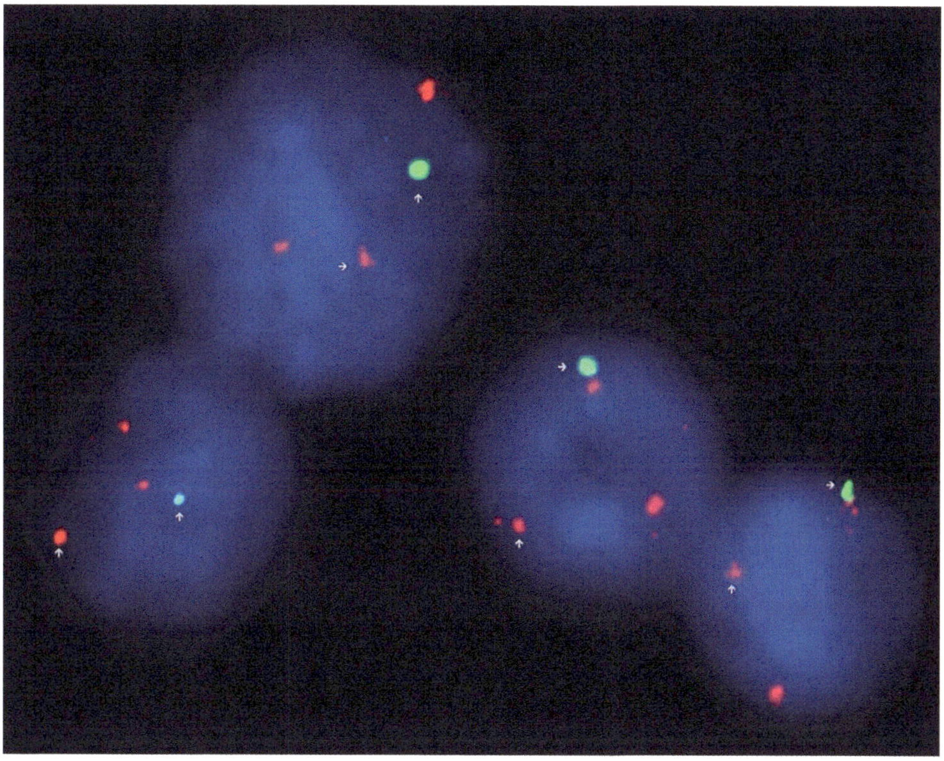

Case Study 11.9 (1) Interphase nuclei showing 1G3R signal pattern using 1p/1q LSI probe indicating 1q gain along with 1p deletion

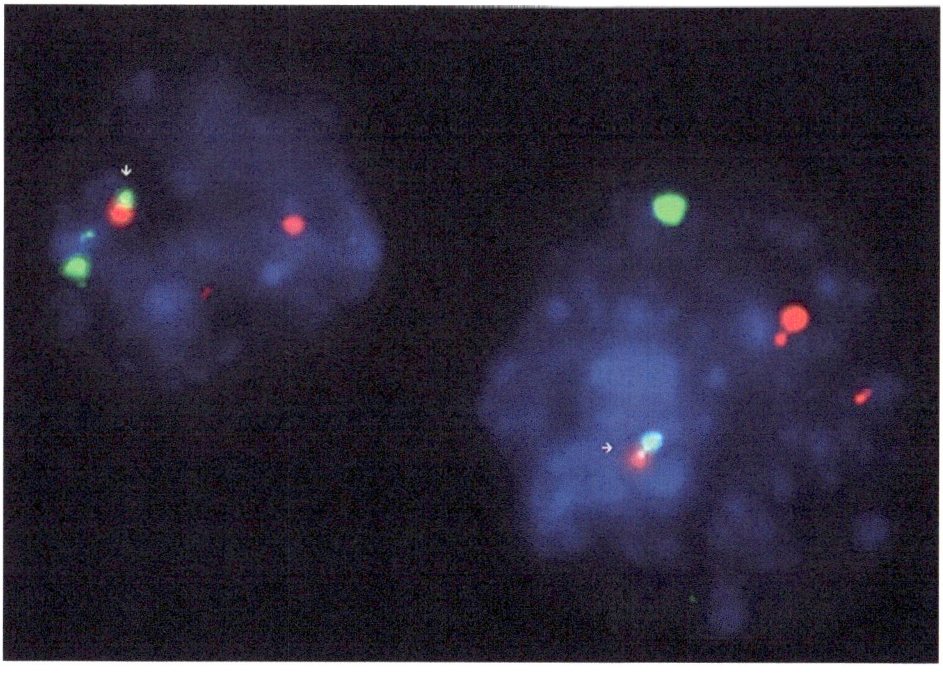

Case Study 11.9 (2) Interphase nuclei showing 1F1G2R aberrant signal pattern using *IGH/FGFR3* LSI probe indicating t(4;14)

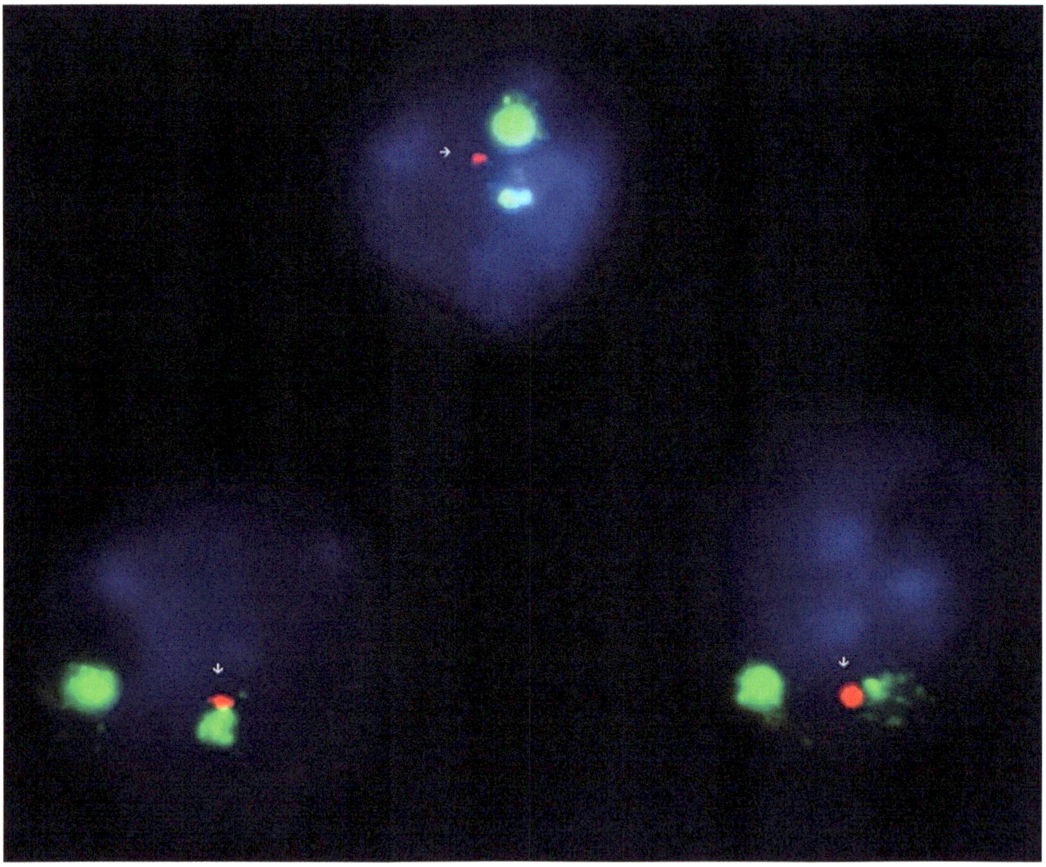

Case Study 11.9 (3) Interphase nuclei showing 2G1R signal pattern using *TP53* LSI probe indicating deletion of *TP53* gene

Discussion

As in another case of triple-hit MM described here in Triple-Hit Multiple Myeloma with TP53 Gene Deletion, t(4;14) with Gain of 1q and Deletion 1p, this patient too presented with multiple high-risk factors with a difference that there were no monosomies and there was a deletion of 1p in addition to gain of 1q. In one of these studies, 159 patients classified as per M-SMART guidelines of Mayo Clinic [7], 24 patients were observed to have 1 high-risk, 7 with 2 high-risk, and 2 with 3 high-risk determinants. Overall survival with 1 high-risk factor was 32 months, with 2 high-risk factors 6 months. Significantly, in double-hit myeloma patients, 4 had co-occurrence of 17p deletion and t(4;14) (as in this patient), one patient had gain of 1q and 17p deletion, one had t(14;16) 17p deletion. In triple-hit myeloma, one patient had 17p del, t(4;14), and t(14;16) and one had 17p del, t(4;14), and t(14;20) showing multiplicity of high-risk factors [19].

11.10 Case Study 11.10: Triple-Hit Multiple Myeloma with IgH Gene Rearrangement t(4;14), TP53 Gene Deletion and Monosomy Chromosome 17, Monosomy 13 and Gain of 1q

Brief Clinical History

A 63-year-old female was admitted to the ICU in a critical condition. She had undergone PTCA and diagnosed to have multiple myeloma as well. Evaluation done for multiple myeloma post one dose of Bortezomib.

CBC: *Hb*: 9.8 g/dl, TLC: 6050 /cu mm and Platelets: 66,000 /cu mm.

DLC: Plasma cells: 2.0%, Neutrophils: 67.0%, Lymphocytes: 27.0%, Eosinophils: 0.0%,

Monocytes: 4.0%, Basophils: 0.0%.

Peripheral Blood Smear and Bone Marrow

Smear showed blood counts within normal limits. Two percent circulating plasma cells were seen. Bone marrow aspirate showed plasma cells 64% of all nucleated cells including abnormal forms. Picture consistent with plasma cell myeloma. Biopsy showed a similar picture with IHC showing CD138+, CD56+, kappa light chain−, and lambda light chain+.

Serum-free Light Chain Assay Post One Dose of Bortezomib

Kappa light chain 15.7 (Ref. Range 3.3–19.4), Lambda light chain 24.4 (Ref. Range 5.7–26.3).

Kappa/lambda ratio 0.643 (Ref. Range 0.26–1.65). The picture was consistent with lambda light chain restricted multiple myeloma.

FISH: Multiple Myeloma Panel on Bone Marrow

FISH with IGH breakapart probe showed IGH rearrangement. Further characterization showed t(4;14) in 85% cells and TP53 gene deletion in 97% cells, Monosomy 13 in 97% cells, and Monosomy 17 in 5% cells analyzed. Additional probe using CKS1B/CDKN2C Dual-Color Probe for the detection of gains/amplifications affecting the chromosomal region 1q21.3-q22 (CKS1B) and/or deletions of the chromosomal region 1p32.2 (CDKN2C) showed there were three copies of 1q, i.e., 1q gain in 90% cells.

Monosomy 13 and Monosomy 17 revealed Hypodiploidy. Hypodiploidy, del(17)(p13.1) (TP53 deletion) and 1q gain are associated with poor prognosis in MM.

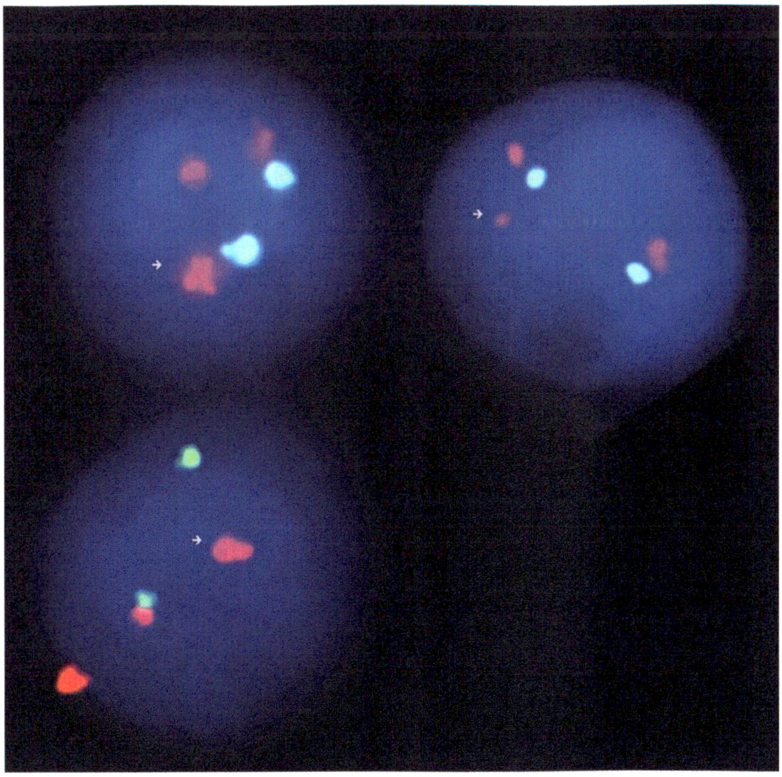

Case Study 11.10 (1) Interphase nuclei showing 2G3R signal pattern using 1p/1q LSI probe indicating 1q gain

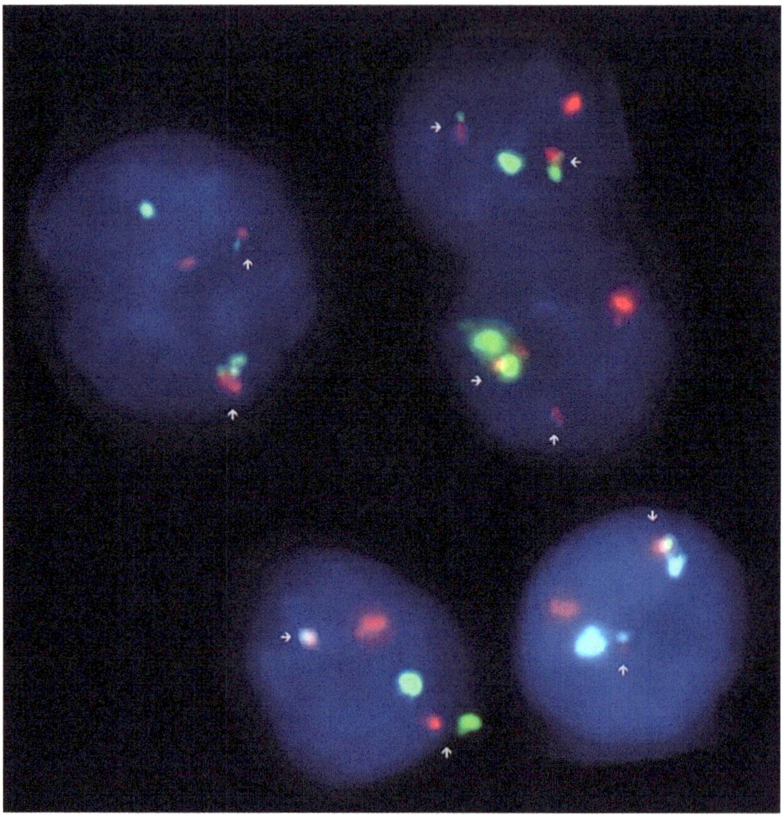

Case Study 11.10 (2) Interphase nuclei showing 2F1G1R signal pattern using *IGH/FGFR3* LSI probe indicating *IGH/FGFR3* fusion, t(4;14)

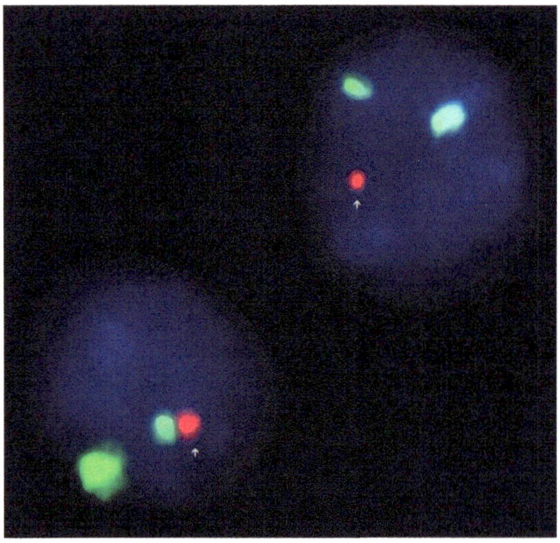

Case Study 11.10 (3) Interphase nuclei showing 2G1R signal pattern using *TP53* LSI probe indicating deletion of *TP53* gene

Discussion

With the utilization of FISH analysis, high-risk multiple myeloma (MM) is defined as having at least one of the mutations signifying poor prognosis including t(4;14), t(14:16), t(14;20), del 17p, p53 mutation, gain 1q, and del 1p [7]. Having two high-risk factors were defined as a double-hit Multiple myeloma and having any three as triple-hit MM. As per these definitions, this patient had triple-hit multiple myeloma. She presented with multiple high-risk factors, three of which being t(4;14), TP53 gene deletion, and gain of 1q, making it a triple-hit multiple myeloma. The concept of double-hit and triple-hit MM have defined groups requiring a different approach to management in these cases. In a retrospective analysis of 159 newly diagnosed MM (NDMM) patients, 7 and 2 patients were defined as double-hit and triple-hit MM, respectively. Both were predictive of lower overall survival [19]. The presence of monosomy 13 and 17 in this patient added further to a poor prognosis.

Annexure

- Case Study 11.1: Multiple Myeloma with IgH Gene Rearrangement, t(11;14)
- Case Study 11.2: Multiple Myeloma with partial deletion of IgH Gene Rearrangement, Monosomy Chromosome 13, Trisomy Chromosome 11 and gain of Chromosome 1q
- Case Study 11.3: Multiple Myeloma with IgH Gene Rearrangement with t(14;16), and Monosomies of Chromosome 13 and 16
- Case Study 11.4: Multiple Myeloma with Biallelic IgH Gene Rearrangement, t(4;14)
- Case Study 11.5: Multiple Myeloma with del (17) (p13.1) and Trisomy of Chromosome 11
- Case Study 11.6: Multiple Myeloma with Trisomy Chromosome 17 and 1q Amplification
- Case Study 11.7: Multiple Myeloma (Plasmablastic) with TP53 Gene Deletion and Tetrasomies of Chromosomes 4,11,13,14, and 17
- Case Study 11.8: Double Hit Multiple Myeloma with TP53 Gene Deletion, Monosomy 13 and gain of 1 q
- Case Study 11.9: Triple Hit Multiple Myeloma with TP53 Gene Deletion, t(4;14) with Gain of 1q and Deletion 1p
- Case Study 11.10: Triple Hit Multiple Myeloma with IgH Gene Rearrangement t(4;14), TP53 Gene Deletion and Monosomy Chromosome 17, Monosomy 13 and gain of 1q

References

1. Rajan AM, Rajkumar SV. Interpretation of cytogenetic results in multiple myeloma per clinical practice. Blood Cancer J. 2015;5(10):e365.
2. Gimondi S, Cavane A, Vendramin A. Identification of clonal IgH gene rearrangement by high through put sequencing of cell free DNA in multiple myeloma patients. Blood. 2015;126(23):2987.
3. Rajkumar SV. Multiple myeloma: 2020 update on diagnosis, risk stratification and management. Annual clinical updates in hematological malignancies. Am J Hematol. 2020;95:548–67.
4. Huh J, Mun YC, Yoo ES. Submicroscopic deletions of immunoglobulin heavy chain gene (IGH) in precursor B acute lymphoblastic leukemia with IGH rearrangements. Ann Lab Med. 2015 Jan;35(1):128–31.
5. Hwang Y, Lee JY, Mun YC. Various patters of IgH deletion identified by FISH using combined IgH and IgH/CCND 1 probes in multiple myeloma and chronic lymphocytic leukemia Int. J Lab Hemat. 2011;33(3):299–304.
6. Shin SY, Eom HS, Sohn JY, et al. Prognostic implications of monosomies in patients with multiple myeloma. Clin Lymphoma, Myeloma Leukemia. 2017;17(3):159–64.
7. Rajkumar SV. Multiple myeloma 2020 update on diagnosis, risk stratification and management. Am J Hematol. 2020;95:548–67.
8. Kumar S, Fonseca R, Ketterling RP. Trisomies in multiple, myeloma impact on survival in patients with high risk cytogenetic. Blood. 2012;119(9):2100–5.
9. Gonzalez D, Burg V, Sanz RG, et al. Immunoglobulin gene rearrangement and the pathogenesis of multiple myeloma. Blood. 2007;110(9):3112.
10. Baker BW, Deane M, Gilleece MH, et al. Distinctive features of immunoglobulin wavy chain variable region rearrangement in multiple myeloma. Leuk Lymphoma. 1994;14:291 30.
11. Kumar S, Fonseca R, Ketterling RP, et al. Trisomies in multiple myeloma: impact on survival in patients with high risk cytogenetics. Blood. 2012;119:2100–5.
12. Hebraud B. Role of additional chromosomal changes in the prognostic value of t(4;14) and del(17p) in multiple myeloma: the IFM experience. Blood. 2015;125:2095–100.
13. Lakshman A, Painuly U, Kumar SK. Natural history of multiple myeloma with de novo del(17p). Blood Cancer Journal. 2019;9:32.
14. Shah GL, Landau H, Londono D, et al. Gain of chromosome 1q portends worse prognosis in multiple myeloma despite novel agent-based induction regimens and autologous transplantation. Leuk Lymphoma. 2017;58(8):1823–31.
15. Sidana S, Jevremovic D, Ketterling RP, et al. Tetraploidy is associated with poor prognosis at diagnosis in multiple myeloma. Am J Hematol. 2019;94(5):E117–20.
16. Yuan J, Shah R, Kulharya A, et al. Near-tetraploidy clone can evolve from a hyperdiploidy clone and cause resistance to *Lenalidomide and dexamethasone* in a multiple myeloma. Leuk Res. 2010;34(7):954–7.

17. Walker BA, Mavrommatis K, Morgan G, et al. A high-risk double hit group of newly diagnosed myeloma identified by genomic analysis. Leukemia. 2019;33:159–70.
18. Binder M, Rajkumar SV, Ketterling RV, et al. Prognostic implications of abnormalities of chromosome 13 and the presence of multiple cytogenetic high risk abnormalities in newly diagnosed multiple myeloma. Blood Cancer J. 2017;7(9):e600. https://doi.org/10.1038/bcj.2017.83.
19. Baysal M, Demirci U, Umit E, et al. Concepts of double hit and triple hit disease in multiple myeloma. Entity and prognostic significance. Sci Rep, Nat Res. 2020;10:5991. https://doi.org/10.1038/s41598-020-62885-0.

GPSR Compliance

The European Union's (EU) General Product Safety Regulation (GPSR) is a set of rules that requires consumer products to be safe and our obligations to ensure this.

If you have any concerns about our products, you can contact us on ProductSafety@springernature.com

In case Publisher is established outside the EU, the EU authorized representative is:

Springer Nature Customer Service Center GmbH
Europaplatz 3
69115 Heidelberg, Germany

Batch number: 09651077

Printed by Printforce, the Netherlands